T0275929

Lecture Notes in Physics

Volume 915

The Lecture Notes in Physics

The series Lecture Notes in Physics (LNP), founded in 1969, reports new developments in physics research and teaching-quickly and informally, but with a high quality and the explicit aim to summarize and communicate current knowledge in an accessible way. Books published in this series are conceived as bridging material between advanced graduate textbooks and the forefront of research and to serve three purposes:

- to be a compact and modern up-to-date source of reference on a well-defined topic
- to serve as an accessible introduction to the field to postgraduate students and nonspecialist researchers from related areas
- to be a source of advanced teaching material for specialized seminars, courses and schools

Both monographs and multi-author volumes will be considered for publication. Edited volumes should, however, consist of a very limited number of contributions only. Proceedings will not be considered for LNP.

Volumes published in LNP are disseminated both in print and in electronic formats, the electronic archive being available at springerlink.com. The series content is indexed, abstracted and referenced by many abstracting and information services, bibliographic networks, subscription agencies, library networks, and consortia.

Proposals should be sent to a member of the Editorial Board, or directly to the managing editor at Springer:

Christian Caron
Springer Heidelberg
Physics Editorial Department I
Tiergartenstrasse 17
69121 Heidelberg/Germany
christian.caron@springer.com

More information about this series at http://www.springer.com/series/5304

Charalampos (Haris) Skokos • Georg A. Gottwald •
Jacques Laskar

Editors

Chaos Detection
and Predictability

 Springer

Editors

Charalampos (Haris) Skokos
Department of Mathematics and Applied
    Mathematics
University of Cape Town
Rondebosch, South Africa

Georg A. Gottwald
School of Mathematics and Statistics
University of Sydney
Sydney, Australia

Jacques Laskar
Observatoire de Paris
IMCCE
Paris, France

ISSN 0075-8450   ISSN 1616-6361   (electronic)
Lecture Notes in Physics
ISBN 978-3-662-48408-1   ISBN 978-3-662-48410-4   (eBook)
DOI 10.1007/978-3-662-48410-4

Library of Congress Control Number: 2015959798

Springer Heidelberg New York Dordrecht London

Printed on acid-free paper

Springer International Publishing AG Switzerland is part of Springer Science+Business Media
(www.springer.com)

Preface

Being able to distinguish chaoticity from regularity in deterministic dynamical systems, as well as to specify the subspace of the phase space in which instabilities are expected to occur, is of utmost importance in as disparate areas as astronomy, particle physics, and climate dynamics. The presence of chaos introduces limitations in our ability to accurately predict the evolution of a dynamical system at scales of different sizes. In many practical applications it is of great importance to determine the significance of this effect to the overall dynamics of the system. For this reason, the development of precise and efficient numerical tools for distinguishing between order and chaos, both locally and globally, becomes imperative, especially in the case of multidimensional systems, whose phase space is not easily visualized. Nowadays there exists a plethora of such methods.

The workshop "Methods of Chaos Detection and Predictability: Theory and Applications" which was held in June 2013 at the Max Planck Institute for the Physics of Complex Systems, in Dresden, Germany, brought together specialists who have developed such methods, as well as researchers applying those techniques to a variety of problems in the natural sciences. This book reviews the theory and numerical implementation of several of the existing methods of chaos detection and predictability and presents the current state of the art. Its chapters are written by the creators of these methods and/or by well-established experts included in the workshop's list of invited speakers.

The most commonly employed method for investigating chaotic dynamics is the computation of the Lyapunov Exponents (LEs). These are asymptotic measures characterizing the average rate of growth (or shrinking) of small perturbations to the orbits of a dynamical system, with the positivity of the maximum LE (mLE) indicating chaoticity. The basic concepts of LEs are presented in the first chapter of the book written by U. Parlitz, where the particular case of the LEs' estimation for time series is discussed and analyzed in depth.

As successful and illuminating LEs have been to characterize chaoticity in deterministic dynamical systems, they suffer in certain situations from serious drawbacks: For example, their computed values can vary significantly in time and may only be used in the long time limit when the exponents have converged with satisfactory accuracy. Furthermore, in the case of (noisy) experimental data, they rely on phase space reconstruction methods, whose inattentive implementation might produce unreliable results.

In the last two decades, several methods have been developed for the fast and reliable determination of the regular or chaotic nature of orbits which were aimed to surmount the shortcomings of the traditional methods involving LEs and phase space reconstruction. These methods can be divided in two broad categories: those which are based on the study of the evolution of deviation vectors from a given orbit, like the computation of the mLE, and those which rely on the analysis of the particular orbit itself.

A technique closely related to the computation of the mLE, which exploits the information provided by the short time evolution of a deviation vector, is the Fast Lyapunov Indicator (FLI) discussed in the second chapter of the book by E. Lega, M. Guzzo, and C. Froeschlé. The next chapter by R. Barrio deals with some variants of the FLI method, namely, the Orthogonal Fast Lyapunov Indicator (OFLI and OFLI2). The method of the Mean Exponential Growth factor of Nearby Orbits (MEGNO), which again is based on the evolution of one deviation vector from the reference orbit, is presented in the next chapter by P. Cincotta and M. Giordano.

The utilization of more than one deviation vector for the characterization of chaos is considered in the next chapter by Ch. Skokos and Th. Manos where the methods of the Smaller (SALI) and the Generalized Alignment Index (GALI) are presented. The method of the Relative Lyapunov Indicator (RLI) where the differences of the finite-time estimators of the mLE of two nearby orbits are used to characterize chaos is the content of the next chapter by Z. Sándor and N. Maffione.

In the following chapter by G. Gottwald and I. Melbourne, the "0-1" test for chaos is discussed in detail. Contrary to the five previous chapters, the analysis in the "0-1" test for chaos is performed directly on the actual orbit (or time series).

The presence of chaos and a positive mLE is often seen as a limitation to the predictability time of the underlying system, which is crudely estimated to be inversely proportional to the mLE (the so-called Lyapunov time). The situation in complex systems evolving on several temporal scales, like for example, in weather forecasting models, can be, however, much more intricate as is shown in the last chapter of the book by S. Siegert and H. Kantz: reliable predictions can be made for times much longer than suggested by the predictability horizon implied by the Lyapunov time.

We hope that this book will be useful both for young scholars, like graduate students, Ph.D. candidates, and postdocs, and for specialists aiming at an up-to-date review of some of the most widely used techniques of chaos detection and predictability.

We thank the Springer Editorial Board, the authors, as well as the reviewers of all chapters, for their work and effort which made possible the publication of this volume.

Cape Town, South Africa   Ch. Skokos
Sydney, Australia   G. Gottwald
June 2015

The original version of the book was revised: The first editor's name was corrected.

The publisher apologizes for having published an early draft edition of the preface.
This has been updated to the final version. The Erratum to the book is available at
DOI 10.1007/978-3-662-48410-4_9

Contents

Chapter 1
Estimating Lyapunov Exponents from Time Series

Ulrich Parlitz

Abstract Lyapunov exponents are important statistics for quantifying stability and deterministic chaos in dynamical systems. In this review article, we first revisit the computation of the Lyapunov spectrum using model equations. Then, employing state space reconstruction (delay coordinates), two approaches for estimating Lyapunov exponents from time series are presented: methods based on approximations of Jacobian matrices of the reconstructed flow and so-called direct methods evaluating the evolution of the distances of neighbouring orbits. Most direct methods estimate the largest Lyapunov exponent, only, but as an advantage they give graphical feedback to the user to confirm exponential divergence. This feedback provides valuable information concerning the validity and accuracy of the estimation results. Therefore, we focus on this type of algorithms for estimating Lyapunov exponents from time series and illustrate its features by the (iterated) Hénon map, the hyper chaotic folded-towel map, the well known chaotic Lorenz-63 system, and a time continuous 6-dimensional Lorenz-96 model. These examples show that the largest Lyapunov exponent from a time series of a low-dimensional chaotic system can be successfully estimated using direct methods. With increasing attractor dimension, however, much longer time series are required and it turns out to be crucial to take into account only those neighbouring trajectory segments in delay coordinates space which are located sufficiently close together.

1.1 Introduction

Lyapunov exponents are a fundamental concept of nonlinear dynamics. They quantify local stability features of attractors and other invariant sets in state space. Positive Lyapunov exponents indicate exponential divergence of neighbouring trajectories and are the most important attribute of chaotic attractors. While the computation of Lyapunov exponents for given dynamical equations is straight

U. Parlitz (✉)
Max Planck Institute for Dynamics and Self-Organization, Am Faßberg 17, 37077 Göttingen, Germany
e-mail: ulrich.parlitz@ds.mpg.de

© Springer-Verlag Berlin Heidelberg 2016
Ch. Skokos et al. (eds.), *Chaos Detection and Predictability*, Lecture Notes in Physics 915, DOI 10.1007/978-3-662-48410-4_1

1

forward, their estimation from time series remains a delicate task. Given a univariate (scalar) time series the first step is to use delay coordinates to reconstruct the state space dynamics. Using the reconstructed states there are basically two approaches to solve the estimation problem: With Jacobian matrix based methods a (local) mathematical model is fitted to the temporal evolution of the states that can then be used like any other dynamical equation. Using this approach in principle all Lyapunov exponents can be estimated if the chosen black-box model is in very good agreement with the underlying dynamics. In practical applications such a high level of fidelity is often difficult to achieve, in particular since the time series typically contain only limited information about contracting directions in state space. With the second approach for estimating (at least the largest) Lyapunov exponents the local divergence of trajectory segments in reconstructed state space is assessed directly. Advantage of this kind of direct methods is their low number of estimation parameters, easy implementation, and last but not least, direct graphical feedback about the (non-) existence of exponential divergence in the given time series.

The following presentation is organised as follows: In Sect. 1.2 the standard algorithm for computing Lyapunov exponents using dynamical model equations is revisited. Methods for computing Lyapunov exponents from time series are presented in Sect. 1.3. In Sect. 1.4 four dynamical systems are introduced to generate time series which are then in Sect. 1.5 used as examples for illustrating and evaluating features of direct estimation of the largest Lyapunov exponent. The examples are: the Hénon map, the hyper chaotic folded-towel map, the Lorenz-63 system, and a 6-dimensional Lorenz-96 model. These time discrete and time continuous models exhibit deterministic chaos of different dimensionality and complexity. In Sect. 1.6 a summary is given and the Appendix contains some information for those readers who are interested in implementing Jacobian based estimation algorithms.

1.2   Computing Lyapunov Exponents Using Model Equations

Lyapunov exponents characterize and quantify the dynamics of (infinitesimally) small perturbations of a state or trajectory in state space. Let the dynamical model be a M-dimensional discrete

$$\mathbf{x}(n + 1) = \mathbf{g}(\mathbf{x}(n)) \tag{1.1}$$

or a continuous

$$\dot{\mathbf{x}} = \frac{d\mathbf{x}}{dt} = \mathbf{f}(\mathbf{x}) \tag{1.2}$$

dynamical system generating a flow

$$\phi^t : \mathbb{R}^M \to \mathbb{R}^M \tag{1.3}$$

with discrete $t = n \in \mathbb{Z}$ or continuous $t \in \mathbb{R}$ time. The temporal evolution of an infinitesimally small perturbation $\mathbf{y}$ of the state $\mathbf{x}$

$$D\phi^t(\mathbf{x}) \cdot \mathbf{y} \tag{1.4}$$

is governed by the linearized dynamics where $D\phi^t(\mathbf{x})$ denotes the Jacobian matrix of the flow ϕ^t. For discrete systems this Jacobian can be computed using the recursion scheme

$$D_x\phi^{n+1}(\mathbf{x}) = D_x\mathbf{g}(\phi^n(\mathbf{x})) \cdot D_x\phi^n(\mathbf{x}) \tag{1.5}$$

with initial value $D_x\phi^0(\mathbf{x}) = I_M$ where I_M denotes the $M \times M$ identity matrix. For continuous systems (1.2) additional linearized ordinary differential equations (ODEs)

$$\frac{d}{dt}Y = D_x\mathbf{f}(\phi^t(\mathbf{x})) \cdot Y \tag{1.6}$$

have to be solved where $\phi^t(\mathbf{x})$ is a solution of Eq. (1.2) with initial value $\mathbf{x}$ and Y is a $M \times M$ matrix that is initialized as $Y(0) = I_M$. The solution $Y(t)$ provides the Jacobian of the flow $D\phi^t(\mathbf{x})$ that describes the local dynamics along the trajectory given by the temporal evolution $\phi^t(\mathbf{x})$ of the initial state $\mathbf{x}$. Since Eq. (1.6) is a linear ODE its solutions consist of exponential functions and the Jacobian of the flow $D\phi^t(\mathbf{x})$ maps a sphere of initial values close to $\mathbf{x}$ to an ellipsoid centered at $\phi^t(\mathbf{x})$ as illustrated in Fig. 1.1. This evolution of the tangent space dynamics can be analyzed using a singular value decomposition (SVD) of the Jacobian of the flow $D\phi^t(\mathbf{x})$

$$D\phi^t(\mathbf{x}) = U \cdot S \cdot V^{tr} \tag{1.7}$$

Fig. 1.1 Temporal evolution of an infinitesimally small sphere in state space

where $S = \text{diag}(\sigma_1, \ldots, \sigma_M)$ is a $M \times M$ diagonal matrix containing the singular values $\sigma_1 \geq \sigma_2 \geq \ldots \geq \sigma_M \geq 0$ and $U = (\mathbf{u}^{(1)}, \ldots, \mathbf{u}^{(M)})$ and $V = (\mathbf{v}^{(1)}, \ldots, \mathbf{v}^{(M)})$ are orthogonal matrices, represented by orthonormal column vectors $\mathbf{u}^{(i)} \in \mathbb{R}^M$ and $\mathbf{v}^{(i)} \in \mathbb{R}^M$, respectively. V^{tr} is the transposed of V coinciding with the inverse $V^{-1} = V^{tr}$, because V is orthogonal. For the same reason $U^{tr} = U^{-1}$ and by multiplying by V from the right we obtain $D\phi^t(\mathbf{x}) \cdot V = U \cdot S$ or

$$D\phi^t(\mathbf{x})\mathbf{v}^{(m)} = \sigma_m \mathbf{u}^{(m)} \quad (m = 1, \ldots, M). \tag{1.8}$$

The column vectors of the matrices V and U span the initial sphere and the ellipsoid, as illustrated in Fig. 1.1, where the singular values $\sigma_m(t)$ give the lengths of the principal axes of the ellipsoid at time t. On average $\sigma_m(t)$ increases or decreases exponentially during the temporal evolution and the *Lyapunov exponents* λ_m are the mean logarithmic growth rates of the lengths of the principal axes

$$\lambda_m = \lim_{t \to \infty} \frac{1}{t} \ln \sigma_m(t). \tag{1.9}$$

The existence of the limit in Eq. (1.9) is guaranteed by the *Theorem of Oseledec* [33] stating that the Oseledec matrix

$$\Lambda(\mathbf{x}) = \lim_{t \to \infty} \left([D\phi^t(\mathbf{x})]^{tr} \cdot D\phi^t(\mathbf{x}) \right)^{\frac{1}{2t}} \tag{1.10}$$

exists. For dissipative systems one set of exponents is associated with each attractor and for almost all initial states $\mathbf{x}$ from each attractor $\Lambda(\mathbf{x})$ takes the same value.

The logarithms of the eigenvalues μ_m of this symmetric positive definite $M \times M$ matrix are the Lyapunov exponents of the attractor or invariant set the initial state $\mathbf{x}$ belongs to

$$\lambda_m = \ln \mu_m \quad (m = 1, \ldots, M). \tag{1.11}$$

Using the SVD of the Jacobian matrix of the flow the Oseledec matrix for finite time t can be written

$$\left(V \cdot S \cdot U^{tr} \cdot U \cdot S \cdot V^{tr} \right)^{\frac{1}{2t}} = \left(V \cdot S^2 \cdot V^{tr} \right)^{\frac{1}{2t}} \tag{1.12}$$

with eigenvalues $\sigma_m^{1/t}$. Taking the logarithm $\frac{1}{t} \ln \sigma_m$ and performing the limit $t \to \infty$ we obtain the Lyapunov exponents (1.11). Unfortunately, this definition and illustration of the Lyapunov exponents cannot be used directly for their numerical computation, because the Jacobian matrix $D\phi^t(\mathbf{x})$ consists of elements that are exponentially increasing or decreasing in time resulting in values beyond the numerical resolution and representation of variables. To avoid these severe numerical problems in 1979 Shimada and Nagashima [45] and in 1980 and Benettin et al. [3] suggested algorithms that exploit the fact that the growth rate of k-dimensional volumes $\lambda^{(k)}$

(in the M-dimensional state space) is given by the sum of the largest k Lyapunov exponents

$$\lambda^{(k)} = \sum_{m=1}^{k} \lambda_m \qquad (1.13)$$

and the Lyapunov exponents can by computed from the volume growth rates as $\lambda_1 = \lambda^{(1)}$, $\lambda_2 = \lambda^{(2)} - \lambda_1$, $\lambda_3 = \lambda^{(3)} - \lambda_2 - \lambda_1$, etc. The volume growth rates $\lambda^{(k)}$ can be computed using a QR decomposition of the Jacobian of the flow $D\phi^t(\mathbf{x})$. Let $O^{(k)} = (\mathbf{o}^{(1)}, \ldots, \mathbf{o}^{(k)})$ be an orthogonal matrix whose column vectors $\mathbf{o}^j$ span a k-dimensional infinitesimal volume with k ranging from 1 to M. After time t this volume is transformed by the Jacobian matrix into a parallelepiped $P^{(k)}(t) = D\phi^t(\mathbf{x}) \cdot O^{(k)}$. To computed the volume spanned by the column vectors of $P^{(k)}(t)$ we perform a QR-decomposition of $P^{(k)}(t)$

$$P^{(k)}(t) = D\phi^t(\mathbf{x}) \cdot O^{(k)} = Q^{(k)}(t) \cdot R^{(k)}(t) \qquad (1.14)$$

where $Q^{(k)}(t)$ is a matrix with k orthonormal columns and $R^{(k)}(t)$ is an upper triangular matrix with non-negative diagonal elements. The volume $V^{(k)}(t)$ of $P^{(k)}(t)$ at time t is given by the product of the diagonal elements $R_{ii}^{(k)}(t)$ of $R^{(k)}(t)$

$$V^{(k)}(t) = R_{11}^{(k)}(t) \cdot \ldots \cdot R_{kk}^{(k)}(t) = \prod_{i=1}^{k} R_{ii}^{(k)}(t). \qquad (1.15)$$

The mean logarithmic growth rate of the k-dimensional volume is thus given by

$$\lambda^{(k)} = \lim_{t \to \infty} \frac{1}{t} \ln V^{(k)}(t) = \lim_{t \to \infty} \frac{1}{t} \sum_{i=1}^{k} \ln R_{ii}^{(k)}(t). \qquad (1.16)$$

Using this relation and Eq. (1.13) we can conclude that the first k Lyapunov exponents $\lambda_1, \ldots, \lambda_k$ are given by

$$\lambda_i = \lim_{t \to \infty} \frac{1}{t} \ln R_{ii}^{(k)}(t). \qquad (1.17)$$

If one would perform the QR-decomposition (1.14) of the Jacobian $D\phi^t(\mathbf{x})$ after a very long period of time (to approximate the limit $t \to \infty$) then one would be faced with the same numerical problems that were mentioned above in the context of the Oseledec matrix. The advantage of the volume approach via QR-decomposition is, however, that this decomposition can be computed recursively for small time intervals avoiding any numerical over or underflow. To exploit this feature the period of time $[0, t]$ is divided into N time intervals of length $T = t/N$ and the Jacobian matrices $D\phi^T(\phi^{t_n}(\mathbf{x}))$ are computed at times $t_n = nT$ ($n = 0, \ldots, N-1$) along the orbit. Employing the chain rule the Jacobian matrix $D\phi^t(\mathbf{x})$ can be written as a product of Jacobian matrices $D\phi^T(\phi^{t_n}(\mathbf{x}))$

$$D\phi^t(\mathbf{x}) = D\phi^T(\phi^{t_{N-1}}(\mathbf{x})) \cdot \ldots \cdot D\phi^T(\phi^{t_0}(\mathbf{x})) = \prod_{n=0}^{N-1} D\phi^T(\phi^{t_n}(\mathbf{x})). \qquad (1.18)$$

Using QR-decompositions

$$\hat{Q}^{(k)}(t_{n+1}) \cdot \hat{R}^{(k)}(t_{n+1}) = D\phi^T(\phi^{t_n}(\mathbf{x})) \cdot \hat{Q}^{(k)}(t_n) \qquad (1.19)$$

the full period of time $[0, t]$ used for averaging the local expansion rates can be decomposed into a sequence of relatively short intervals $[0, T]$ with Jacobian matrices $D\phi^T(\phi^{t_n}(\mathbf{x}))$ that are not suffering from numerical difficulties. Applying the QR-decompositions (1.19) recursively we obtain a scheme for computing the QR-decomposition of the Jacobian matrix of the full time step

$$\begin{aligned}
D\phi^t(\mathbf{x}) \cdot O^{(k)} &= D\phi^T(\phi^{t_{N-1}}(\mathbf{x})) \cdot \ldots \cdot D\phi^T(\phi^{t_0}(\mathbf{x})) \cdot O^{(k)} \\
&= D\phi^T(\phi^{t_{N-1}}(\mathbf{x})) \cdot \ldots \cdot D\phi^T(\phi^{t_1}(\mathbf{x})) \cdot \hat{Q}^{(k)}(t_1) \cdot \hat{R}^{(k)}(t_1) \\
&= D\phi^T(\phi^{t_{N-1}}(\mathbf{x})) \cdot \ldots \cdot D\phi^T(\phi^{t_2}(\mathbf{x})) \cdot \hat{Q}^{(k)}(t_2) \cdot \hat{R}^{(k)}(t_2) \cdot \hat{R}^{(k)}(t_1) \\
&\vdots \\
&= D\phi^T(\phi^{t_{N-1}}(\mathbf{x})) \cdot \hat{Q}^{(k)}(t_{N-1}) \cdot \hat{R}^{(k)}(t_{N-1}) \cdot \ldots \cdot \hat{R}^{(k)}(t_1) \\
&= \hat{Q}^{(k)}(t_N) \cdot \hat{R}^{(k)}(t_N) \cdot \ldots \cdot \hat{R}^{(k)}(t_1)
\end{aligned}$$

which provides $Q^{(k)}(t) = \hat{Q}^{(k)}(t_N)$ and the required matrix $R^{(k)}(t)$ as a product

$$R^{(k)}(t) = \hat{R}^{(k)}(t_N) \cdot \ldots \cdot \hat{R}^{(k)}(t_1). \qquad (1.20)$$

For the diagonal elements of the upper triangular matrices holds the relation

$$R_{ii}^{(k)}(t) = \prod_{n=1}^{N} \hat{R}_{ii}^{(k)}(t_n) \qquad (1.21)$$

and substituting $R_{ii}^{(k)}(t)$ in Eq. (1.17) (with $t = NT$) we obtain the following expression for the ith Lyapunov exponent (with $i \le k \le M$)

$$\lambda_i = \lim_{N \to \infty} \frac{1}{NT} \sum_{n=1}^{N} \ln \hat{R}_{ii}^{(k)}(t_n). \qquad (1.22)$$

Using this approach the computation of all Lyapunov exponents of a given dynamical system became a standard procedure [12, 19, 46, 51] providing the full set of Lyapunov exponents constituting the *Lyapunov spectrum* which is an ordered set of real numbers $\{\lambda_1, \lambda_2, \ldots, \lambda_m\}$. If the system undergoes aperiodic oscillations (after transients decayed) and if the largest Lyapunov exponent λ_1 is positive, then the corresponding attractor is said to be *chaotic* and to show *sensitive dependence on initial conditions*. If more than one Lyapunov exponent is positive the underlying dynamics is called *hyper chaotic*.

The (ordered) spectrum $\lambda_1 \geq \lambda_2 \geq \ldots \geq \lambda_m$ can be used to compute the *Kaplan–Yorke dimension* (also called *Lyapunov dimension*) [34]

$$D_{KY} = k + \frac{\sum_{i=1}^{k} \lambda_i}{|\lambda_{k+1}|} \tag{1.23}$$

where k is the maximum integer such that the sum of the k largest exponents is still non-negative. D_{KY} is an upper bound for the information dimension of the underlying attractor.

1.3 Estimating Lyapunov Exponents from Time Series

All methods for computing Lyapunov exponents are based on state space reconstruction from some observed (univariate) time series [11, 42, 43, 49]. For reconstructing the dynamics most often delay coordinates are used due to their efficacy and robustness.

To reconstruct the multi-dimensional dynamics from an observed (univariate) time series $\{s_n\}$ sampled at times $t_n = n\Delta t$ we use delay coordinates providing the $N \times D$ trajectory matrix

$$X = (\mathbf{x}_1, \mathbf{x}_2, \ldots, \mathbf{x}_N)^{tr} \tag{1.24}$$

where each row is a reconstructed state vector[1]

$$\mathbf{x}_n = (s_n, s_{n+L}, \ldots, s_{n+(D-1)L}) \tag{1.25}$$

at time n (with lag L and dimension D). From a time series $\{s_n\}$ of length N_d a total number of $N = N_d - (D-1)L$ states can be reconstructed. To achieve useful (nondistorted) reconstructions the time window length $(D-1)L$ of the delay vector should cover typical time scales of the dynamics like natural periods or the first zero or minimum of the autocorrelation function or the (auto) mutual information [1, 26].

Since Lyapunov exponents are invariant with respect to diffeomorphic changes of the coordinate system the Lyapunov exponents estimated for the reconstructed flow

---

[1]We use forward delay coordinates here. Delay reconstruction backward in time provides equivalent results.

will coincide with those of the original system. Technically, different approaches exist for computing Lyapunov exponents from embedded time series. *Jacobian-based methods* employ the standard algorithm outlined in Sect. 1.2 except for the computation of the Jacobian matrix $D\phi^t(\mathbf{x})$ which is now based on approximations of the flow in the reconstructed state space. This class of methods will be briefly presented in Sect. 1.3.1. In particular with noisy data reliable estimation of the Jacobian matrix may be a delicate task. This is one of the reasons why several authors proposed methods for estimating the largest Lyapunov exponent directly from diverging trajectories in reconstructed state space. Such *direct methods* will be discussed in detail in Sect. 1.3.2 and will be illustrated and evaluated in Sect. 1.5. They do not require Jacobian matrices but are mostly used to compute the largest Lyapunov exponent, only. A major advantage of direct methods, however, is the fact that they provide direct visual feedback to the user whether the available time series really exhibits exponential divergence on small scales. Therefore, we shall focus on this class of methods in the following.

1.3.1 Jacobian-Based Methods

With *Jacobian methods*, first a model is fitted to the data and then the Jacobian matrices of the model equations are used to compute the Lyapunov exponents using standard algorithms (see Sect. 1.2) which have been developed for the case when the equations of the dynamical system are known [3, 12, 19, 45]. In this context usually local linear approximations are used for modeling the flow in reconstructed state space [14, 22, 29, 36, 40, 47, 48, 53, 54]. An investigation of the data requirements for Jacobian-based methods may be found in [13, 15]. Technical details and more information about the implementation of Jacobian-based methods are given in the Appendix.

To employ the standard algorithm for computing Lyapunov exponents (Sect. 1.2) also for time series analysis the Jacobian matrices along the orbit in reconstruction space are required and have to be estimated from the temporal evolution of reconstructed states. Here two major challenges occur:

(a) The Jacobian matrices (derivatives) have to be estimated using reconstructed states that are scattered along the unstable direction(s) of the attractor but *not* in transversal directions (governed by contracting dynamics). This may result in ill-posed estimation problems and is a major obstacle for estimating negative Lyapunov exponents. Furthermore, the estimation problem is often even more delicate because we aim at approximating (partial) derivatives (the elements of the Jacobian matrix) from typically noisy data where estimating derivatives is a notoriously difficult problem.

(b) To properly unfold the attractor and the dynamics in reconstruction space the embedding dimension D has in general to be larger than the dimension of the original state space M (see Sect. 1.3). Therefore, a straightforward computation of Lyapunov exponents using (estimated) $D \times D$ Jacobian matrices in reconstruction space and QR-decomposition (see Sect. 1.2) will provide D Lyapunov exponents, although the underlying M-dimensional system possesses $M < D$ exponents, only. The additional $D - M$ Lyapunov exponents are called *parasitic* or *spurious* exponents and they have to be identified (or avoided), because their values are *not* related to the dynamics to be characterized.

Spurious Lyapunov exponents can take any values [8], depending on details of the approximation scheme used to estimate the Jacobian matrices, the local curvature of the reconstructed attractor, and perturbations of the time series (e.g., noise). Therefore, without taking precautions spurious Lyapunov exponents can occur between "true" exponents and may spoil in this sense the observed spectrum (resulting in false conclusion about the number of positive exponents or the Kaplan–Yorke dimension, for example). To cope with this problem many authors presented different approaches for avoiding spurious Lyapunov exponents or for reliably detecting them [44].

To identify spurious Lyapunov exponents one can estimate the local thickness of the attractor along the directions associated with the different Lyapunov exponents [6, 7] or compare the exponents obtained with those computed for the time reversed series [35, 36], because spurious exponents correspond to directions where the attractor is very thin and because in general they do not change their signs upon time reversal (in contrast to the true exponents). The latter method, however, works only for data of very high quality that enable also a correct estimation of negative Lyapunov exponents which in most practical situations is not the case. Furthermore, in some cases also spurious Lyapunov exponents may change signs and can then not be distinguished from true exponents. Another method for identifying spurious Lyapunov exponents employing covariant Lyapunov vectors been suggested in [27, 52].

Spurious Lyapunov exponents can be avoided by globally unfolding the dynamics in a D-dimensional reconstruction space and locally approximating the (tangent space) dynamics in a lower dimensional d-dimensional space (with $d \leq M$). This can be done using two different delay coordinates where the set of indices of neighbouring points of a reference point is identified using a D-dimensional delay reconstruction and then these indices are used to reconstruct states representing "proper" neighbours in a d-dimensional delay reconstruction (with $d < D$) which is used for subsequent modeling of the dynamics (flow and its Jacobian matrices) [6, 7, 14]. To cover relevant time scales it is recommended [14] to use for both delay reconstructions different lags L_D and L_d so that the delay vectors span the same or similar windows in time (i.e., $(D - 1)L_D \approx (d - 1)L_d$). An alternative approach for evaluating the dynamics in a lower dimensional space employs local projections into d-dimensional subspaces of the D-dimensional delay embedding space given by singular value decompositions of local trajectory matrices [10, 48].

For evaluating the uncertainty in Lyapunov exponent computations from time series employing Jacobian based algorithms bootstrapping methods have been suggested [28].

1.3.2 Direct Methods

There are (slightly) different ways to implement a direct method for estimating the largest Lyapunov exponent and they all rely on the fact that almost all tangent vectors (or perturbations) converge to the subspace spanned by the first Lyapunov vector(s) with an asymptotic growth rate given by the largest Lyapunov exponent λ_1 (see Sect. 1.5.1). In practice, however, from a time series of finite length only a finite number of reconstructed states is available with a finite lower bound for their mutual distances. If the nearest neighbour $\mathbf{x}_{m(n)}$ of a reference point $\mathbf{x}_n$ is chosen from the set of reconstructed states the trajectory segments emerging from both states will (on average) diverge exponentially until the distance $\|\mathbf{x}_{m(n)+k} - \mathbf{x}_{n+k}\|$ exceeds a certain threshold and ceases to grow but oscillates bounded by the size of the attractor. For direct methods it is crucial that the reorientation towards the most expanding direction takes place and is finished *before* the distance between the states saturates. Then the period of exponential growth characterised by the largest Lyapunov exponent can be detected and estimated for some period of time as a linear segment in a suitable semi-logarithmic plot. This feature is illustrated in Fig. 1.2a showing the average of the logarithms of distances of neighbouring trajectories vs. time on a semi-logarithmic scale. In phase I the difference vector between states from both trajectories converges towards the most expanding direction. Then in phase II exponential divergence results in a linear segment until in phase III states from both trajectory segments are so far away from each other that nonlinear folding occurs and the mean distance converges to a constant value (which is related to the diameter of the attractor).

Different implementations of the direct approach have been suggested in the past 25 years [18, 25, 30, 38, 41] that are based on the following considerations.

Let $\mathbf{x}(m(n))$ be a neighbour of the reference state $\mathbf{x}(n)$ (with respect to the Euclidean norm or any other norm) and let both states be *temporally separated*

$$|m(n) - n| > w \qquad (1.26)$$

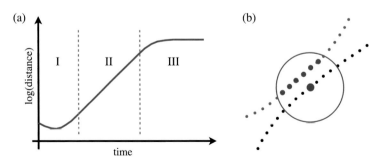

Fig. 1.2 (**a**) Sketch showing the mean logarithmic distance of neighbouring states on different trajectory segments vs. time. (**b**) Illustration motivating the exclusion of temporal neighbours (Theiler window)

where w is a characteristic time scale (e.g., a mean period) of the time series. The temporal separation (also called *Theiler window w* [50]) is necessary to make sure that this pair of neighbouring states can be considered as initial conditions of *different* trajectory segments and the largest Lyapunov exponent can be estimated from the mean rate of separation of states along these two orbits. Figure 1.2b shows an illustration of a case where a Theiler window of $w = 3$ would be necessary to exclude neighbours of the state marked by a big dot to avoid temporal neighbours on the same trajectory segment (small dots preceding and succeeding the reference state within the circle, indicating the search radius).

We shall quantify the separation of states by the distance

$$d(m(n), n, k) = \|\mathbf{x}(m(n) + k) - \mathbf{x}(n + k)\| \qquad (1.27)$$

of the neighbouring states after k time steps (i.e. a period of time $T = k\Delta t$). Most often [18, 30, 38, 41] the Euclidean norm

$$d_E(m(n), n, k) = \|\mathbf{x}(m(n) + k) - \mathbf{x}(n + k)\|_2 \qquad (1.28)$$

is used to define this distance, although Kantz [25] pointed out that it is sufficient to consider the difference

$$d_L(m(n), n, k) = |x(m(n) + (D - 1)L + k) - x(n + (D - 1)L + k)| \qquad (1.29)$$

of the last components of both reconstructed states, because these projections also grow exponentially with the largest Lyapunov exponent. Here L denotes again the time lag used for delay reconstruction. With the same argument, one can also consider the difference of the first component

$$d_F(m(n), n, k) = |x(m(n) + k) - x(n + k)| \qquad (1.30)$$

and in the following we shall compare all three choices. Within the linear approximation (very small $d(m(n), n, k)$) the temporal evolution of the distance $d(m(n), n, k)$ is given by

$$d(m(n), n, k) \approx d(m(n), n, 0)\, e^{\hat{\lambda}_1(n)k\Delta t} \qquad (1.31)$$

where $d(m(n), n, 0)$ stands for the initial separation of both orbits and $\hat{\lambda}_1(n)$ denotes the (largest) local expansion rate of orbits starting at $\mathbf{x}(n)$. Taking the logarithm we obtain

$$\hat{\lambda}_1(n) \approx \frac{1}{k\Delta t}\left[\ln(d(m(n), n, k)) - \ln(d(m(n), n, 0))\right] \qquad (1.32)$$

$$= \frac{1}{k\Delta t}\ln\left(\frac{d(m(n), n, k)}{d(m(n), n, 0)}\right).$$

Here and in the following expansion rates and Lyapunov exponents are computed using the natural logarithm $\ln(\cdot)$.

Since expansion rates vary on the attractor we have to average along the available trajectory by choosing for each *reference state* $\mathbf{x}(n)$ some *neighbouring states* $\{\mathbf{x}(m(n)) : m(n) \in \mathcal{U}_n\}$ where $\mathcal{U}_n$ defines the chosen neighbourhood of $\mathbf{x}(n)$ that can be of *fixed mass* (a fixed number K of nearest neighbours of $\mathbf{x}(n)$) or of *fixed size* (all points with distance smaller than a given bound ϵ). Often a fixed mass with $K = 1$ (i.e., using only the nearest neighbour) is used [18, 38, 41], but a fixed size may in some cases be more appropriate to avoid mixing of scales [25, 30]. In the following $|\mathcal{U}_n|$ denotes the number of neighbours of $\mathbf{x}(n)$.

Furthermore, it may be appropriate to use not all available reconstructed states $\mathbf{x}(n)$ $(n = 1, \ldots, N)$ as reference points but only a subset $\mathcal{R}$ consisting of $N_r = |\mathcal{R}|$ points. This speeds up computations and may even result in better results if $\mathcal{R}$ contains only those reconstructed states that possess very close neighbours (where $d(m(n), n, 0)$ is very small). This issue will be discussed and demonstrated in the results section.

With averaged logarithmic distances

$$E(k) = \frac{1}{N_r} \sum_{n \in \mathcal{R}} \frac{1}{|\mathcal{U}_n|} \sum_{m \in \mathcal{U}_n} \ln(d(m, n, k)) \tag{1.33}$$

and

$$S(k) = \frac{1}{N_r} \sum_{n \in \mathcal{R}} \frac{1}{|\mathcal{U}_n|} \sum_{m \in \mathcal{U}_n} \ln\left(\frac{d(m, n, k)}{d(m, n, 0)}\right) = E(k) - E(0) \tag{1.34}$$

and the local expansion rates (1.32) the averaged growth rate can be expressed as

$$\bar{\lambda}_1 = \frac{1}{N_r} \sum_{n \in \mathcal{R}} \hat{\lambda}_1(n) \approx \frac{1}{N_r} \frac{1}{k \Delta t} \sum_{n \in \mathcal{R}} \frac{1}{|\mathcal{U}_n|} \sum_{m \in \mathcal{U}_n} \ln\left(\frac{d(m, n, k)}{d(m, n, 0)}\right) \tag{1.35}$$

$$= \frac{1}{k \Delta t} S(k) = \frac{1}{k \Delta t} [E(k) - E(0)] \tag{1.36}$$

providing the relations

$$S(k) \approx k \Delta t \bar{\lambda}_1 \tag{1.37}$$

and

$$E(k) \approx k \Delta t \bar{\lambda}_1 + E(0). \tag{1.38}$$

Here $E(k)$ stands for $E_E(k)$, $E_F(k)$, or $E_L(k)$ depending on the distance measure d_E, d_F, or d_L used when computing E in Eq. (1.33).

In 1987 Sato et al. [41] suggested to estimated the largest Lyapunov exponent by the slope of a linear segment of the graph obtained when plotting $S(k)$ vs. $k\Delta t$. The same approach was suggested later in 1993 by Gao and Zheng [18]. The same year Rosenstein et al. [38] recommended to avoid the normalization by the initial distance $d(m(n), n, 0)$ in Eqs. (1.32) and (1.34) and to plot $E(k)$ vs. $k\Delta t$. As can be seen from Eqs. (1.37) and (1.38) both procedures are equivalent, because both graphs differ only be a constant shift $E(0)$. Instead of estimating the slope in $S(k)$ vs. $k\Delta t$ Sato et al. [41] and Kurths and Herzel [30] independently suggested in 1987 to consider

$$\bar{\lambda}_1 \approx \frac{E(k+l) - E(k)}{l\Delta t} = \frac{S(k+l) - S(k)}{l\Delta t} \tag{1.39}$$

and to identify a plateau in the graph $[E(k+l) - E(k)]$ vs. k (that should occur for the same range of k values where the linear segment occurs with the previous methods). This is basically a finite differences approximation of the slope of the graph $E(k)$ vs. $k\Delta t$. To obtain best results the time interval $l\Delta t$ should be large but $(l+k)\Delta t$ must not exceed the linear scaling region(s) where distances grow exponentially (and this range is in general not known a priori).

In 1985 Wolf et al. [51] suggested a method to estimate the largest Lyapunov exponent(s) which avoids the saturation of mutual distances of reference states and local neighbours due to nonlinear folding. The main idea is to monitor the distance between the reference orbit and the neighbouring orbit and to replace (once a threshold is exceeded) the neighbouring state by another neighbouring state that is closer to the reference orbit *and* which lies on or near the line from the current reference state to the last point of the previous neighbouring orbit in order to preserve the (local) direction corresponding to the largest Lyapunov exponent. Criteria for the replacement threshold and other details of the algorithm are given in [51], including a FORTRAN program. In principle, it is possible to use this strategy also for computing the second largest Lyapunov exponent [51], but this turns out to be quite difficult. When applied to stochastic time series the Wolf algorithm yields inconclusive results and may provide any value for the Lyapunov exponent depending on computational parameters and pre-filtering [9]. Due to its robustness the Wolf-algorithm is often used for the analysis of experimental data (see, for example, [16, 17]). A drawback of this method (similar to Jacobian based algorithms) is the fact that the user has no possibility to check whether exponential growth underlies the estimated values or not. Even if the amount of data available or the type and quality of the time series would not be sufficient to quantify exponential divergence the algorithm would provide a number that might be misinterpreted as the largest Lyapunov exponent of the underlying process. Therefore, we do not consider this method in more detail in the following.

1.4 Example Time Series

To illustrate and evaluate the direct method for estimating the largest Lyapunov exponent, time series generated by four different chaotic dynamical systems are used that will be introduced in the following subsections.

1.4.1 The Hénon Map

The first system is the *Hénon map* [21]

$$x_1(n+1) = 1 - ax_1^2(n) + bx_2(n) \qquad (1.40a)$$

$$x_2(n+1) = x_1(n) \qquad (1.40b)$$

with parameters $a = 1.4$ and $b = 0.3$. The Lyapunov exponents of this system are $\lambda_1 = 0.420$ and $\lambda_2 = -1.624$ (computed with the natural logarithm $\ln(\cdot)$, note that for the Hénon map $\lambda_1 + \lambda_2 = \ln(b) = -1.204$). In the following we shall assume that a x_1 time series of length $N_d = 4096$ is given.[2] A special feature of the Hénon map is that its original coordinates $(x_1(n), x_2(n))$ coincide with 2-dimensional delay coordinates $(x_1(n), x_1(n-1)) = (x_1(n), x_2(n))$. Figure 1.3a shows the Hénon attractor reconstructed from a clean $\{x_1(n)\}$ time series and in Fig. 1.3b a reconstruction is given based on a time series with additive measurement noise of signal-to-noise ration (SNR) of 30 dB (generated by adding normally distributed random numbers).

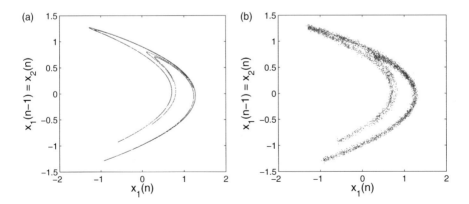

Fig. 1.3 Attractor of the Hénon map (1.40) (**a**) without noise and (**b**) with noise (SNR = 30 dB)

---

[2]Since $x_2(n+1) = x_1(n)$ any x_2 time series will give the same results.

1.4.2 The Folded-Towel Map

The second system is the *folded-towel map* introduced in 1979 by Rössler [39]

$$x(n + 1) = 3.8x(n)(1 - x(n)) - 0.05(y(n) + 0.35)(1 - 2z(n)) \quad (1.41a)$$

$$y(n + 1) = 0.1[(y(n) + 0.35)(1 - 2z(n)) - 1](1 - 1.9x(n)) \quad (1.41b)$$

$$z(n + 1) = 3.78z(n)(1 - z(n)) + 0.2y(n) \quad (1.41c)$$

which generates the chaotic attractor shown in Fig. 1.4. The folded-towel map has two positive Lyapunov exponents $\lambda_1 = 0.427$, $\lambda_2 = 0.378$, and a negative exponent $\lambda_3 = -3.30$. The Kaplan–Yorke dimension of this attractor equals $D_{KY} = 2.24$.

Figure 1.5 shows delay reconstructions based on x, y, and z time series of length $N_d = 65{,}536$ (i.e. 64k). In the first row (Fig. 1.5a–c) clean data are used while for the reconstructions shown in the second row (Fig. 1.5d–f) noisy data (64k) with signal-to-noise ratio (SNR) of 30 dB are used that were obtained by adding normally distributed random numbers to the clean data shown in the first row. These noisy time series will be used to evaluate the robustness of methods for estimating Lyapunov exponents.

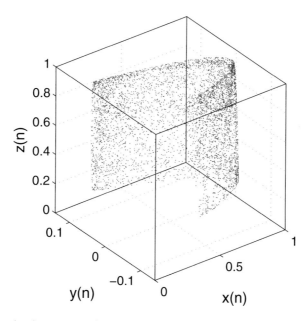

Fig. 1.4 Hyperchaotic attractor of the folded-towel map (1.41), original coordinates (x, y, z) (length $N_d = 65{,}536$)

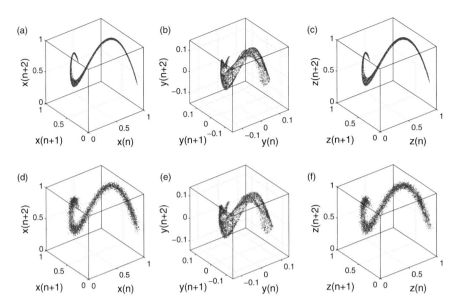

Fig. 1.5 Delay reconstructions of the attractor of the folded-towel map (1.41). (**a**)–(**c**) Without and (**d**)–(**f**) with additive (measurement) noise of SNR 30 dB. Reconstruction from (**a**), (**d**) $\{x(n)\}$ time series, (**b**), (**e**) $\{y(n)\}$ time series, and (**c**), (**f**) $\{z(n)\}$ time series

1.4.3 Lorenz-63 System

As an example of a low dimensional continuous time system we use the Lorenz-63 system [31] given by the following set of ordinary differential equations (ODEs)

$$\dot{x}_1 = \sigma(x_2 - x_1) \tag{1.42a}$$

$$\dot{x}_2 = x_1(R - x_3) - x_2 \tag{1.42b}$$

$$\dot{x}_3 = x_1 x_2 - b x_3. \tag{1.42c}$$

With parameter values $\sigma = 16$, $R = 45.92$, and $b = 4$ this systems generates a chaotic attractor with Lyapunov exponents $\lambda_1 = 1.51$, $\lambda_2 = 0$, and $\lambda_3 = -22.5$.

1.4.4 Lorenz-96 System

As an example of a continuous time system exhibiting complex dynamics we shall employ a 6-dimensional Lorenz-96 system [32] describing a ring of 1-dimensional dynamical elements. The differential equations for the model read

$$\frac{dx_i(t)}{dt} = x_{i-1}(t)(x_{i+1}(t) - x_{i-2}(t)) - x_i(t) + f \tag{1.43}$$

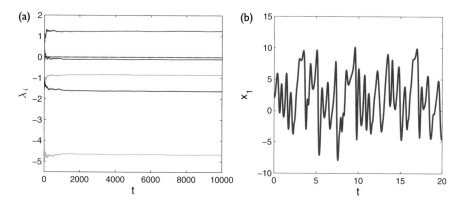

Fig. 1.6 (**a**) Convergence of Lyapunov exponents of the 6-dimensional Lorenz-96 system (1.43) generated with parameter value $f = 10$. (**b**) Typical oscillation

with $i = 1, 2, \ldots, 6$, $x_{-1}(t) = x_5(t), x_0(t) = x_6(t)$, and $x_7(t) = x_1(t)$. With a forcing parameter $f = 10$ the system generates a chaotic attractor characterized by a Lyapunov spectrum $\{1.249, 0.000, -0.098, -0.853, -1.629, -4.670\}$ and a resulting Kaplan–Yorke dimension of $D_{KY} = 4.18$. Figure 1.6a shows the convergence of the six Lyapunov exponents upon their computation using the full model equations (1.43) and in Fig. 1.6b, a typical time series of the Lorenz-96 system is plotted.

1.5   Estimation of Largest Lyapunov Exponents Using Direct Methods

1.5.1   Convergence of Small Perturbations

As illustrated in Fig. 1.2a any (random) perturbation first undergoes a transient phase I and converges to the direction of the Lyapunov vector(s) corresponding to the largest Lyapunov exponent. Then, in phase II, it grows linearly (on a logarithmic scale) until the perturbation exceeds the linear range (phase III). In the following we shall study this convergence process for the four example systems which were introduced in the previous section. The asymptotic average stretching of almost any initial perturbation (tangent vector) $\mathbf{z}(0)$ is given by the largest Lyapunov exponent λ_1. Using the tangent space basis $\{\mathbf{v}^{(1)}, \ldots, \mathbf{v}^{(m)}\}$ provided by the SVD (1.7) the initial tangent vector can be written as

$$\mathbf{z}(0) = \sum_{m=1}^{M} c_i \mathbf{v}^{(m)} = V \cdot \mathbf{c} \tag{1.44}$$

where $\mathbf{c} = (c_1, \ldots, c_M)$ denotes the vector of projection coefficients. Its temporal evolution is thus given by

$$\mathbf{z}(t) = D\phi^t(\mathbf{x}) \cdot \mathbf{z}(0) = U \cdot S \cdot V^{tr} \cdot \mathbf{z}(0) = U \cdot S \cdot \mathbf{c} = \sum_{m=1}^{M} c_m \sigma_m \mathbf{u}^{(m)}. \tag{1.45}$$

If we approximate the singular values σ_m by $e^{\lambda_m t}$ we obtain for (the square of) the Euclidean norm of $\mathbf{z}(t)$

$$Z^2(t) = \|\mathbf{z}(t)\|^2 = \sum_{m=1}^{M} c_m^2 e^{2\lambda_m t} \tag{1.46}$$

and this yields

$$\lim_{t \to \infty} \frac{1}{t} \ln\left(Z(t)\right) = \lim_{t \to \infty} \frac{1}{t} \ln\left(\|\mathbf{z}(t)\|\right) = \lim_{t \to \infty} \frac{1}{2} \ln\left(\sum_{m=1}^{M} c_m^2 e^{2\lambda_m t}\right)^{1/t} = \lambda_1 \tag{1.47}$$

because the term $e^{\lambda_m t}$ with the largest λ_m dominates the sum as time t goes to infinity. The speed of convergence depends on the full Lyapunov spectrum. Figure 1.6 shows $\ln(Z(t)) = \ln(\|\mathbf{z}(t)\|)$ (as defined in Eq. (1.47)) vs. t for the Hénon map (1.40), the folded towel map (1.41), the Lorenz-63 system (1.42), and the Lorenz-96 system (1.43). While the local slopes of the Hénon map and the folded towel map reach the value of the largest Lyapunov exponent after a period of time of about $t \approx 1$ the random initial tangent vectors $\mathbf{z}(0)$ of the Lorenz-96 system need about twice the time and converge to λ_1 only after $t \approx 2$ (Fig. 1.7).

1.5.2 Hénon Map

Figure 1.8 shows an application of the direct estimation method to a $\{x_1(n)\}$ time series of the Hénon map (1.40). The time series has a length of $N = 4096$ samples, the lag equals $L = 1$, and different reconstruction dimensions $D = 2$, $D = 4$, and $D = 6$ are used. Figure 1.8a shows $E_E(k)$ vs. $k\Delta t$ where $\Delta t = 1$ denotes the

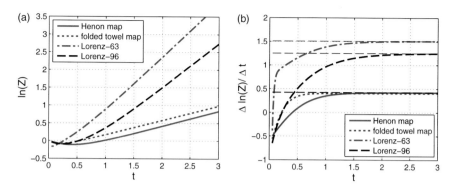

Fig. 1.7 (a) Average values of $\ln(Z(t)) = \ln(\|z(t)\|)$ vs. t (see Eq. (1.47)) for the Hénon map (*red*), the folded towel map (*blue, dotted line*), The Lorenz-63 system (*green, dashed-dotted line*) and the Lorenz-96 system (*dashed line*). The curves are computed by averaging 2000 realizations with randomly chosen initial vectors $z(0)$ with $\|z(0)\| = 1$. (b) Local slopes of curves shown in (a) indicating the convergence to the value of the corresponding largest Lyapunov exponent (given by *horizontal dashed lines*)

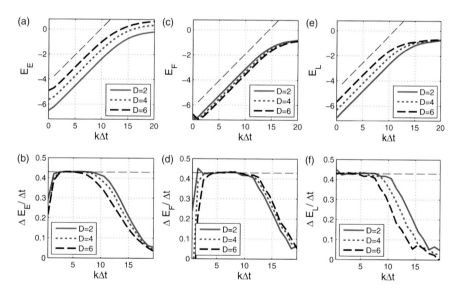

Fig. 1.8 Direct estimation of the largest Lyapunov exponent from a Hénon time series for different reconstruction dimensions $D = 2, D = 4, D = 6$ using a lag of $L = 1$. The diagrams (**a**), (**c**), and (**e**) show E_E, E_L and E_F vs. $k\Delta t$ with $\Delta t = 1$ for different measures of distance (1.28), (1.29), and (1.30). In (**b**), (**d**), and (**f**) the corresponding slopes $\Delta E/\Delta t$ vs. $k\Delta t$ (Eq. (1.48)) are shown. The *dashed lines* indicate the true result $\lambda_1 = 0.42$

sampling time and E_E (1.33) is computed with the Euclidean distance d_E (1.28). In Fig. 1.8b the slope

$$\frac{dE_E}{dt}((k + 0.5)\Delta t) \approx \frac{E_E(k + 1) - E_E(k)}{\Delta t} = \frac{\Delta E_E}{\Delta t} \tag{1.48}$$

is plotted. Figure 1.8c, d and e, f show the corresponding diagrams obtained with the distance measures d_F (1.30) and d_L (1.29), respectively. In all diagrams three phases occur (see also Fig. 1.2):

- First the difference vector $\mathbf{x}(m(n) + k) - \mathbf{x}(n + k)$ converges for increasing k to the subspace spanned by the first Lyapunov vector(s). The slope increases.
- Then the difference vector experiences the expansion rate given by the largest Lyapunov exponent. The slope is constant indicating exponential divergence.
- Finally, the lengths of the difference vector exceeds the range of the linearized dynamics and its length saturates due to nonlinear folding in the (reconstructed) state space. The slope decreases.

The lengths of the linear scaling regions in Fig. 1.8a, c, e and of the plateaus in Fig. 1.8b, d, f shrink with increasing embedding dimension. They also shrink, if the length of the time series is reduced or the number of nearest neighbours K is increased.

The results shown in Fig. 1.8 are computed by using each reconstructed state as a reference point. Reliable estimates may be obtained, however, already with a subset of reference points which reduces computation time almost linearly. This subset can be randomly selected from all reconstructed states or it can be chosen to include only those reconstructed states that possess the nearest neighbours. The latter choice has the advantage that more steps of the diverging neighbouring trajectory segments are governed by the linearised flow and exhibit exponential growths resulting in longer scaling regions. Figure 1.9 shows results based on those 25 % of the total number N of reference points that possess the closest neighbours (i.e., where the chosen distance measure d_E, d_F, or d_L (see Eqs. (1.28)–(1.30)) takes the smallest values). The scaling regions are extended compared to Fig. 1.8 but the local slopes plotted in Fig. 1.9c, d, f show more statistical fluctuations due to the smaller number of reference points (for $D = 6$ we have $N_r = 1018$ reference points in Fig. 1.9e, f compared to $N_r = 4070$ in Fig. 1.8e, f).

To illustrate the impact of (additive) measurement noise Fig. 1.10 shows results obtained with a noisy Hénon time series (compare Fig. 1.2b). As can be seen in all diagrams noise leads to shorter scaling intervals and a bias towards smaller values underestimating the largest Lyapunov exponent. Decreasing the number N_r of reference points (with nearest neighbours) reduces the bias but increases statistical fluctuations.

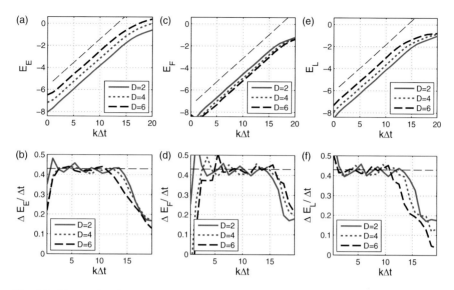

Fig. 1.9 Direct estimation of the largest Lyapunov exponent from a Hénon time series for different reconstruction dimensions $D = 2, D = 4, D = 6$ using a lag of $L = 1$. The diagrams (**a**), (**c**), and (**e**) show E_E, E_L and E_F vs. $k\Delta t$ with $\Delta t = 1$ for different measures of distance (1.28), (1.29), and (1.30). In (**b**), (**d**), and (**f**) the corresponding slopes $\Delta E/\Delta t$ vs. $k\Delta t$ (Eq. (1.48)) are shown. The *dashed lines* indicate the true result $\lambda_1 = 0.42$. In contrast to Fig. 1.8 only those 25 % of the reconstructed states with closest neighbours have been used as reference points

1.5.3 Folded Towel Map

To address the question whether the direct methods also work with hyper-chaotic dynamics we shall now analyze time series generated by the folded-towel map (1.41). Figure 1.11 shows results obtained from a $\{x(n)\}$ time series of length $N_d = 65{,}536$ using all N reconstructed states as reference points. As can be seen no linear scaling region exists, because this time series provides poor reconstructions of the underlying attractor (compare the reconstruction shown in Fig. 1.4a). Results can be improved by using a longer time series and only those reference points with very close neighbours. Alternatively, one may consider reconstructions based on a $\{y(n)\}$ time series which provide better unfolding of the chaotic attractor (compare Fig. 1.4b). Figure 1.12 shows results computed using a $\{y(n)\}$ time series from the folded-towel map (1.41) with length $N_d = 65{,}536$, where only 10 % of the reconstructed states (with closest neighbours) are used for estimating exponential divergence. As can be seen the $\{y(n)\}$ time series is more suited for estimating the largest Lyapunov exponent of the folded towel map and exhibits for reconstruction dimensions $D = 4$ and $D = 6$ the expected scaling behaviour. For $D = 2$ no clear scaling occurs and results differ significantly from those obtained with $D = 4$ and $D = 6$, because 2-dimensional delay coordinates are not

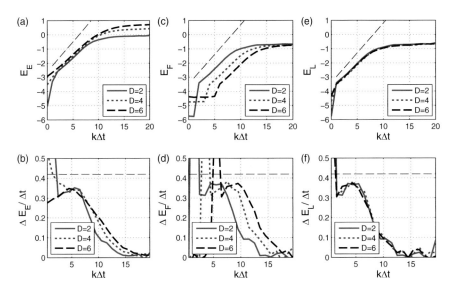

Fig. 1.10 Direct estimation of the largest Lyapunov exponent from a noisy Hénon time series (SNR 30 dB) for different reconstruction dimensions $D = 2$, $D = 4$, $D = 6$ using a lag of $L = 1$. The diagrams (**a**), (**c**), and (**e**) show E_E, E_L and E_F vs. $k\Delta t$ with $\Delta t = 1$ for different measures of distance (1.28), (1.29), and (1.30). In (**b**), (**d**), and (**f**) the corresponding slopes $\Delta E/\Delta t$ vs. $k\Delta t$ (Eq. (1.48)) are shown. All reconstructed states are used as reference points and the *dashed lines* indicate the true result $\lambda_1 = 0.42$

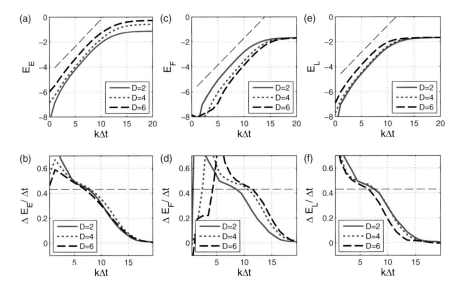

Fig. 1.11 Direct estimation of the largest Lyapunov exponent from a $\{x(n)\}$ time series of the folded-towel map of length $N_d = 65{,}536$ for different embedding dimensions $D = 2$, $D = 4$, $D = 6$ using a lag of $L = 1$. The diagrams (**a**), (**c**), and (**e**) show E vs. $k\Delta t$ with $\Delta t = 1$ for the Euclidean norm. In (**b**), (**d**), and (**f**) the corresponding slopes $\Delta E/\Delta t$ vs. k (Eq. (1.48)) are shown. The *dashed lines* indicate the true result $\lambda_1 = 0.43$

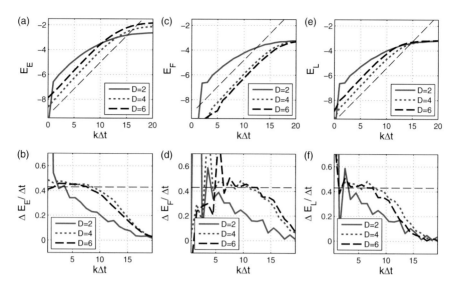

Fig. 1.12 Direct estimation of the largest Lyapunov exponent from a $\{y(n)\}$ time series of the folded-towel map of length $N_d = 65,536$ for different embedding dimensions $D = 2$, $D = 4$, $D = 6$ using a lag of $L = 1$. The diagrams (**a**), (**c**), and (**e**) show E vs. $k\Delta t$ with $\Delta t = 1$ for the Euclidean norm. In (**b**), (**d**), and (**f**) the corresponding slopes $\Delta E/\Delta t$ vs. k (Eq. (1.48)) are shown. The *dashed lines* indicate the true result $\lambda_1 = 0.43$. Only those 10 % of the reconstructed states possessing the most nearest neighbours are used as reference points for estimating exponential divergence ($N_r = 6551$ for $D = 6$)

sufficient for reconstructing this chaotic attractor (with Kaplan–Yorke dimension $D_{KY} = 2.24$).

Figure 1.13 shows results obtained with a noisy $\{y(n)\}$ time series (SNR 30 dB) generated by the folded-towel map (compare Fig. 1.4e) with reconstruction dimension $D = 4$, $D = 6$, and $D = 8$ and 10 % reference points. Scaling intervals are barely visible due to the added measurement noise.

1.5.4 Lorenz-63

We shall now use as data source the Lorenz-63 system which is an example of a low dimensional continuous system exhibiting deterministic chaos. Figure 1.14 shows results for a x_1 time series of length $N_d = 65,536$ sampled with $\Delta t = 0.025$ for reconstruction dimensions $D = 4$, $D = 12$, and $D = 21$ using a delay of $L = 1$. The resulting time windows $(D - 1)L$ covered by the delay vectors are 3, 11, and 20, respectively, where the latter corresponds to a typical oscillation period of the Lorenz-63 system. Here the sampling time $\Delta t = 0.025$ is much smaller compared to

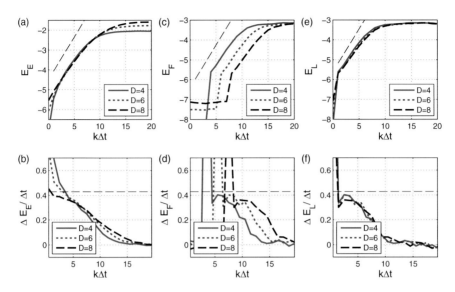

Fig. 1.13 Direct estimation of the largest Lyapunov exponent from a noisy $\{y(n)\}$ time series ($SNR = 30\,\text{dB}$) of the folded-towel map of length $N_d = 65{,}536$ for different embedding dimensions $D = 4$, $D = 6$, $D = 8$ using a lag of $L = 1$. The diagrams (**a**), (**c**), and (**e**) show E vs. $k\Delta t$ with $\Delta t = 1$ for the Euclidean norm. In (**b**), (**d**), and (**f**) the corresponding slopes $\Delta E/\Delta t$ vs. k (Eq. (1.48)) are shown. The *dashed lines* indicate the true result $\lambda_1 = 0.43$. Only 10 % of the reconstructed states with the smallest distances to their neighbours are used for estimating (exponential) growth rates

the iterated maps considered so far. To avoid strong fluctuations of the slope values the derivative $\Delta E/\Delta t$ is estimated by

$$\frac{\Delta E}{\Delta t}(t) \approx \frac{E(t + 3\Delta t) - E(t - 3\Delta t)}{6\Delta t} \qquad (1.49)$$

where E-values at $t \pm 3\Delta$ are used when estimating $\Delta E/\Delta t$ at time t. Note that the oscillations are less pronounced for higher reconstruction dimensions. Only 20 % of the reconstructed states are used as reference points (those which possess the closest neighbours). The linear scaling regions are clearly visible in the semi-logarithmic diagrams.

Figure 1.15 shows diagrams with reconstruction dimensions $D = 6$, $D = 11$, and $D = 21$ and corresponding lags $L = 4$, $L = 2$, and $L = 1$, respectively. In this case all reconstructed states represent the same windows in time with a length of $(D-1)L = 5 \cdot 4 = 10 \cdot 2 = 20 \cdot 1 = 20$ time steps of size $\Delta t = 0.025$, i.e. a period of time of length $20 \cdot 0.025 = 0.5$ which is close to the period of the natural oscillations of the Lorenz-63 system. The results for all three state space

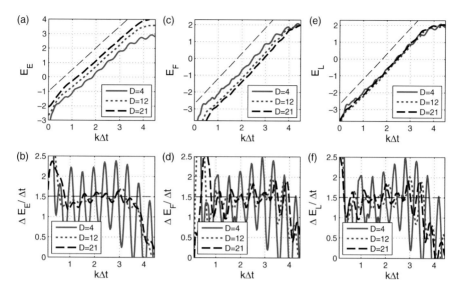

Fig. 1.14 Direct estimation of the largest Lyapunov exponent from a $\{x(n))\}$ time series of the Lorenz-63 system of length $N_d = 65{,}536$ for different reconstruction dimensions $D = 4$, $D = 12$, and $D = 21$, all with a lag of $L = 1$. As reference points only those 20 % of the reconstructed states are used that possess the nearest neighbours. The diagrams (**a**), (**c**), and (**e**) show E vs. $k\Delta t$ with $\Delta t = 0.025$ for the Euclidean norm. In (**b**), (**d**), and (**f**) the corresponding slopes $\Delta E/\Delta t$ vs. k (Eq. (1.48)) are shown. The *dashed lines* indicate the true result $\lambda_1 = 1.51$

reconstruction coincide very well and the amplitude of oscillations of the slope is much smaller compared to the results shown in Fig. 1.14.

1.5.5 Lorenz-96

Although it possesses only a single positive Lyapunov exponent the 6-dimensional Lorenz-96 systems turns out to be a surprisingly challenging case for estimating the largest Lyapunov exponent from time series. Figure 1.16 shows estimation results for time series of different lengths (first column: $N_d = 10{,}000$, second column $N_d = 100{,}000$, third column $N_d = 1{,}000{,}000$) and a different number of reference points given by those reconstructed states with closest neighbours (first row: 1 %, second row: 10 %). All examples employing 10 % of the reconstructed states as reference points provide diagrams where no suitable scaling region exists (even with $N_d = 1{,}000{,}000$ data points, see Fig. 1.16F, f). If only 1 % of the reconstructed states is used, the diagram based on $N_d = 100{,}000$ samples (Fig. 1.16B, b) gives a rough estimate of λ_1 and with $N_d = 1{,}000{,}000$ data points a linear scaling

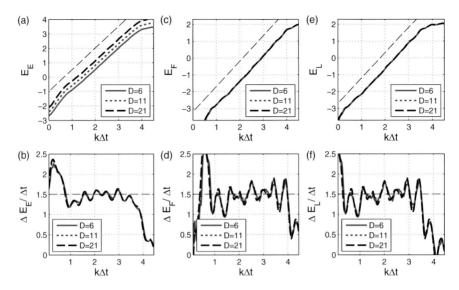

Fig. 1.15 Direct estimation of the largest Lyapunov exponent from a $\{x(n)\}$ time series of the Lorenz-63 system of length $N_d = 65{,}536$ for different reconstruction dimensions $D = 6$, $D = 11$, and $D = 21$, with lags $L = 4$, $L = 2$, and $L = 1$, respectively. As reference points only those 20 % of the reconstructed states are use that possess the nearest neighbours. The diagrams (**a**), (**c**), and (**e**) show E vs. $k\Delta t$ with $\Delta t = 0.025$ for the Euclidean norm. In (**b**), (**d**), and (**f**) the corresponding slopes $\Delta E / \Delta t$ vs. k (Eq. (1.48)) are shown. The *dashed lines* indicate the true result $\lambda_1 = 1.51$

regime (with the correct slope) is clearly visible in Fig. 1.16C, c. The reconstruction dimensions used here are $D = 9, 18$, and 36 with lags $L = 4, 2$, and 1, respectively, resulting in window lengths $8 \cdot 4 = 32$, $17 \cdot 2 = 34$, and $35 \cdot 1 = 35$. The slopes given in Fig. 1.16 were computed with Eq. (1.48) and only the case of the Euclidean norm E_E is shown here, because E_F and E_L show very similar results. The observation that a time series of length $N_d = 1{,}000{,}000$ (at least) is required to obtain reliable and correct results is consistent with the results of Eckmann and Ruelle [13] who estimated that the amount of required data points increases as a power of the attractor dimension. For comparison, the Kaplan–Yorke dimension of the Lorenz-96 attractor ($D_{KY} = 4.18$) is more than twice as large as the dimension of the Lorenz-63 model and so instead of 64k data a time series of length longer than $64^2\mathrm{k} = 4\mathrm{M}$ would be necessary to obtain comparable results.[3]

---

[3]This is just a rough estimate, because the choice of the sampling time Δt and the resulting distribution of reconstructed states on the attractor have also to be taken into account when estimating the required length of the time series.

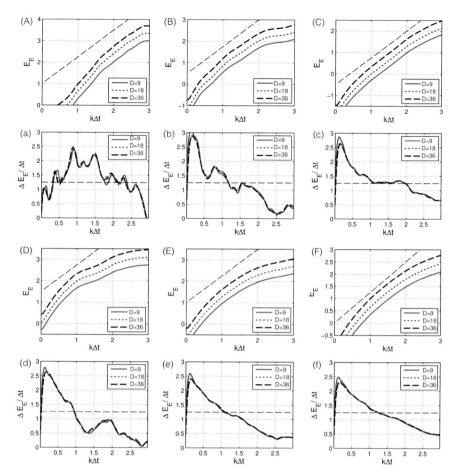

Fig. 1.16 Direct estimation of the largest Lyapunov exponent from a $\{x_1(n)\}$ time series of the Lorenz-96 system for different reconstruction dimensions $D = 9$, $D = 18$, and $D = 36$ with corresponding lags $L = 4$, $L = 2$, and $L = 1$, respectively. Diagrams (**A**)–(**F**) show E_E vs. $k\Delta t$ with $\Delta t = 0.025$ and diagrams (**a**)–(**f**) give the corresponding local slopes $\Delta E / \Delta t$ vs. k (Eq. (1.48)) (E_E is the error with respect to the Euclidean norm). The *dashed lines* indicate the true result $\lambda_1 = 1.249$. In diagrams (**A**)–(**C**), (**a**)–(**c**) only 1 % of the reconstructed states with the smallest distances to their neighbours are selected for estimating (exponential) growth rates, while in (**D**)–(**F**), (**d**)–(**f**) 10 % are used. Diagrams (**A**), (**a**) and (**D**), (**d**) are generated using $N_d = 10,000$ samples, figures (**B**), (**b**) and (**E**), (**e**) are computed from $N_d = 100,000$ data points, and diagrams (**C**), (**c**) and (**F**), (**f**) show results obtained from a time series of length $N_d = 1,000,000$

1.6   Conclusion

Estimating Lyapunov exponents from time series is a challenging task and since any algorithm provides "results" (i.e., numbers) some error control is very important to avoid misleading interpretation of the values obtained. From the variety of estimation methods, currently only the direct methods provide some feedback to the user whether local exponential divergence is properly identified or not. The presented examples included cases where this was *not* the case, due to:

(a) a too short time series (compared to the dimension of the underlying attractor), resulting in neighbouring reconstructed states whose distances exceed the range of validity of locally linearized dynamics, see for example Fig. 1.16A, B
(b) (measurement) noise, see for example Fig. 1.13, or
(c) an observable which is not suitable (to faithfully unfold the dynamics in reconstruction space), see for example Fig. 1.11.

This failure was in all cases directly visible in the semi-logarithmic diagrams showing the average growth of mutual distances of neighbouring states vs. time, where no linear scaling region could be identified. If, on the contrary, such a linear scaling region exists then it provides strong evidence for deterministic chaos and the estimated slope can be trusted to be a good estimate of the largest Lyapunov exponent. The choice of the norm for quantifying the divergence of trajectories turned out to be noncritical because all three norms used (E_E, E_F, and E_L, see Sect. 1.3.2) used exhibited equivalent performance.

A particular challenge are time series from high dimensional chaotic attractors. Eckmann and Ruelle [13] estimated that the number of required data points N_d exponentially grows with the attractor dimension D_a as $N_d \approx const^{D_a}$. The results obtained for the folded-towel map (Sect. 1.5.3) and the 6-dimensional Lorenz-96 model (Sect. 1.5.5) confirmed this ('pessimistic') prediction. Although the 6-dimensional Lorenz-96 model possesses a chaotic attractor with a single positive Lyapunov exponent it possesses a Kaplan–Yorke dimension of $D_{KY} = 4.18$. Due to this relatively high attractor dimension, satisfying estimates of the largest Lyapunov exponent were obtained only from very long time series (Fig. 1.16C) *and* if only those trajectory segments are used for estimating local divergence which started from very closely neighbouring reconstructed states (1 % in Fig. 1.16C). This selection of suitable reference points is very similar to a fixed size approach (see Sect. 1.3.2) using a relatively small radius ϵ and the results obtained for the folded-towel map and the Lorenz-96 model indicate its importance for coping with high dimensional chaos. On the other hand, these examples clearly show that data requirements (and practical difficulties) increase exponentially with the dimension of the underlying attractor (at least for the direct estimation methods employed here) and this fact imposes fundamental bounds for estimating Lyapunov exponents from time series generated by processes of medium or even high complexity.

Acknowledgements Inspiring scientific discussions with S. Luther and all members of the Biomedical Physics Research Group and financial support from the German Federal Ministry of Education and Research (BMBF) (project FKZ 031A147, GO-Bio) and the German Research Foundation (DFG) (Collaborative Research Centre SFB 937 Project A18) are gratefully acknowledged.

Appendix

Let $\varphi^k : \mathbb{R}^D \rightarrow \mathbb{R}^D$ be the induced flow in reconstruction space mapping reconstructed states $\mathbf{x}_n = (s_n, s_{n+L}, \ldots, s_{n+(D-1)L})$ to their future values $\varphi^k(\mathbf{x}_n) = \mathbf{x}_{n+k}$. To estimate the $D \times D$ Jacobian matrices $D\varphi^k(\mathbf{x})$ from the temporal evolution of the reconstructed states $\{\mathbf{x}_n\}_{n=1}^N$ the flow φ^k has to be approximated by a general ansatz (black-box model) like a neural network [20] or a superposition of I basis functions $b_i : \mathbb{R}^D \rightarrow \mathbb{R}$ providing an approximating function

$$\psi^k(\mathbf{x}) = \left(\psi_1^k(\mathbf{x}), \ldots, \psi_D^k(\mathbf{x})\right) = \left(\sum_{i=1}^I c_{i1} b_i(\mathbf{x}), \ldots, \sum_{i=1}^I c_{iD} b_i(\mathbf{x})\right) = \mathbf{b}(\mathbf{x}) \cdot C$$

(1.50)

where $C = (c_{ij})$ denotes a $I \times D$ matrix of coefficients with columns $\mathbf{c}^{(j)}$ ($j = 1, \ldots, D$) that have to be estimated and $\mathbf{b}(\mathbf{x}) = (b_1(\mathbf{x}), \ldots, b_I(\mathbf{x}))$ is a row vector consisting of the values of all basis functions evaluated at the state $\mathbf{x}$.[4]

For the special choice $k = L$ (evolution time step equals the lag of the delay coordinates) the first $D - 1$ components of the map $\psi^k(\mathbf{x}_n)$ are known (due to the delay reconstruction) and only for the last component an approximation is required

$$\psi^L(\mathbf{x}) = \left(s_{n+L}, s_{n+2L}, \ldots, s_{n+(D-1)L}, \sum_{i=1}^I c_i b_i(\mathbf{x})\right)$$

(1.51)

$$= (x_{n2}, \ldots, x_{nD}, \mathbf{b}(\mathbf{x}) \cdot \mathbf{c}).$$

(1.52)

With this notation the approximation $D\psi^k(\mathbf{x})$ of the desired Jacobian matrix $D\varphi^k(\mathbf{x})$ of the (induced) flow $\varphi(\mathbf{x})$ in embedding space can be written as

$$D\psi^k(\mathbf{x}) = \begin{pmatrix} \frac{\partial b_1}{\partial x_1} & \cdots & \frac{\partial b_I}{\partial x_1} \\ \vdots & \ddots & \vdots \\ \frac{\partial b_1}{\partial x_D} & \cdots & \frac{\partial b_I}{\partial x_D} \end{pmatrix} \cdot C = G \cdot C$$

(1.53)

where G will be called *derivative matrix* in the following.

For $k = L$ the Jacobian matrix of the approximating function ψ^k is given as

$$D\psi^L(\mathbf{x}) = \begin{pmatrix} 0 & 1 & 0 & \cdots & 0 \\ 0 & 0 & 1 & \cdots & 0 \\ \vdots & \vdots & \vdots & \ddots & \vdots \\ 0 & 0 & 0 & \cdots & 1 \\ \sum_{i=1}^I c_i \frac{\partial b_i}{\partial x_1} & \cdots & \cdots & \cdots & \sum_{i=1}^I c_i \frac{\partial b_i}{\partial x_D} \end{pmatrix}$$

(1.54)

---

[4]The matrix C and its column vectors $\mathbf{c}^{(j)}$ depend on the time step k. To avoid clumsy notation this dependance is not explicitly indicated.

Linear basis functions $b_i(\mathbf{x})$ can be used to model the (linearized) flow (very) close to the reference points $\mathbf{x}_n$ along the orbits. To approximate the flow in a larger neighbourhood of $\mathbf{x}_n$ or even globally, nonlinear basis functions are required, like multidimensional polynomials [2, 4–7], or radial basis functions [23, 24, 35].

To estimate the coefficient matrix C in Eq. (1.50) or the coefficient vector $\mathbf{c}$ in Eq. (1.51) we select a set of representative states $\{\mathbf{z}^j\}$ whose temporal evolution $\varphi^k(\mathbf{z}^j)$ is known. For local modeling this set of states consists of nearest neighbours $\{\mathbf{x}_{m(n)} : m(n) \in \mathcal{U}_n\}$ of the reference point $\mathbf{x}_n$ where $\mathcal{U}_n$ defines the chosen neighbourhood that can be of *fixed mass* (a fixed number K of nearest neighbours of $\mathbf{x}_n$) or of *fixed size* (all points with distance smaller than a given bound ϵ). For global modeling of the flow the set $\{\mathbf{z}^j\}$ is usually a (randomly sampled) subset of all reconstructed states. Let

$$Y = \begin{pmatrix} \varphi_1^k(\mathbf{z}^1) \ldots \varphi_D^k(\mathbf{z}^1) \\ \vdots \quad \vdots \quad \vdots \\ \varphi_1^k(\mathbf{z}^J) \ldots \varphi_D^k(\mathbf{z}^J) \end{pmatrix} \tag{1.55}$$

be a $J \times D$ matrix whose rows are components the (known) future values $\varphi^k(\mathbf{z}^j)$ of the J states $\{\mathbf{z}^j\}$ and let

$$B = \begin{pmatrix} b_1(\mathbf{z}^1) \ldots b_I(\mathbf{z}^1) \\ \vdots \quad \vdots \quad \vdots \\ b_1(\mathbf{z}^J) \ldots b_I(\mathbf{z}^J) \end{pmatrix} \tag{1.56}$$

be the $J \times I$ (design) matrix [37] whose rows are the basis functions $b_i(\cdot)$ evaluated at the selected states $\{\mathbf{z}^j\}$. Using this notation the approximation task can be stated as a minimization problem with a cost function

$$g(\mathbf{c}^{(j)}) = \|B \cdot \mathbf{c}^{(j)} - \mathbf{y}^{(j)}\|^2 \tag{1.57}$$

where $\mathbf{y}^{(j)}$ denotes the j-th column of the matrix Y (given in Eq. (1.55)), or

$$g(C) = \|B \cdot C - Y\|_F^2 \tag{1.58}$$

where $\|\cdot\|_F$ = denotes the *Frobenius matrix norm* (also called *Schur norm*).

The solution of this optimization problem may suffer from the fact that typically the states $\{\mathbf{z}^j\}$ cover only some subspace of the reconstructed state space. Therefore, in particular for local modeling ill-posed optimization problems may occur with many almost equivalent solutions. For estimating Lyapunov exponents we prefer to select solutions for the coefficient matrix C that provide partial derivatives (elements of the Jacobian matrix) with small magnitudes, because in this way spurious Lyapunov exponents are shifted towards $-\infty$. This goal can be achieved by *Tikhonov–Philips regularization* where the cost function of the optimization problem (1.58) is extended by a term $\rho\|A \cdot C\|$ resulting in

$$g(\mathbf{c}^{(j)}) = \|B \cdot \mathbf{c}^{(j)} - \mathbf{y}^{(j)}\|^2 + \rho^2\|A \cdot \mathbf{c}^{(j)}\|^2 \tag{1.59}$$

where A denotes a so-called stabilizer matrix and $\rho \in \mathbb{R}$ is the *regularization parameter* that is used to control the impact of the regularization term on the solution of the minimization problem. If the identity matrix is used as stabilizer $A = I$ then $\|\mathbf{c}(j)\|$ is minimized and the solution with the smallest coefficients is selected (also called Tikhonov stabilization). Another possible choice is the derivative matrix (1.53) $A = G$. In this case we minimize the sum of all squared singular values σ_i of $D\psi^k(\mathbf{x}) = U \cdot S \cdot V^{tr}$, because

$$\|G \cdot C\|_F^2 = \|D\psi^k(\mathbf{x})\|_F^2 = \text{trace}\left([D\psi^k(\mathbf{x})]^{tr} \cdot D\psi^k(\mathbf{x})\right) \tag{1.60}$$

$$= \text{trace}\left(V \cdot S^2 \cdot V^{tr}\right) = \text{trace}(S^2) = \sum_{i=1}^{D} \sigma_i^2 \tag{1.61}$$

and so we minimize Lyapunov exponents by maximizing contraction rates.

To solve the optimization problem (1.59) we rewrite it as an augmented least squares problem with a cost function

$$g(\mathbf{c}^{(j)}) = \|\begin{pmatrix} B \\ \rho A \end{pmatrix} \cdot \mathbf{c}^{(j)} - \begin{pmatrix} \mathbf{y}^{(j)} \\ 0 \end{pmatrix}\|^2 = \|\hat{B} \cdot \mathbf{c}^{(j)} - \hat{\mathbf{y}}^{(j)}\|^2 \tag{1.62}$$

that can be minimized by a solution of the corresponding normal equations

$$\left(B^{tr} \cdot B + \rho^2 A^{tr} \cdot A\right) \cdot \mathbf{c}^{(j)} = B^{tr} \cdot \mathbf{y}^{(j)} \tag{1.63}$$

using a sequence of Householder transformations [35] or by employing the singular value decomposition of the matrix $\hat{B} = U_{\hat{B}} \cdot S_{\hat{B}} \cdot V_{\hat{B}}^{tr}$ providing the minimal solution [37]

$$\mathbf{c}^{(j)} = V_{\hat{B}} \cdot S_{\hat{B}}^{-1} \cdot U_{\hat{B}}^{tr} \cdot \hat{\mathbf{y}}^{(j)}. \tag{1.64}$$

for each column $\hat{\mathbf{c}}^{(j)}$ or

$$C = V_{\hat{B}} \cdot S_{\hat{B}}^{-1} \cdot U_{\hat{B}}^{tr} \cdot \hat{Y} \tag{1.65}$$

for the full coefficient matrix C where $\hat{Y} = \begin{pmatrix} Y \\ 0 \end{pmatrix}$.

For the stabilizer $A = I$ the elements of the diagonal matrix $S_{\hat{B}}^{-1}$ are given by

$$\frac{\hat{\sigma}_i}{\hat{\sigma}_i^2 + \rho^2} \tag{1.66}$$

where $\hat{\sigma}_i$ are the diagonal elements of $S_{\hat{B}}$ (i.e., the singular values of $\hat{B}$).

References

1. Abarbanel, H.D.I.: Analysis of Observed Chaotic Data. Springer, New York (1996)
2. Abarbanel, H.D.I., Brown, R., Kennel, M.B.: Lyapunov exponents in chaotic systems: their importance and their evaluation using observed data. Int. J. Mod. Phys. B **5**, 1347–1375 (1991)
3. Benettin, G., Galgani, L., Giorgilli, A., Strelcyn, J.M.: Lyapunov characteristic exponents for smooth dynamical systems and for hamiltonian systems; a method for computing all of them. Part II: Numerical application. Meccanica **15**, 21–30 (1980)
4. Briggs, K.: An improved method for estimating Liapunov exponents of chaotic time series. Phys. Lett. A **151**, 27–32 (1990)
5. Brown, R.: Calculating Lyapunov exponents for short and/or noisy data sets. Phys. Rev. E **47**(6), 3962–3969 (1993)
6. Brown, R., Bryant, P., Abarbanel, H.D.I.: Computing the Lyapunov spectrum of a dynamical system from an observed time series. Phys. Rev. A **43**, 2787–2806 (1991)
7. Bryant, P., Brown, R., Abarbanel, H.D.I.: Lyapunov exponents from observed time series. Phys. Rev. Lett. **65**, 1523–1526 (1990)
8. Čenys, A.: Lyapunov spectrum of the maps generating identical attractors. Europhys. Lett. **21**(4), 407–411 (1993)
9. Dämmig, M., Mitschke, F.: Estimation of Lyapunov exponents from time series: the stochastic case. Phys. Lett. A **178**, 385–394 (1993)
10. Darbyshire, A.G., Broomhead, D.S.: Robust estimation of tangent maps and Liapunov spectra. Physica D **89**(3–4), 287–305 (1996)
11. Dechert, W.D., Gençay, R.: The topological invariance of Lyapunov exponents in embedded dynamics. Physica D **90**, 40–55 (1996)
12. Eckmann, J.-P., Ruelle, D.: Ergodic theory of chaos and strange attractors. Rev. Mod. Phys. **57**, 617–656 (1985)
13. Eckmann, J.-P., Ruelle, D.: Fundamental limitations for estimating dimensions and Lyapunov exponents in dynamical systems. Physica D **56**, 185–187 (1992)
14. Eckmann, J.-P., Kamphorst, S.O., Ruelle, D., Ciliberto, S.: Lyapunov exponents from time series. Phys. Rev. A **34**, 4971–4979 (1986)
15. Ellner, S., Gallant, A.R., McCaffrey, D., Nychka, D.: Convergence rates and data requirements for Jacobian-based estimates of Lyapunov exponents from data. Phys. Lett. A **153**, 357–363 (1991)
16. Fell, J., Beckmann, P.: Resonance-like phenomena in Lyapunov calculations from data reconstructed by the time-delay method. Phys. Lett. A **190**, 172–176 (1994)
17. Fell, J., Röschke, j., Beckmann, P.: Deterministic chaos and the first positive Lyapunov exponent: a nonlinear analysis of the human electroencephalogram during sleep. Biol. Cybern. **69**, 139–146 (1993)
18. Gao, J., Zheng, Z.: Local exponential divergence plot and optimal embedding of a chaotic time series. Phys. Lett. A **181**, 153–158 (1993)
19. Geist, K., Parlitz, U., Lauterborn, W.: Comparison of different methods for computing Lyapunov exponents. Prog. Theor. Phys. **83**, 875–893 (1980)

20. Gencay, R., Dechert, W.D.: An algorithm for the n Lyapunov exponents of an n-dimensional unknown dynamical system. Physica D **59**, 142–157 (1992)
21. Hénon, M.: A two-dimensional mapping with a strange attractor. Commun. Math. Phys. **50**(1), 69–77 (1976)
22. Holzfuss, J., Lauterborn, W.: Liapunov exponents from a time series of acoustic chaos. Phys. Rev. A **39**, 2146–2152 (1989)
23. Holzfuss, J., Parlitz, U.: Lyapunov exponents from time series. In: Arnold, L., Crauel, H., Eckmann, J.-P. (eds.) Proceedings of the Conference *Lyapunov Exponents*, Oberwolfach 1990. Lecture Notes in Mathematics, vol. 1486, pp. 263–270. Springer, Berlin (1991)
24. Kadtke, J.B., Brush, J., Holzfuss, J.: Global dynamical equations and Lyapunov exponents from noisy chaotic time series. Int. J. Bifurcat. Chaos **3**, 607–616 (1993)
25. Kantz, H.: A robust method to estimate the maximal Lyapunov exponent of a time series. Phys. Lett. A **185**, 77–87 (1994)
26. Kantz, H., Schreiber, T.: Nonlinear Time Series Analysis. Cambridge University Press, Cambridge (2004)
27. Kantz, H., Radons, G., Yang, H.: The problem of spurious Lyapunov exponents in time series analysis and its solution by covariant Lyapunov vectors. J. Phys. A: Math. Theor. **46**, 254009 (2013)
28. Kostelich, E.: Bootstrap estimates of chaotic dynamics. Phys. Rev. E **64**, 016213 (2001)
29. Kruel, Th.M., Eiswirth, M., Schneider, F.W.: Computation of Lyapunov spectra: effect of interactive noise and application to a chemical oscillator. Physica D **63**, 117–137 (1993)
30. Kurths, J., Herzel, H.: An attractor in solar time series. Physica D **25**, 165–172 (1987)
31. Lorenz, E.N.: Deterministic nonperiodic flow. J. Atmos. Sci. **20**(2), 130–141 (1963)
32. Lorenz, E.N.: Predictability a problem partly solved. In: Proceedings of the Seminar on Predictability, vol. 1, pp. 1–18. ECMWF, Reading (1996)
33. Oseledec, V.I.: A multiplicative ergodic theorem. Lyapunov characteristic numbers for dynamical systems. Trans. Moscow Math. Soc. **19**, 197–231 (1968)
34. Ott, E.: Chaos in Dynamical Systems. Cambridge University Press, Cambridge (1993)
35. Parlitz, U.: Identification of true and spurious Lyapunov exponents from time series. Int. J. Bifurcat. Chaos **2**, 155–165 (1992)
36. Parlitz, U.: Lyapunov exponents from Chua's circuit. J. Circuits Syst. Comput. **3**, 507–523 (1993)
37. Press, W.H., Teukolsky, S.A., Vetterling, W.T., Flannery, B.P.: Numerical Recipes: The Art of Scientific Computing, 3rd edn. Cambridge University Press, Cambridge (2007)
38. Rosenstein, M.T., Collins, J.J., de Luca, C.J.: A practical method for calculating largest Lyapunov exponents from small data sets. Physica D **65**, 117–134 (1993)
39. Rössler, O.E.: An equation for hyperchaos. Phys. Lett. A **71**, 155–157 (1979)
40. Sano, M., Sawada, Y.: Measurement of the Lyapunov spectrum from a chaotic time series. Phys. Rev. Lett. **55**, 1082–1085 (1985)
41. Sato, S., Sano, M., Sawada Y.: Practical methods of measuring the generalized dimension and largest Lyapunov exponent in high dimensional chaotic systems. Prog. Theor. Phys. **77**, 1–5 (1987)
42. Sauer, T., Yorke, J.A.: How many delay coordinates do you need? Int. J. Bifurcat. Chaos **3**, 737–744 (1993)
43. Sauer, T., Yorke, J., Casdagli, M.: Embedology. J. Stat. Phys. **65**, 579–616 (1991)
44. Sauer, T.D., Tempkin, J.A., Yorke, J.A.: Spurious Lyapunov exponents in attractor reconstruction. Phys. Rev. Lett. **81**, 4341–4344 (1998)
45. Shimada, I., Nagashima, T.: A numerical approach to ergodic problems of dissipative dynamical systems. Prog. Theor. Phys. **61**, 1605–1616 (1979)
46. Skokos, Ch.: The Lyapunov characteristic exponents and their computation. In: Lecture Notes in Physics, vol. 790, pp. 63–135. Springer, Berlin (2010)
47. Stoop, R., Meier, P.F.: Evaluation of Lyapunov exponents and scaling functions from time series. J. Opt. Soc. Am. B **5**, 1037–1045 (1988)

48. Stoop, R., Parisi, J.: Calculation of Lyapunov exponents avoiding spurious elements. Physica D **50**, 89–94 (1991)
49. Takens, F.: Detecting strange attractors in turbulence. In: Lecture Notes in Mathematics, vol. 898, pp. 366–381. Springer, Berlin (1981)
50. Theiler, J.: Estimating fractal dimension. J. Opt. Soc. Am. A **7**, 1055–1073 (1990)
51. Wolf, A., Swift, J.B., Swinney, L., Vastano, J.A.: Determining Lyapunov exponents from a time series. Physica D **16**, 285–317 (1985)
52. Yang, H.-L., Radons, G., Kantz, H.: Covariant Lyapunov vectors from reconstructed dynamics: the geometry behind true and spurious Lyapunov exponents. Phys. Rev. Lett. **109**(24), 244101 (2012)
53. Zeng, X., Eykholt, R., Pielke, R.A.: Estimating the Lyapunov-exponent spectrum from short time series of low precision. Phys. Rev. Lett. **66**, 3229–3232 (1991)
54. Zeng, X., Pielke, R.A., Eykholt, R.: Extracting Lyapunov exponents from short time series of low precision. Mod. Phys. Lett. B **6**, 55–75 (1992)

Chapter 2
Theory and Applications of the Fast Lyapunov Indicator (FLI) Method

Elena Lega, Massimiliano Guzzo, and Claude Froeschlé

Abstract In the last 20 years numerical experiments have allowed to study dynamical systems in a new way providing interesting results. The development of tools for the detection of regular and chaotic orbits has been one of the key points to access the global properties of dynamical systems. In many cases the visualization of suitably chosen sections of the phase space has been determinant for the comprehension of the fascinating and complex interplay between order and chaos. The Fast Lyapunov Indicator introduced in Froeschlé et al. (Celest Mech Dyn Astron 67:41–62, 1997) and further developed in Guzzo et al. (Physica D 163(1–2):1–25, 2002), is an easy to implement and sensitive tool for the detection of order and chaos in dynamical systems. Closely related to the computation of the Largest Lyapunov Exponent, the Fast Lyapunov Indicator relies on the idea that the computation of tangent vectors contains a lot of information even on short integration times, while for the Largest Lyapunov Indicator large integration times are required in order to accurately approximate a limit value. The aim of this Chapter is to provide the definition of the Fast Lyapunov indicator and some simple examples of applications for readers that would like to implement and use the indicator for the first time. We associate to each example of application the references to more specific papers that we have published during these years.

E. Lega (✉)
Université de Nice Sophia Antipolis, CNRS UMR 7293, Observatoire de la Côte d'Azur, Bv. de l'Observatoire, CS 34229, 06304 Nice cedex 4, France
e-mail: elena@oca.eu

M. Guzzo
Dipartimento di Matematica, Università degli studi di Padova, Via Trieste, 63 - 35121 Padova, Italy
e-mail: guzzo@math.unipd.it

C. Froeschlé
Université de Nice Sophia Antipolis, CNRS UMR 7293, Observatoire de la Côte d'Azur, Bv. de l'Observatoire, B.P. 4229, 06304 Nice cedex 4, France
e-mail: claude@gmail.com

© Springer-Verlag Berlin Heidelberg 2016
Ch. Skokos et al. (eds.), *Chaos Detection and Predictability*, Lecture Notes in Physics 915, DOI 10.1007/978-3-662-48410-4_2

2.1 Introduction

The study of the interplay between order and chaos is one of the keys for under-standing the behaviour of complex systems. Since the pioneering work of Hénon and Heiles [27] the use of numerical simulations together with the development of different tools for the detection of chaos has provided interesting results in different domains of physics (celestial mechanics, particle accelerators, dynamical astronomy, statistical physics, plasma physics).

In their study, Hénon and Heiles, searching for the existence of a third integral of motion in a galactic potential, were surprised by finding that ordered and chaotic motions co-existed for some values of the total energy of the system. As usual in numerical experiments the authors searched for eventual numerical errors. Listening to a seminar by Arnold about new theoretical results on stability of quasi-integrable Hamiltonian systems (the nowadays celebrated KAM theorem [1, 28, 43]), M. Hénon understood that order and chaos are complementary rather than antagonist dynamical behaviours and got convinced on the numerical results he had obtained with C. Heiles in their study of the galactic potential. Since then, numerical experiments have become a sort of laboratory, very often used to extend the domain of validity of theorems and therefore showing their interest for physical problems.

As an example, we consider the problem of the long term stability properties of a dynamical system, problem of particular interest in the domain of celestial mechanics. During the last decade the numerical detection of the resonances of a system using dynamical indicators has been one of the major tools for studying the long-term stability in the specific case of celestial mechanics (for recent examples, see [15, 16, 29, 42, 49–51, 55]). The reason is that many problems of interest for celestial mechanics can be studied with KAM [1, 28, 43] and Nekhoroshev theorems [48]. For small values of the perturbation parameters the KAM theorem leaves the possibility of large instabilities only on a peculiar subset of the phase space, the so-called Arnold web. According to the Nekhoroshev theorem, on the Arnold web the diffusion times are expected to increase at least exponentially with an inverse power of the norm of the perturbation. This phenomenon of extremely slow diffusion was introduced by Arnold [2] on an ad-hoc model well suited to the mathematical demonstration rather than for numerical experiments.

We recall that, for many years, researchers were convinced that Arnold's diffusion could not be detected numerically, and therefore, in some sense, the phenomenon was not interesting for the study of physical systems.

The Fast Lyapunov Indicator (FLI hereafter), introduced in [12] and further developed in [21], is an easy (to implement) and sensitive tool for the detection of the Arnold web of a system. The FLI method was first tested by comparing results obtained with other chaos indicators. A detailed comparison with the frequency analysis application on two and four dimensional mapping [30, 31] can be found in [10, 32]. The comparison with other chaos indicators was presented in [33]. Without entering in the details, we can say that the FLI belongs to the class of the so called finite time chaos indicators (such as the Finite Time Lyapunov Exponent [52], the MEGNO [8, 9] as well as OFLI and OFLI2 [3]) which are able to discriminate between regular orbits and chaotic orbits on times significantly smaller than the time required for a reliable estimation of the largest characteristic Lyapunov exponent or of the frequency.

The detailed detection of the resonances obtained with the FLI on models which satisfy the hypothesis of both KAM and Nekhoroshev theorems allowed us to measure directly the quantitative features of the Arnold's diffusion [14, 18, 22–26, 35, 36, 53] showing its interest for physical systems. Later in [17, 24, 37–40] we have used the FLI for the detection of the stable and unstable manifolds. More recently the FLI has been applied to the planar circular restricted three body problem for the detection and characterization of close encounters and resonances [19, 41]; more precisely, we have formulated the FLI method using the Levi–Civita regularization in order to handle the singularity of the gravitational potential.

The majority of our studies concern conservative systems, however, we have used the FLI for studying the dynamics of dissipative systems in [5–7]; more recently we have provided an application of the FLI to track the diffusion of orbits of a quasi integrable Hamiltonian system perturbed with a very small non-Hamiltonian perturbation [18].

In this Chapter, rather than providing a review of the results obtained with the FLI, we present the indicator for readers that would like to implement it for the first time. At this purpose we provide in Sect. 2.2 the definition and use of the FLI on a simple 2-dimensional discrete model: the standard map. On this model we try to answer to some frequently asked questions about the implementation and use of the method. In Sect. 2.3 we show the use of the FLI for the computation of the stable and unstable manifolds. In Sect. 2.4 we provide an application on a generic Hamiltonian model. In Sect. 2.5 we show an application of the FLI for the detection of the resonances of a quasi-integrable Hamiltonian system with 3 degrees of freedom and we show how to use the FLI to follow the diffusion of orbits along resonant lines. Conclusions are provided in Sect. 2.6.

2.2   Definition of the Fast Lyapunov Indicator

Given a set of differential equations:

$$\frac{dx}{dt} = F(x) \ , \ x = (x_1, x_2, ...x_n) \tag{2.1}$$

for any solution $x(t)$ with initial condition $x(0)$ the evolution $v(t)$ of any tangent vector with initial value $v(0)$ is obtained by integrating the variational equations:

$$\begin{cases} \frac{dx}{dt} = F(x) \\ \frac{dv}{dt} = \frac{\partial F}{\partial x} v. \end{cases} \tag{2.2}$$

If instead one considers the discrete-time dynamics defined by the map

$$x(t + 1) = \psi(x(t)), \tag{2.3}$$

the evolution of the tangent vector is defined by:

$$\begin{cases} x(t + 1) = & \psi(x(t)) \\ v(t + 1) = & \frac{\partial \psi}{\partial x}(x(t))v(t). \end{cases} \tag{2.4}$$

With this setting, for both systems (2.1) and (2.3), the simplest definition of the fast Lyapunov indicator of a point $x(0)$ and of a tangent vector $v(0)$, at time t, is:

$$\text{FLI}_t(x(0), v(0)) = \log \frac{\|v(t)\|}{\|v(0)\|} . \tag{2.5}$$

The FLI is defined in such a way that, unless $v(0)$ belongs to some lower dimensional linear spaces, the quantity $\text{FLI}_t(x(0), v(0))/t$ tends to the largest Lyapunov exponent as t goes to infinity. If Eq. (2.1) is Hamiltonian and if the motion is regular (except for some peculiar hyperbolic structures, such as whiskered tori) then the largest Lyapunov exponent is zero, otherwise it is positive. This property has been largely used to discriminate between chaotic and ordered motions. However, among regular motions the Lyapunov exponent does not distinguish between circulation and libration orbits. In contrast, the FLI distinguishes between them ([13, 34], see Sect. 2.1).

Therefore, the computation of $\text{FLI}_t(x, v)$ on grids of initial conditions x and for the same fixed tangent vector v allows one to detect the distribution of invariant tori and resonances (i.e. circulation and libration orbits) in relatively short CPU times [11, 13].

We remark that the FLI depends parametrically on the initial vector $v(0)$ and on the integration time t. A frequently asked question concerns the choice of $v(0)$ for the practical implementation of the method. As for the computation of the largest Lyapunov exponent, one has in principle to avoid special choices of $v(0)$. In order to reduce the dependence of the computation on the choice of the initial tangent vector we suggested in [18] to compute the average (or alternatively the maximum) of the FLIs obtained for an orthonormal basis of tangent vectors. It happens that any orthonormal basis is suitable to detect the dynamics of the system. A second frequently asked question concerns the choice of the integration time t. We answer to both questions in the following using the standard map as a model problem.

2.2.1 The Standard Map as a Model Problem

We consider as a model problem the two-dimensional standard map, whose phase space variables are denoted by $(I, \varphi) \in \mathbb{R} \times \mathbb{S}^1$, and whose dynamics is defined by

$$(I(t + 1), \varphi(t + 1)) = \psi(I(t), \varphi(t)) \tag{2.6}$$

with

$$\psi(I, \varphi) = (I + \epsilon \sin(\varphi + I), \varphi + I),$$

and ϵ is a parameter. As it is well known, the map has interesting dynamics for $\epsilon \neq 0$. As an example, in Fig. 2.1, we report the phase-portrait of the map for $\epsilon = 0.3$: we can appreciate the presence of invariant curves, as well as of a small chaotic zone around the hyperbolic fixed point $(0, 0)$. We select three initial conditions on the phase-portrait of Fig. 2.1, according to different dynamical features: we select an initial condition in the small chaotic region around the origin, a second one corresponding to a resonant libration and finally a third one corresponding to a circulation curve (see Fig. 2.1). In Fig. 2.2 we report the time evolution of the FLI for the three orbits. We remark that about 10 iterations of the map are enough to differentiate the chaotic orbit, whose tangent vector growths approximately exponentially with time, from the regular libration and circulation, whose tangent vectors increase almost linearly with time (correspondingly, the FLI increases almost linearly with time for the chaotic orbit, and approximately logarithmically for the regular motions).

In order to reduce the fluctuations that appear on Fig. 2.2 one can conveniently compute, instead of the indicator defined in (2.5), the indicator:

$$\text{FLI}(x(0), v(0), t) = \sup_{0 \leq k \leq t} \log \|v(k)\| \tag{2.7}$$

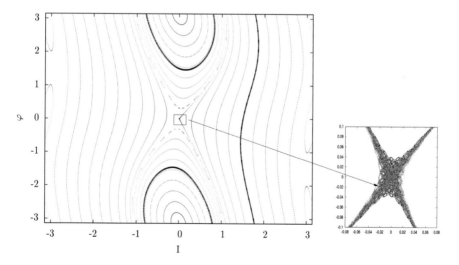

Fig. 2.1 *On the left*: a set of orbits of the standard mapping for $\epsilon = 0.3$. The *black points* correspond to a resonant libration orbit of initial conditions $(I(0), \varphi(0)) = (0, 1.5)$ and to a circulation orbit of initial conditions $(I(0), \varphi(0)) = (1.5, 0)$. *On the right*: enlargement around the hyperbolic fixed point at $I(0) = 0$, $\varphi(0) = 0$

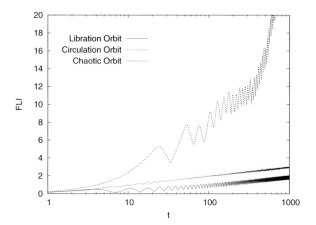

Fig. 2.2 Evolution with time of the FLI for the standard map of Eq. (2.6) for $\epsilon = 0.3$ for 3 orbits of initial conditions $(10^{-5}, 0)$ corresponding to the small chaotic region around the hyperbolic fixed point at the origin (right panel of Fig. 2.1), $(0, 1.5)$ and $(1.5, 0)$ corresponding respectively to the libration orbit and to the circulation curve marked with points in Fig. 2.1

Figure 2.3 shows that computing the FLI as in (2.7) the fluctuations become negligible. We remark that, while the Largest Lyapunov exponent is zero for both libration and circulation orbits, their corresponding FLI are different. In fact, using a refined perturbation theory we have shown in [21] that the value of the FLI differs at order 0 in ϵ, between libration and circulation motions even for more general systems. In Fig. 2.4 we show the FLI value at $t = 1000$, obtained for a set of 900×900 orbits of the standard map with $\epsilon = 0.3$ and with $I(0)$ and $\varphi(0)$ regularly spaced in the interval $[-\pi : \pi]$. We have considered 2 orthogonal initial vectors $v(0) = (1, 0)$ and $w(0) = (0, 1)$ and we have computed the FLI value using Eq. (2.7) on both vectors; we plot the largest between the two FLI values. When compared with Fig. 2.1 we clearly see that the three different dynamics are well distinguished: the largest FLI values corresponding to chaotic motions, the intermediate values to circulation orbits and the lower values to libration orbits. In [4] it was shown on a pendulum problem that some spurious pattern appear when using FLI (their Fig. 3). We notice that considering the largest between the two FLI values obtained on orthogonal initial vectors, there are no spurious structures in the FLI computation shown in Fig. 2.4.

2.2.2 The Choice of the Integration Time

A second frequently asked question concerns the choice of the integration time. A practical way to choose a suitable integration time is to compute the FLI for different time values and see for which time the picture gets stable. For example,

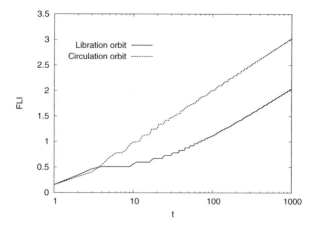

Fig. 2.3 Time evolution of the FLI for the circulation orbit and the libration orbit of Fig. 2.2. The FLI is computed as in (2.7). When considering the supremum of $\log \|v(t)\|$ the fluctuations, which are due to the geometry of the orbits, become negligible and libration motion is well distinguished from circulation

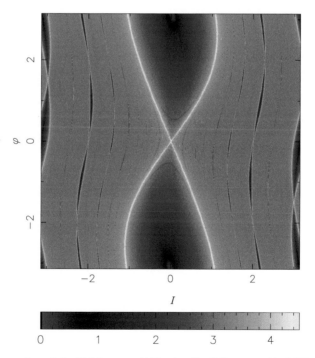

Fig. 2.4 Computation of the FLI for $t = 1000$ using Eq. (2.7) on a grid of 900×900 initial conditions regularly spaced in the interval $[-\pi : \pi]$. Precisely, two FLIs have been computed on 2 orthogonal initial vectors $v(0) = (1, 0)$ and $w(0) = (0, 1)$, the largest FLI value is plotted

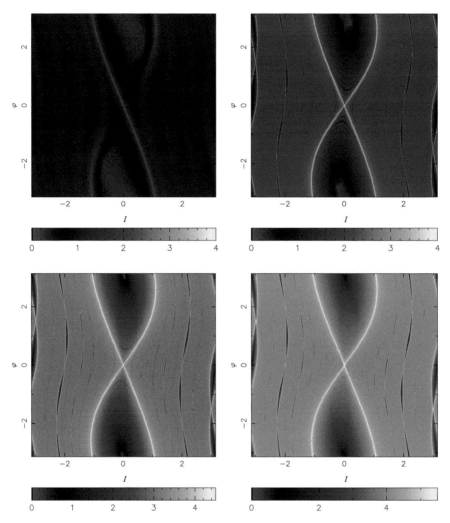

Fig. 2.5 Computation of the FLI as in Fig. 2.4 at $t = 10$ (*top left*), $t = 100$ (*top right*), $t = 1000$ (*bottom left*), $t = 10{,}000$ (*bottom right*)

let's consider the FLI computation shown in Fig. 2.4 for different times, say $t = 10, 100, 1000, 10{,}000$. We see clearly on Fig. 2.5 that $t = 10$ is a too short time to distinguish the dynamics while already at $t = 100$ we clearly distinguish between the different motions. Few more weakly chaotic orbits are detected at $t = 1000$ and no difference appears between $t = 1000$ and $t = 10{,}000$. For this case, $t = 1000$ has to be considered a suitable integration time.

We remark that, the FLI chart obtained in [14] on the relatively short time $t \simeq 1000$ provided a representation of the geometry of the resonances which allowed to follow the diffusion of orbits up to the very long times $t \simeq 10^{11}$.

For further details about the sensitivity of the method in detecting high order resonances we refer to [21].

2.3 The FLI for the Computation of the Stable and Unstable Manifolds

Since the work of Poincaré it is well known that the complexity of chaotic motions in deterministic systems can be appreciated from the analysis of the stable and unstable manifolds associated to hyperbolic orbits. Different methods can be found in the literature for the detection of hyperbolic manifolds. The FLI method for the computation of the hyperbolic manifolds has been introduced in [13, 17, 54] and used in [24, 25, 39, 40] to investigate the relation between the topology of hyperbolic manifolds and diffusion. Recently, the method has been used for the detection of the tube manifolds related to the Lyapunov periodic orbits [41] and for the detection of multiple close encounters [19] in the case of the restricted planar three body problem. We do not enter in the details here, we just recall that we have recently provided [20] an analytic description of the growth of tangent vectors for orbits with initial conditions which are close to the stable-unstable manifolds of a hyperbolic saddle point; as a matter of fact, we explain why the Fast Lyapunov Indicator detects the stable-unstable manifolds of all fixed points which satisfy a certain condition and we provide a suitably modified Fast Lyapunov Indicator if the condition is not satisfied.

Here we illustrate the use of the FLI in detecting hyperbolic manifolds associated to the hyperbolic point $(0, 0)$ on the standard map of Eq. (2.6). Let us recall that the for a two-dimensional standard map the unstable manifold $W_u(I_h, \varphi_h)$ of an hyperbolic point $(I_h, \varphi_h) \in \mathbb{R} \times \mathbb{S}^1$ is the set of (I, φ) such that:

$$W_u(I_h, \varphi_h) = \{(I(0), \varphi(0)) : \lim_{t \to \infty} d((I_h, \varphi_h), (I(-t), \varphi(-t))) = 0\},$$

the stable manifold $W_s(I_h, \varphi_h)$, is the set of (I, φ) such that:

$$W_s(I_h, \varphi_h) = \{(I(0), \varphi(0)) : \lim_{t \to \infty} d((I_h, \varphi_h), (I(t), \varphi(t))) = 0\}.$$

The numerical localization of the unstable manifold of an hyperbolic fixed point can be obtained by propagating a small neighborhood of initial conditions up to a time T of the order of some Lyapunov times of the fixed point (see [40] and references therein). In such a way, one directly constructs a neighborhood of a finite piece of the unstable manifold (for the stable manifold one repeats the construction for the inverse flow). This method gives very good results for fixed points of two dimensional maps, because the neighborhoods of the fixed points are two dimensional and can be propagated with reasonable CPU times. Figure 2.6, left panel shows the detection of a piece of the stable and of the unstable manifolds of the standard map of Eq. (2.6) for $\epsilon = 0.3$. Figure 2.6, right panel shows the value of the FLI obtained on a two dimensional grid of regularly spaced initial conditions

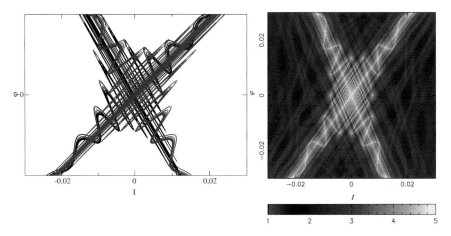

Fig. 2.6 *Left panel*: Detection of a piece of the stable and unstable manifold of the standard map computed with the usual method of set propagation (see [40] and references therein). *Right panel*: Representation of the FLI for the standard map (2.6) for $t = 50$. Precisely, we plot the average of two FLIs obtained on the direct and on the inverse map. We can appreciate the details of the lobes associated to the hyperbolic manifolds and the agreement with the results of the usual method of set propagation

for a short integration time $t = 50$. Precisely, two FLIs have been computed one on the direct and one on the inverse map, and the average of the two FLIs is plotted. We can appreciate the details of the lobes associated to the hyperbolic manifolds. The comparison with Fig. 2.6, left panel shows the quality of the detection of pieces of hyperbolic manifolds as obtained with the FLI computation. The advantage is that the use of the FLI method easily extends to higher dimensional systems and moreover one does not need to know in advance the local approximations of the hyperbolic manifolds.

2.4 Application to a Continuous System

The FLI is easily implemented also for generic continuous dynamical systems. We consider here, as an example, the computation of the FLI for a particle in an accelerated logarithmic potential which models the mean motion of stars in a flat rotation curve galaxy that sustains an asymmetric jet, whose dynamics has been previously studied in [47] using the traditional method of Poincaré surface of sections. The problem of the influence of stellar jets on the dynamics of protoplanetary discs was studied in [44–46]. In [47] the motion of stars in a flat rotation curve galaxy that sustains wind episodes was modeled by:

$$H(\rho, z, p_\rho, p_z) = \frac{1}{2}(p_\rho^2 + p_z^2) + \frac{h_z^2}{2\rho^2} + \frac{1}{2}\log(\rho^2 + z^2) - z \qquad (2.8)$$

where ρ and z are the cylindrical coordinates of the star in a reference frame with origin on the galactic center, p_ρ and p_z are the corresponding conjugate momenta, h_z is the projection of the angular momentum along the direction of acceleration (the z-axis). H and h_z are constants of motion of the system. In order to numerically compute the FLI, we integrate the Hamilton equations of (2.8) with a symplectic integrator and we write the variational equations of the map representing the numerical integrator. When dealing with multi-dimensional systems it is evident that we can't visualize the whole phase space as we did for the 2-dimensional standard map. However, we can still provide a global view of the dynamics on suitably chosen 2-dimensional sections. Precisely, the surface of constant energy H is three dimensional, and we can further reduce the study to a two dimensional space by fixing the value of one of the three independent variables.

Figure 2.7 shows the FLI computed on a bidimensional grid of 1000×1000 initial conditions regularly spaced in ρ and z for $E = -0.75$ and $h_z^2 = 0.05831$ with integration time $t = 200$. The other initial conditions are $p_z = 0$ and p_ρ is

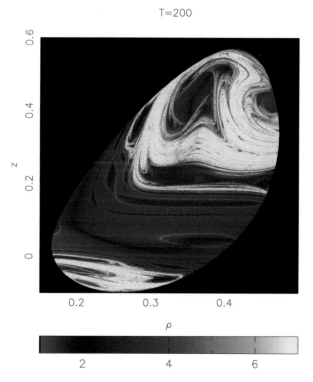

Fig. 2.7 FLI computation of 1000×1000 initial conditions regularly spaced in ρ and z for $H = -0.75$ and $h_z^2 = 0.05831$ with integration time $t = 200$. The other initial conditions are $p_z = 0$ and p_ρ is obtained from the energy equation. For all initial conditions we have chosen $v(0) = (1, 1, 0.5(\sqrt{5} - 1), 1)$

T=200

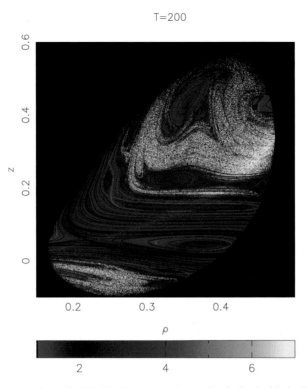

Fig. 2.8 FLI computation as in Fig. 2.7. To compare the results obtained with the FLI to those of the traditional method of surface of section we have drawn with *black points* the intersections of a set of orbits with the plane ρ, z obtained setting $p_z = 0$

obtained from (2.8). For all initial conditions we have chosen $v(0) = (1, 1, 0.5$ $(\sqrt{5} - 1), 1)$.

We now compare the result of the FLI computation with the results obtained with the traditional method of surface of section. In Fig. 2.8 we plot with black points the intersections of a set of orbits with the plane ρ, z obtained setting $p_z = 0$. We can observe that, as usual, the larger FLI values correspond to chaotic orbits (dispersed points on the surface of section) while intermediate and lower FLI values provide regular motions (closed curves on the surface of section). For the specific case of $H = -0.75$ and $h_z^2 = 0.05831$ the system has a large chaotic region for smaller and larger elevations z. Moreover, using the FLI, we do not only recover the results in [47] concerning the integrability, but we easily obtain much more details in the dynamics. This appears clearly in Fig. 2.9 where the FLI is computed zooming out Fig. 2.7 and we can see the complexity of the chaotic structures. In the bottom part of the figure (for ρ close to 0.25) we recognize the typical lobes related to the hyperbolic manifold of hyperbolic periodic orbits.

T=100

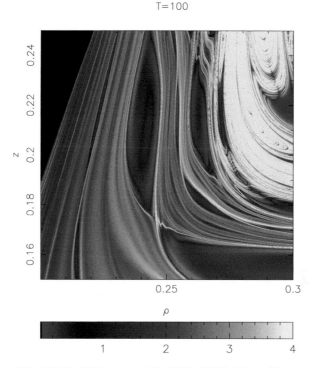

Fig. 2.9 Zoom of Fig. 2.7. The FLI is computed for 1000×1000 initial conditions regularly spaced in ρ and z for $H = -0.75$ and $h_z^2 = 0.05831$ with integration time $t = 100$

2.5 The FLI for Detecting the Geography of Resonances

The problem of long term stability of an Hamiltonian system is strongly related to the famous KAM [1, 28, 43] and Nekhoroshev [48] theorems which leave the possibility of a slow drift of the orbits on a peculiar subset of the phase space, the so called Arnold's web. The detection of the Arnold's web is therefore the first step to achieve if we are interested in studying the long term stability properties of a system.

We consider the following quasi-integrable hamiltonian (that we studied in several papers, see for example [22, 25, 35]):

$$H_\epsilon = \frac{I_1^2}{2} + \frac{I_2^2}{2} + I_3 + \epsilon f \ , \quad f = \frac{1}{\cos(\varphi_1) + \cos(\varphi_2) + \cos(\varphi_3) + 3 + c} \ , \quad (2.9)$$

where ϵ is a small parameter, $c > 0$, and (I, φ) are action angle variables. In the integrable case (defined by $\epsilon = 0$) the actions are constants of motion while the angles $\varphi_1(t) = \varphi_1(0) + I_1 t$, $\varphi_2(t) = \varphi_2(0) + I_2 t$, $\varphi_3(t) = \varphi_3(0) + t$ rotate with frequencies $\omega_1 = I_1$, $\omega_2 = I_2$, $\omega_3 = 1$. Therefore, a couple of actions I_1, I_2

characterizes an invariant torus $\mathbb{T}^3$. For any small ϵ different from zero, H_ϵ is not expected to be integrable. However, if ϵ is sufficiently small, the KAM theorem applies[1]: for any invariant torus of the original system with Diophantine non-resonant frequencies there exists an invariant torus in the perturbed system which is a small deformation of the unperturbed one.

The phase space has dimension 6, therefore, as for the system discussed in Sect. 2.4, we need to properly choose sections in order to provide a visual representation of the dynamics. Since the action I_3 does not enter the equations of motion of all the other variables, we can consider the time evolution in the reduced phase-space $(I_1, I_2, \varphi_1, \varphi_2, \varphi_3)$, and then in this space we consider various sections by fixing the values of some of the variables, for example, we fix the angles and consider the section:

$$S_0 = \{(I_1, I_2, \varphi_1, \varphi_2, \varphi_3) \in \mathbb{R}^2 \times \mathbb{T}^3 : \quad \varphi_1, \varphi_2, \varphi_3 = 0\}. \tag{2.10}$$

or, alternatively, we fix one action and two angles and consider the section:

$$S_1 = \{(I_1, I_2, \varphi_1, \varphi_2, \varphi_3) \in \mathbb{R}^2 \times \mathbb{T}^3 : \quad I_2 = I_2(0), \varphi_2, \varphi_3 = 0\} \tag{2.11}$$

or:

$$S_2 = \{(I_1, I_2, \varphi_1, \varphi_2, \varphi_3) \in \mathbb{R}^2 \times \mathbb{T}^3 : \quad I_1 = I_1(0), \varphi_2, \varphi_3 = 0\}. \tag{2.12}$$

In Fig. 2.10 we show the FLI computed for the three different sections S_0, S_1, S_2 represented in the three dimensional space (I_1, I_2, φ_1). Precisely, in the horizontal plane we have represented the FLI computed on section S_0 using a grid of 500×500 initial conditions regularly spaced in (I_1, I_2) in the interval $[-0.5, 1.5]$; on the vertical plane on the left (right) we have represented the FLI computed on section S_1 (S_2) using a grid of 500×500 initial conditions regularly spaced in (I_1, φ_1) (respectively (I_2, φ_1)) in the intervals $[-0.5, 1.5]$ and $[0, 2\pi]$, with respectively $I_2(0) = 1.5$, $I_1(0) = 1.5$.

In the horizontal plane we clearly see a web of resonance, located near the straight lines defined by: $k_1\omega_1 + k_2\omega_2 + k_3 \equiv k_1 I_1 + k_2 I_2 + k_3 = 0$, with $k_1, k_2, k_3 \in \mathbb{Z}\backslash 0$. For examples the resonances $I_1 = 0$ and $I_2 = 0$ appear as large lines in the horizontal plane. They both have a chaotic boundary (shown in yellow). When looking at the vertical panels it appears clearly that the amplitude of the resonances change with values of the angles.

We now study the evolution of a set of $N = 100$ chaotic orbits with initial conditions $I_1(i) = 1.5$, $I_2(i) = 0$, $\varphi_2(i) = \varphi_3(i) = 0$, $\varphi_1 = \pi + 10^{-6}i$, $i = 1, ...N$. The orbits evolve in a multi dimensional space, therefore it is useful to consider the points of the orbits intersecting two dimensional sections. On the FLI map of Fig. 2.11 we plot as black dots the points of the orbits which have returned after

---

[1]H_ϵ is real analytic and H_0 is isoenergetically non-degenerate.

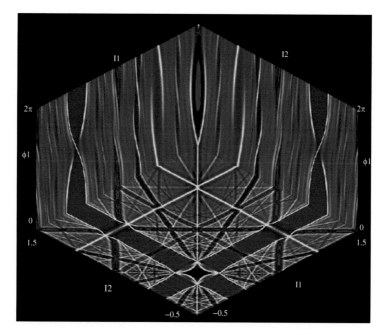

Fig. 2.10 FLI computation for the Hamiltonian of Eq. (2.9) for a value of the perturbing parameter $\epsilon = 0.04$. The integration time is $t = 100$. The initial conditions are regularly spaced on grids of 500×500 points on three different sections of the phase space defined in the text as S_0 (*horizontal plane*), S_1 (*vertical left plane*) and S_2 (*vertical right plane*)

some time on the sections S_2 and S_0. Of course, since computed orbits are discrete, we represent the points that return in a small neighbourhood of S_0 defined by

$$\tilde{S}_0 = \{(I_1, I_2, \varphi_1, \varphi_2, \varphi_3) : \ |\varphi_1| \leq 0.05, |\varphi_2| \leq 0.05, \varphi_3 = 0\}$$

and those that return in a small neighbourhood of S_2 defined by

$$\tilde{S}_2 = \{(I_1, I_2, \varphi_1, \varphi_2, \varphi_3) : \ |\varphi_1| \leq 0.05, |I_1 - 1.5| < 0.005, \varphi_3 = 0\}.$$

Reducing the size of the neighborhood of the sections S_0 and S_2 reduce the number of points but doesn't change the results. In Fig. 2.11 we consider three different integration times: $t = 2\,10^5$ (left panel), $t = 2\,10^7$ (middle panel), $t = 2\,10^8$ (right panel).

On the vertical plane the black points cover the whole chaotic region associated to the chaotic border of the resonance $I_2 = 0$ already on short times. As we can observe in Fig. 2.11 the number of points returning in $\tilde{S}_2$ increases with time but the region covered by the orbits doesn't change.

On the horizontal plane, it is interesting to observe that the orbits slowly diffuse along the resonance $I_2 = 0$, precisely the points that return in $\tilde{S}_0$ cover a portion

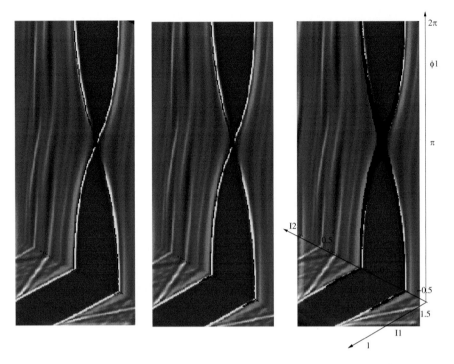

Fig. 2.11 Zoom of Fig. 2.10 around the resonance $I_2 = 0$. We plot as *black dots* the points of a set of $N = 100$ chaotic orbits which have returned after some time on the sections S_2 and S_0. Precisely, we represent the points that return in a small neighbourhood of S_0 defined by $\tilde{S}_0 = \{(I_1, I_2, \varphi_1, \varphi_2, \varphi_3) : |\varphi_1| \leq 0.05, |\varphi_2| \leq 0.05, \varphi_3 = 0\}$ and those that return in a small neighbourhood of S_2 defined by $\tilde{S}_2 = \{(I_1, I_2, \varphi_1, \varphi_2, \varphi_3) : |\varphi_1| \leq 0.05, |I_1 - 1.5| < 0.005, \varphi_3 = 0\}$. The three panels correspond to different integrations times: $t = 2\,10^5$ (*left panel*), $t = 2\,10^7$ (*middle panel*) and $t = 2\,10^8$ (*right panel*). The initial conditions are chosen in the neighbourhood of the hyperbolic point: $I_1(i) = 1.5, I_2(i) = 0, \varphi_2(i) = \varphi_3(i) = 0, \varphi_1 = \pi + 10^{-6}i, i = 1, \ldots N$

of the chaotic border of the resonance that increases for increasing values of the integration time. Let us remark that the FLI, computed on a short total time $t = 100$, allows to properly follow orbits on a much larger integration time ($t = 2\,10^8$ on the right panel of Fig. 2.11).

Let us remark that the section S_0 is particularly suited to the detection of the slow diffusion since the large oscillations of the action I_2 that we observe on section S_2 are filtered when considering the points that return in $\tilde{S}_0$. Moreover, the FLI chart allows us to check that the orbits really diffuse along the chaotic border of a resonance.

We have quantified the diffusion by measuring a diffusion coefficient using the points of sets of orbits returning in the section $\tilde{S}_0$ using various Hamiltonian or discrete mapping models in [14, 18, 22, 24–26, 35, 36]. The computation of the diffusion coefficient has many technical difficulties that we do not recall here. The interested reader can refer to [25] for a numerical characterization of the statistical properties of the diffusion and to [18] for the technical aspects of the computation of the diffusion coefficient.

In [22] we have shown that the diffusion of orbits occurring along the peculiar set of resonances has a global character and in [26] we measured a diffusion coefficient decreasing exponentially through 40 orders of magnitude thus showing that Arnold's diffusion concerns and can be measured on systems of physical interest.

2.6 Conclusions

Since the pioneering work of Hénon and Heiles [27] it appeared clearly that understanding the dynamical behaviours of a system required global studies of the phase space. Different tools for the detection of chaotic and ordered motions have been developed since then, providing interesting results in different domains of physics (celestial mechanics, particle accelerators, dynamical astronomy, statistical physics, plasma physics). The Fast Lyapunov Indicator was introduced in [12] as an easy (to implement) and sensitive tool for distinguish between ordered and chaotic motion. The method was used in [13] for the detection of the Arnold web of a system and further developed in [21] using a refined perturbation theory which provided the behaviour of the FLI for different orbits.

In the last 10 years we have used the FLI for studies for which a global visualization of the phase space was one of the key ingredients for understanding the problem. For example, the Arnold web computed with short integration time allowed us to choose possibly diffusing initial conditions and to follow their evolution on much longer integration time. With a rather technical method we have measured diffusion coefficients decreasing faster than a power low and possibly exponentially through many orders of magnitude [14, 18, 22, 24–26, 35, 36] showing the interest of Arnold's diffusion for physical systems.

Later in [17, 24, 38–40] we have used the FLI for the detection of the stable and unstable manifolds. This is in general a difficult task that requires sophisticated methods. We have recently provided [20] an analytic description of the growth of tangent vectors for orbits with initial conditions which are close to the stable-unstable manifolds of a hyperbolic saddle point and we have explained the condition for the detection the stable-unstable manifolds with the FLI.

Moreover, the FLI has been applied to the planar circular restricted three body problem for the detection and characterization of close encounters and resonances [19, 41].

In this chapter we have we presented the indicator for readers that would like to implement it for the first time. At this purpose we have chosen simple discrete and continuous model problems giving the element to reproduce this cases. As examples of applications we have shown (1) the use of the FLI for the detection of the stable/unstable manifold of a two dimensional model, (2) the FLI for the detection of the resonances of continuous systems and (3) we have explained how to use the indicator to follow the diffusion of orbits along resonant lines.

References

1. Arnold, V.I.: Proof of a theorem by A.N. Kolmogorov on the invariance of quasi-periodic motions under small perturbations of the Hamiltonian. Russ. Math. Surv. **18**, 9 (1963)
2. Arnold, V.I.: Instability of dynamical systems with several degrees of freedom. Sov. Math. Dokl. **6**, 581–585 (1964)
3. Barrio, R.: Sensitivity tools vs. Poincaré sections. Chaos Solitons Fractals **25**, 711–726 (2005)
4. Barrio, R., Borczyk, W., Breiter, S.: Spurious structures in chaos indicators maps. Chaos Solitons Fractals **40**, 1697–1714 (2009)
5. Celletti, A., Froeschlé, C., Lega, E.: Dissipative and weakly-dissipative regimes in nearly-integrable mappings. Discrete Contin. Dyn. Syst. Ser. A **16**(4), 757–781 (2006)
6. Celletti, A., Froeschlé, C., Lega, E.: Dynamics of the conservative and dissipative spin–orbit problem. Planet. Space Sci. **55**, 889–899 (2007)
7. Celletti, A., Stefanelli, L., Lega, E., Froeschlé, C.: Global dynamics of the regularized restricted three-body problem with dissipation. Celest. Mech. Dyn. Astron. **109**, 265–284 (2011)
8. Cincotta, P.M., Simó, C.: Simple tools to study global dynamics in non-axisymmetric galactic potentials - I. Astron. Astrophys. Suppl. Ser. **147**, 205 (2000)
9. Cincotta, P.M., Giordano, C.M., Simó, C.: Phase space structure of multi-dimensional systems by means of the mean exponential growth factor of nearby orbits. Physica D **182**(3–4), 151–178 (2003)
10. Froeschlé, C., Lega, E.: Weak chaos and diffusion in Hamiltonian systems. From Nekhoroshev to Kirkwood. In: Roy, A.E. (eds.) The Dynamics of Small Bodies in the Solar System: A Major Key to Solar System Studies. NATO/ASI. Kluwer Academic, Boston (1998)
11. Froeschlé, C., Lega, E.: On the structure of symplectic mappings. The fast Lyapunov indicator: a very sensitive tool. Celest. Mech. Dyn. Astron. **78**(1/4), 167–195 (2000)
12. Froeschlé, C., Lega, E., Gonczi, R.: Fast Lyapunov indicators. Application to asteroidal motion. Celest. Mech. Dyn. Astron. **67**, 41–62 (1997)
13. Froeschlé, C., Guzzo, M., Lega, E.: Graphical evolution of the Arnold web: from order to chaos. Science **289**(5487), 2108–2110 (2000)
14. Froeschlé, C., Guzzo, M., Lega, E.: Local and global diffusion along resonant lines in discrete quasi-integrable dynamical systems. Celest. Mech. Dyn. Astron. **92**(1–3), 243–255 (2005)
15. Guzzo, M.: The web of three-planets resonances in the outer solar system. Icarus **174**(1), 273–284 (2005)
16. Guzzo, M.: The web of three-planet resonances in the outer solar system II: a source of orbital instability for Uranus and Neptune. Icarus **181**, 475–485 (2006)
17. Guzzo, M.: Chaos and diffusion in dynamical systems through stable–unstable manifolds. In: Perozzi, Mello, F. (eds.) Space Manifolds Dynamics: Novel Spaceways for Science and Exploration. Novel Spaceways for scientific and exploration missions, a dynamical systems approach to affordable and sustainable space applications held in Fucino Space Centre (Avezzano), 15–17 October 2007. Springer, New York/Dordrecht/Heidelberg/London (2010)
18. Guzzo, M., Lega, E.: The numerical detection of the Arnold web and its use for long-term diffusion studies in conservative and weakly dissipative systems. Chaos **23**, 23124 (2013)
19. Guzzo, M., Lega, E.: On the identification of multiple close encounters in the planar circular restricted three-body problem. Mon. Not. R. Astron. Soc. Lett. **428**, 2688–2694 (2013)
20. Guzzo, M., Lega, E.: Evolution of the tangent vectors and localization of the stable and unstable manifolds of hyperbolic orbits by Fast Lyapunov Indicators. SIAM J. Appl. Math. **74**(4), 1058–1086 (2014)

21. Guzzo, M., Lega, E., Froeschlé, C.: On the numerical detection of the effective stability of chaotic motions in quasi-integrable systems. Physica D **163**(1–2), 1–25 (2002)
22. Guzzo, M., Lega, E., Froeschlé, C.: First numerical evidence of Arnold diffusion in quasi-integrable systems. Discrete Continuous Dyn. Syst. Ser. B **5**(3), 687–698 (2005)
23. Guzzo, M., Lega, E., Froeschlé, C.: Diffusion and stability in perturbed non-convex integrable systems. Nonlinearity **19**, 1049–1067 (2006)
24. Guzzo, M., Lega, E., Froeschlé, C.: A numerical study of the topology of normally hyperbolic invariant manifolds supporting Arnold diffusion in quasi-integrable systems. Physica D **238**, 1797–1807 (2009)
25. Guzzo, M., Lega, E., Froeschlé, C.: A numerical study of Arnold diffusion in a priori unstable systems. Commun. Math. Phys. **290**, 557–576 (2009)
26. Guzzo, M., Lega, E., Froeschlé, C.: First numerical investigation of a conjecture by N.N. Nekhoroshev about stability in quasi-integrable systems. Chaos **21**(3), 033101-1–033101-12 (2011)
27. Hénon, M., Heiles, C.: The applicability of the third integral of motion: some numerical experiments. Astron. J. **69**, 73–79 (1964)
28. Kolmogorov, A.N.: On the conservation of conditionally periodic motions under small perturbation of the Hamiltonian. Dokl. Akad. Nauk. SSSR **98**, 527–530 (1954)
29. Laskar, J.: The chaotic motion of the Solar system. A numerical estimate of the size of the chaotic zones. Icarus **88**, 266–291 (1990)
30. Laskar, J.: Frequency analysis for multi-dimensional systems. Global dynamics and diffusion. Physica D **67**, 257–281 (1993)
31. Laskar, J., Froeschlé, C., Celletti, A.: The measure of chaos by the numerical analysis of the fundamental frequencies. Application to the standard mapping. Physica D **56**, 253 (1992)
32. Lega, E., Froeschlé, C.: Fast Lyapunov Indicators. Comparison with other chaos indicators. Application to two and four dimensional maps. In: Henrard, J., Dvorak, R. (eds.) The Dynamical Behaviour of our Planetary System. Springer, The Netherlands (1997)
33. Lega, E., Froeschlé, C.: Comparison of convergence towards invariant distributions for rotation angles, twist angles and local Lyapunov characteristic numbers. Planet. Space Sci. **46**, 1525–1534 (1998)
34. Lega, E., Froeschlé, C.: On the relationship between fast Lyapunov indicator and periodic orbits for symplectic mappings. Celest. Mech. Dyn. Astron. **81**, 129–147 (2001)
35. Lega, E., Guzzo, M., Froeschlé, C.: Physica D **182**, 179–187 (2003)
36. Lega, E., Froeschlé, C., Guzzo, M.: Diffusion in Hamiltonian quasi-integrable systems. In: Benest, D., Froeschlé, C., Lega, E. (eds.) Topics in Gravitational Dynamics. Lecture Notes in Physics, vol. 729. Springer, Berlin (2007)
37. Lega, E., Guzzo, M., Froeschlé, C.: Measure of the exponential splitting of the homoclinic tangle in four dimensional symplectic mappings. Celest. Mech. Dyn. Astron. **104**, 191–204 (2009)
38. Lega, E., Guzzo, M., Froeschlé, C.: A numerical study of the size of the homoclinic tangle of hyperbolic tori and its correlation with Arnold diffusion in Hamiltonian systems. Celest. Mech. Dyn. Astron. **107**, 129–144 (2010)
39. Lega, E., Guzzo, M., Froeschlé, C.: A numerical study of the hyperbolic manifolds in a priori unstable systems. A comparison with Melnikov approximations. Celest. Mech. Dyn. Astron. **107**, 115–127 (2010)
40. Lega, E., Guzzo, M., Froeschlé, C.: Numerical Studies of hyperbolic manifolds supporting diffusion in symplectic mappings. Eur. Phys. J. Spec. Top. **186**, 3–31 (2010)
41. Lega, E., Guzzo, M., Froeschlé, C.: Detection of Close encounters and resonances in three body problems through Levi-Civita regularization. Mon. Not. R. Astron. Soc. Lett. **418**, 107–113 (2011)
42. Mitchenko, T.A., Ferraz–Mello, S.: Astron. J. **122**, 474–481 (2001)
43. Moser, J.: On invariant curves of area-preserving maps of an annulus. Commun. Pure Appl. Math. **11**, 81–114 (1958)
44. Namouni, F.: Astron. J. **130**, 280 (2005)

45. Namouni, F.: LNP **729**, 233 (2007)
46. Namouni, F., Guzzo, M.: Celest. Mech. Dyn. Astron. **99**, 31 (2007)
47. Namouni, F., Guzzo, M., Lega, E.: On the integrability of stellar motion in an accelerated logarithmic potential. Astron. Astrophys. **489**, 1363 (2008)
48. Nekhoroshev, N.N.: Exponential estimates of the stability time of near-integrable Hamiltonian systems. Russ. Math. Surv. **32**, 1–65 (1977)
49. Robutel, P.: Frequency map analysis and quasiperiodic decompositions. In: Benest et al. (eds.) Hamiltonian Systems and Fourier Analysis, pp. 179–198. Taylor and Francis. Adv. Astron. Astrophys., Cambridge Sci. Publ., Cambridge (2005)
50. Robutel, P., Galern, F.: The resonant structure of Jupiter's Trojan asteroids I. Long term stability and diffusion. Mon. Not. R. Astron. Soc. **372**, 1463–1482 (2006)
51. Robutel, P., Laskar, J.: Frequency map and global dynamics in the Solar System I. Icarus **52**(1), 4–28 (2001)
52. Tang, X.Z., Boozer, A.H.: Finite time Lyapunov exponent and advection-diffusion equation. Physica D **95**(3–4), 283–305 (1996)
53. Todorović, N., Guzzo, M., Lega, E., Froeschlé, C.: A numerical study of the stabilization effect of steepness. Celest. Mech. Dyn. Astron. **110**(4), 389–398 (2011)
54. Villac, B.F.: Using FLI maps for preliminary spacecraft trajectory design in multi-body environments. Celest. Mech. Dyn. Astron. **102**, 29–48 (2008)
55. Wayne, B.H., Malykh, A.V., Danforth, C.M.: The interplay of chaos between the terrestrial and giant planets. Mon. Not. R. Astron. Soc. **407**(3), 1859–1865 (2010)

Chapter 3
Theory and Applications of the Orthogonal Fast Lyapunov Indicator (OFLI and OFLI2) Methods

Roberto Barrio

Dedicated to the Memory of Eugenio Barrio (1934–2014)

Abstract During the last decades the Nonlinear Dynamics field has produced a large number of numerical techniques oriented to the analysis of the behavior of the orbits in different systems. These methods are mainly focused to distinguish chaotic from regular behavior. Among the variational methods, based into the variational equations, we discuss in this paper the so-called Orthogonal Fast Lyapunov Indicator (OFLI and OFLI2) methods that are variants of the FLI method but designed to obtain also some information about the periodic orbits of the systems. We review the OFLI and OFLI2 methods and we show several computational aspects related with avoiding the appearance of spurious structures, with their use in the analysis of regular/chaotic behaviors, but also with the analysis of periodic orbits and regular regions, and with the efficient computation of the solution of the variational equations by means of Taylor series methods. Finally, the methods are shown in several Hamiltonian problems, as well as in several classical dissipative systems, as the Lorenz and Rössler models.

3.1 Orthogonal Fast Lyapunov Indicators

When we intend to analyze the behavior of a dynamical system one of the most interesting questions is if it is possible to know if a given initial condition generates a chaotic orbit or not. In fact, this question cannot be answered rigorously without a carefully theoretical study of the particular problem. Therefore, this has been done only for some important problems (note that nowadays the computer assisted proof of chaos is an active research field [1, 2]). Thus, a numerical evidence of the behavior of a dynamical system has become an invaluable tool in the analysis of a problem. One of the most popular techniques is the computation of Poincaré sections,

R. Barrio (✉)
Computational Dynamics Group (CODY), IUMA and Departamento de Matemática Aplicada, Universidad de Zaragoza, E50009 Zaragoza, Spain
e-mail: rbarrio@unizar.es; http://cody.unizar.es

© Springer-Verlag Berlin Heidelberg 2016   55
Ch. Skokos et al. (eds.), *Chaos Detection and Predictability*, Lecture Notes
in Physics 915, DOI 10.1007/978-3-662-48410-4_3

which allow us to distinguish regular from chaotic orbits. However, the Poincaré sections have several drawbacks: they are useful only for two-degrees of freedom Hamiltonian systems, we have to select carefully the two-dimensional surface that is transverse to most of the trajectories for a fixed value of the Hamiltonian and, finally, in the case of chaotic regions it is quite difficult to distinguish among different structures. Another important technique is the Maximum Lyapunov Exponent (MLE) that study the divergence among trajectories of close initial conditions. The definition of the MLE for an initial value problem

$$\frac{d\mathbf{y}}{dt} = \mathbf{f}(t, \mathbf{y}), \qquad \mathbf{y}(t_0) = \mathbf{y}_0 \tag{3.1}$$

is given by

$$\text{MLE} = \lim_{t \to +\infty} \frac{1}{t} \ln \frac{\|\delta \mathbf{y}(t)\|}{\|\delta \mathbf{y}(t_0)\|}$$

being $\delta \mathbf{y}(t)$ the solution of the first order variational equations

$$\frac{d\delta \mathbf{y}}{dt} = \frac{\partial \mathbf{f}(t, \mathbf{y})}{\partial \mathbf{y}} \delta \mathbf{y}, \qquad \delta \mathbf{y}(t_0) = \delta \mathbf{y}_0. \tag{3.2}$$

The value of the MLE gives a way of measuring the degree of sensitivity to initial conditions [3], and so it has been used as an indicator of chaos. The problem of the MLE is that its practical computation is not so simple because as its definition is a limit we have to integrate the system up to a long time and so the computer time is large, being therefore useful just for the analysis of a short number of orbits. It is interesting to note that the equations (3.2) are the first order sensitivity equations with respect to the initial conditions if we take $\delta \mathbf{y}_0 = \mathbb{I}$, being $\mathbb{I}$ the identity matrix (note that in (3.2) we have just selected one particular directional derivative specified by the initial conditions $\delta \mathbf{y}_0$). The last few decades several fast chaos indicators have been designed to overcome the drawbacks of the MLE, among others the frequency map analysis [4, 5], the Mean Exponential Growth factor of Nearby Orbits (MEGNO) [6, 7], the Fast Lyapunov Indicator (FLI) [8, 9], the Smaller ALigment Index (SALI) [10], the 0-1 test [11, 12], spike-counting diagrams [13], kneading invariants [14], and so on (see this volume for more indicators). Among all of the above the FLI seems to be one of the fastest and more efficient ways of studying numerically the dynamical behavior of a set of orbits [15], but it maintains some difficulties in locating periodic orbits and it can generate some spurious structures.

On the present paper we intend to analyze briefly the use of two modifications of the FLI chaos indicator, the OFLI [16] and the OFLI2 [17–19]. The Fast Lyapunov Indicator (FLI) [8] was introduced as the initial part (up to a stopping time t_f) of the computation of the Maximum Lyapunov Exponent (MLE [3]):

$$\text{FLI}(\mathbf{y}(t_0), \delta \mathbf{y}(t_0), t_f) := \sup_{t_0 < t < t_f} \log \|\delta \mathbf{y}(t)\|,$$

where $\mathbf{y}(t)$ and $\delta\mathbf{y}(t)$ are the solutions of the system (3.1) and the first order variational equations (3.2).

In order to detect easily periodic orbits, a variation called OFLI [16] (Orthogonal Fast Lyapunov Indicator) was introduced by Fouchard, Lega, Froeschlé C. and Froeschlé Ch. in 2002 and it is defined by

$$\text{OFLI}(\mathbf{y}(t_0), \delta\mathbf{y}(t_0), t_f) := \sup_{t_0 < t < t_f} \log \|\delta\mathbf{y}^{\perp}(t)\|$$

where $\delta\mathbf{y}^{\perp}$ is the component of $\delta\mathbf{y}$ orthogonal to the flow at that point. The problem is that these indicators (and most of the methods that are on the literature) may exhibit spurious structures [19]. To minimize the appearance of these spurious artifacts and to accelerate the detection of chaos, Barrio in 2005 developed the OFLI2 method [17–19] (also denominated $\text{OFLI}^2_{\text{TT}}$). The method is based on the use of the second order variational equations. We use as numerical ODE integrator a specially developed Taylor method [20, 21] that gives a fast and accurate numerical integration as we will explain below in the Appendix. The OFLI2 looks for detecting the set of initial conditions where we may expect sensitive dependence on initial conditions. The OFLI2 indicator at the final time t_f is given by

$$\text{OFLI2}(\mathbf{y}(t_0), t_f) := \sup_{t_0 < t < t_f} \log \left\| \left\{ \delta\mathbf{y}(t) + \frac{1}{2}\delta^2\mathbf{y}(t) \right\}^{\perp} \right\|, \qquad (3.3)$$

where $\delta\mathbf{y}$ and $\delta^2\mathbf{y}$ are the first and second order sensitivities with respect to carefully chosen initial vectors and $\mathbf{x}^{\perp}$ stands for the orthogonal component to the flow of the vector $\mathbf{x}$. Note that in many practical applications the computation of sensitivities is an important task as it gives the dependence of a system with respect to the initial conditions or parameters of the problem, that is defined by the corresponding partial derivatives and directional derivatives. In our case we need the variational equations up to second order and we fix the initial conditions:

$$\frac{d\mathbf{y}}{dt} = \mathbf{f}(t, \mathbf{y}), \qquad\qquad \mathbf{y}(t_0) = \mathbf{y}_0,$$

$$\frac{d\,\delta\mathbf{y}}{dt} = \frac{\partial\mathbf{f}(t, \mathbf{y})}{\partial\mathbf{y}}\delta\mathbf{y}, \qquad\qquad \delta\mathbf{y}(t_0) = \frac{\mathbf{f}(t_0, \mathbf{y}_0)}{\|\mathbf{f}(t_0, \mathbf{y}_0)\|}, \qquad (3.4)$$

$$\frac{d\,\delta^2\mathbf{y}_j}{dt} = \frac{\partial f_j}{\partial\mathbf{y}}\delta^2\mathbf{y} + \delta\mathbf{y}^{\mathsf{T}}\frac{\partial^2 f_j}{\partial\mathbf{y}^2}\delta\mathbf{y}, \qquad \delta^2\mathbf{y}(t_0) = \mathbf{0}.$$

Note that the last line of Eq. (3.4) is written for a single jth component $\delta^2\mathbf{y}_j$ of the second order variational equations to simplify the notation.

The evolution of the FLI and OFLI indicators is explained in [9] for quasi-integrable systems:

Proposition *Given the Hamiltonian function*

$$\mathscr{H}_\epsilon(I, \phi) = h(I) + \epsilon f(I, \phi), \tag{3.5}$$

with action-angle variables $I_1, \dots, I_n \in \mathbb{R}$ and $\phi_1, \dots, \phi_n \in \mathbb{S}$ and ϵ a small parameter, and suppose that the functions h and f satisfy the hypotheses of both KAM and Nekhoroshev theorems then the following estimates were given:

1. If the initial conditions are on the KAM torus

$$\|(\delta I^\epsilon(t), \delta\phi^\epsilon(t))\| = \left\| \frac{\partial^2 h}{\partial I^2}(I(t_0))\, \delta I(t_0) \right\| t + \mathscr{O}(\epsilon^\alpha t) + \mathscr{O}(1),$$

for some $\alpha > 0$.

2. If the initial conditions are on a regular resonant motion

$$\|(\delta I^\epsilon(t), \delta\phi^\epsilon(t))\| = \|C_\Lambda \Pi_{\Lambda^{\mathrm{ort}}} \delta I(t_0)\| t + \mathscr{O}(\epsilon^\beta t) + t\, \mathscr{O}(\rho^2)$$

$$+ \mathscr{O}(\sqrt{\epsilon}\, t) + \mathscr{O}\left(\frac{1}{\sqrt{\epsilon}}\right),$$

with some $\beta > 0$, $\Lambda \subseteq \mathbb{Z}^n$ a d-dimensional lattice that defines a resonance through the relation $\Pi_\Lambda(\partial h / \partial I) = 0$ where Π_Λ denotes the Euclidean projection of a vector onto the linear space spanned by Λ and C_Λ a linear operator depending on the resonant lattice Λ and the initial action $I(t_0)$.

For the OFLI2 it is possible to obtain similar estimates [17]. Again, if we assume that the Hamiltonian (3.5) satisfies the hypotheses of the KAM theorem then, we obtain the following estimate for some $\alpha > 0$ and for initial conditions on the KAM torus

$$\|(\delta^2 I^\epsilon(t), \delta^2\phi^\epsilon(t))\| = \|(\delta^2 I^0(0), \delta^2\phi^0(0))\| t + \mathscr{O}(\epsilon^\alpha t).$$

Then, we expect the OFLI and OFLI2 behave as $\log t$ [17] for initial conditions on a KAM tori and on a regular resonant motion but with different rate of growing (and so they grow linearly in a logarithmic time scale), tend to a constant value for periodic orbits and grow exponentially (in a logarithmic time scale) for chaotic orbits.

As exemplary problem, we provide some tests on the very classical Hénon–Heiles Hamiltonian system [22] given by the Hamiltonian

$$\mathscr{H}(x, y, X, Y) = \frac{1}{2}(X^2 + Y^2 + x^2 + y^2) + x^2 y - \frac{1}{3}y^3 \tag{3.6}$$

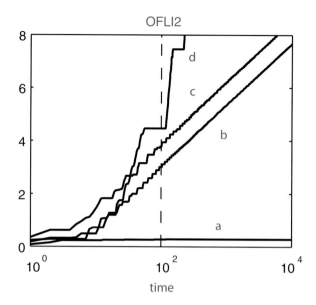

Fig. 3.1 Evolution of the OFLI2 values of four orbits of the Hénon–Heiles Hamiltonian with energy $E = 1/12$: $-a-$ (a periodic orbit), $-b-$ and $-c-$ (orbits on a KAM torus) and $-d-$ (a chaotic orbit close to the hyperbolic point)

and so the differential system is

$$\dot{x} = X, \ \dot{X} = -x - 2xy,$$
$$\dot{y} = Y, \ \dot{Y} = -y - x^2 + y^2.$$

The reason of using this problem is understandable: it was extensively used as a benchmark for chaos indicators.

In Fig. 3.1 we show the time evolution of the OFLI2 values in the time interval [1, 10,000] for four particular orbits of the Hénon–Heiles problem. The orbits are indicated with the letters $-a-$ (a periodic orbit), $-b-$ and $-c-$ (orbits on a KAM torus) and $-d-$ (a chaotic orbit close to the hyperbolic point). From the different behaviour of the curves we observe how the indicator works and how it is easy to classify the orbits comparing one each other.

From now on, all the biparametric figures with the OFLI2 results use the red (white) color to point the chaotic regions and blue (black) for the most regular ones in the color (B&W) plots, and being the intermediate colors the transition from one to the another situation.

3.2   Locating Periodic Orbits and Avoiding Spurious
Structures

In this section we study two of the points where the OFLI2 and OFLI techniques give a better performance than other chaos indicators (in fact two of the reasons for their design): in avoiding the appearance of spurious structures inside the regular regions and in some analysis of periodic orbits and regular regions, besides the identification of regular/chaotic behavior.

3.2.1   Avoiding Spurious Structures

One of the problems of the FLI and OFLI indicators is the dependence of the results on the initial conditions of the variational equations. This fact was already detected in [8], where a set of different initial tangent vectors were considered for a symplectic mapping showing how the magnitude of the FLI indicator changes. In any case, for symplectic mappings the most important fact is the choice of the same tangent vector for the whole set of orbits (see [8]). In the continuous case the dependency on the initial tangent vector is higher that in the symplectic mappings and we have to proceed with more care, especially for short integration times. We have to remark that these indicators give in all cases good results for a long time analysis of an orbit but, depending on the initial conditions, the short time analysis of a region (that is, as a global picture) may give us wrong pictures (although the individual analysis of each orbit is correct due to the local character of the indicators).

We have performed [17] several numerical tests with the Hénon–Heiles system (3.6) to show the temporary dependence on the initial conditions of the variational equations and we have computed the OFLI values for four sets of initial vectors, three of them forming an orthonormal basis of the orthogonal subspace to the flow at the initial time and the unitary initial tangent vector given by

$$
\begin{aligned}
\delta \mathbf{y}_T(t_0)   &= \mathbf{f}(t_0)/\|\mathbf{f}(t_0)\| = (t_1,\, t_2,\, t_3,\, t_4), \\
\delta \mathbf{y}_{O_1}(t_0) &= (\ \ t_2,\, -t_1,\ \ \ t_4,\, -t_3), \\
\delta \mathbf{y}_{O_2}(t_0) &= (-t_4,\, -t_3,\ \ \ t_2,\ \ \ t_1), \\
\delta \mathbf{y}_{O_3}(t_0) &= (-t_3,\ \ \ t_4,\ \ \ t_1,\, -t_2).
\end{aligned}
$$

We use $t_0 = 0$ and $\mathbf{f}(t_0) = (x_0, X_0, y_0, Y_0)$ for the Hénon–Heiles system.

On each picture of Fig. 3.2 we show 400×400 values of the OFLI depending on the initial conditions of the sensitivity vector for the Hénon–Heiles system with energy$= 1/12$ and $x = 0$. The stopping time in the numerical integration is $t_f = 100$. From the pictures we may appreciate as the whole figure presents different configurations depending on the set of initial conditions of the sensitivity vector. The first set, O_1, gives an acceptable figure but the other two orthogonal vectors, O_2 and O_3, generate some spurious structures (although the analysis of each particular orbit is correct, the problem is the different time scales to reach a global picture). The tangent initial conditions, T, gives a correct analysis but the evolution is too slow (the values of the OFLI are of order 10^{-17}) and so it is difficult to distinguish the dynamical structures. Note that with this orthonormal basis one vector gives a good result, another one outlines a good result and two of them give some spurious structures. We have chosen them on purpose, just to show that among the infinity possibilities we may find good and bad ones. On the middle plots of Fig. 3.2 we show on the pictures the evolution of the OFLI on the line $x = Y = 0$ (the discontinuous line on the top figures) for different initial conditions for the sensitivity values (we plot the OFLI value at $t_f = 100$). On the bottom pictures we show the evolution of the OFLI values in the time interval [1, 10000] for four particular orbits (remarked with a dotted vertical line on the pictures on the middle) indicated with the letters a (a periodic orbit), b and c (orbits on a KAM torus) and d (a chaotic orbit close to the hyperbolic point). We observe that the OFLI with tangent initial conditions T gives very small values and therefore, the evolution is much slower than in the other cases (for example the periodic orbit has not reach yet its constant value), the OFLI_{O_2} gives constant values also for non-periodic orbits although it points out clearly the hyperbolic point, the OFLI_{O_3} does not clearly identify the periodic orbit for a low time integration and OFLI_{O_1} seems to give a correct information. Therefore, from these pictures it seems necessary a more detailed study (see [19] for more details) on how to select the initial conditions of the variational equations.

Recalling that variational methods tend to estimate the maximum Lyapunov characteristic exponent, let us first review some results of the Lyapunov exponents theory. In the continuous case, we have a dynamical system [23] on the state space M defined by a diffeomorphic flow map

$$\phi^t : M \rightarrow M,$$
$$\mathbf{y} \mapsto \phi^t(\mathbf{y})$$

given by an ordinary differential equation $\dot{\mathbf{y}} = \mathbf{f}(\mathbf{y})$ with formal solution $\mathbf{y} = \mathbf{y}(t; \mathbf{y}_0) = \phi^t(\mathbf{y}_0) \in M$. The Lyapunov exponents are based on the solution on the standard orthonormal basis of the linearized flow map Y that maps the tangent space $T_{\mathbf{y}_0}M$ into $T_{\phi^t(\mathbf{y}_0)}M$ and is given by the resolvent (or stability matrix) of the matrix linear system

$$\dot{Y}(t) = D\mathbf{f}(t, \mathbf{y}(t))\, Y(t) := \frac{\partial \mathbf{f}(t, \mathbf{y}(t))}{\partial \mathbf{y}}\, Y(t), \qquad Y(t_0) = \mathbb{Y},$$

where $\mathbb{Y}$ denotes any orthonormal basis (usually we take $\mathbb{Y} = \mathbb{I}$, that is, the identity matrix). Once we have the resolvent we have all the solutions of the variational equations on the form $\delta\mathbf{y}(t) = Y(t)\, \delta\mathbf{y}_0$. The resolvent matrix may have complex eigenvalues, so—in order to simplify the stability analysis—it is common to use the singular value decomposition (SVD) of the resolvent [24]. That is, to put $Y = UDV^\top$ with U and V orthogonal matrices and D diagonal. The diagonal

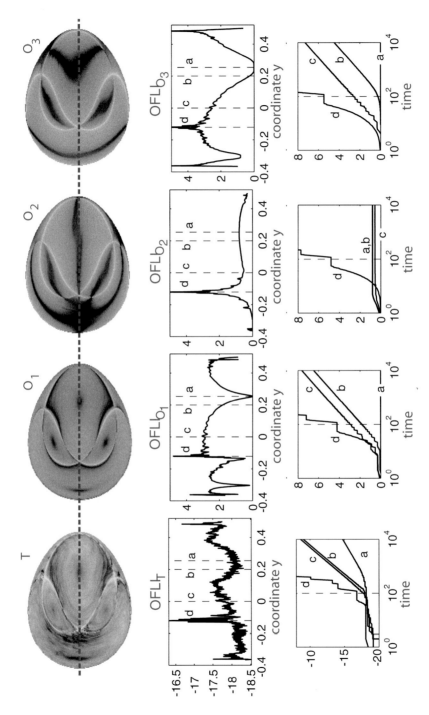

Fig. 3.2 *Top*: OFLI values for different initial conditions for the sensitivity values (T, O_1, O_2 or O_3) for the Hénon–Heiles system

elements of D are the square roots of the eigenvalues of the matrix $Y^\top Y$, which is now a symmetric positive definite matrix, hence its eigenvalues are real and its eigenvectors form an orthogonal basis. We denote by $\Lambda_i(t; \mathbb{Y})$ the eigenvalues of $Y^\top Y$ at time t and using the initial matrix $\mathbb{Y}$. We call *local or finite-time Lyapunov exponents* the real numbers

$$\lambda_i(t; \mathbb{Y}) := \frac{1}{2t} \ln \Lambda_i(t; \mathbb{Y})$$

and *local or finite-time Lyapunov vectors* the eigenvectors of

$$Y^\top(t; \mathbb{Y}) Y(t; \mathbb{Y}).$$

The matrix $Y^\top Y$ may be also interpreted in a more geometrical setting as the flat metric tensor of Eulerian space transformed to Lagrangian coordinates (for details see [25]).

The asymptotic behavior of the finite-time Lyapunov exponents and vectors is governed by the multiplicative ergodic theorem of Osedelec [26] that states that for any ergodic probability measure p on the state space $M = \mathbb{R}^n$ and for any solution $\mathbf{y}(t)$ of the differential equation we have [27]:

1. For p-almost all $\mathbf{v} \in \mathbb{R}^n$, there exists a finite exponent

$$\lambda = \lim_{t \to \infty} \frac{1}{t} \ln \frac{\|Y(t)\mathbf{v}\|}{\|\mathbf{v}\|},$$

   that does not depend on the initial time and takes at most n values $\lambda_1 \geq \lambda_2 \geq \ldots \geq \lambda_n$ (*the Lyapunov exponents*).
2. There exists the limit matrix,

$$L(t_0) = \lim_{t \to \infty} \{Y^\top(t; \mathbb{Y})Y(t; \mathbb{Y})\}^{1/2t}.$$

   The non-integer power of the matrix $Y^\top Y$ is defined by diagonalization. Note that this fact asses that the finite-time Lyapunov exponents $\lambda_i(t; \mathbb{Y})$ will have a limit, the Lyapunov exponents λ_i of the orbit.
3. There exists a sequence of embedded subspaces

$$S_n(t_0) \subset S_{n-1}(t_0) \subset \ldots \subset S_1(t_0) = \mathbb{R}^n,$$

   such that on the complement $S_i(t_0) \backslash S_{i+1}(t_0)$ of $S_{i+1}(t_0)$ in $S_i(t_0)$ the exponential growth (or decay) rate is λ_i.

The eigenvectors $\mathbf{l}_i$ of the limit matrix $L(t_0)$ are called Lyapunov vectors and the ith one belongs to $S_i(t_0) \backslash S_{i+1}(t_0)$. The convergence of the Lyapunov vectors is exponential, so its behaviour at finite time describes well its asymptotic limit, but the convergence of the Lyapunov exponents is very slow. This fact is usually employed to obtain the MLE (the kernel of all the variational indicators). However, problems appear when all $\lambda_i = 0$. Now, following the ergodic theorem it is not easy to compute for all the orbits the same Lyapunov exponent (in fact the same finite-time Lyapunov exponent).

The key point to study some spurious patterns is to study the directions associated with zero Lyapunov exponents.

Proposition *The function $V = \mathbf{f}(t, \mathbf{y})$ is the solution of the variational equation (3.2) with initial conditions $\delta \mathbf{y}_0 = \mathbf{f}(t_0, \mathbf{y}_0)$. Moreover, if the support of the ergodic measure p does not reduce to a fixed point then these initial conditions in the variational equations generate a zero Lyapunov exponent.*

The above Proposition [28] establishes that for any orbit at least one Lyapunov exponent vanishes. We may enforce the above result just pointing that the solution of the variational equations using any vector tangent to the flow will generate a solution tangent to the flow with the same proportionality constant.

But, what happens if we work with Hamiltonian systems? Now the differential system and the Lyapunov spectra possess a specific structure [29]. Given a $2n$ degrees of freedom Hamiltonian function $\mathscr{H}$ the differential system is given by

$$\dot{\mathbf{y}} = J \nabla \mathscr{H},$$

with J the skew-symmetric matrix

$$J = \begin{pmatrix} 0 & \mathbb{I} \\ -\mathbb{I} & 0 \end{pmatrix}.$$

In this case the stability matrix Y is symplectic (that is, $Y J Y^{\top} = J$) and, for conservative Hamiltonians, if λ_i is a Lyapunov exponent then also $-\lambda_i$ is another one: the exponents are grouped in pairs. Therefore, as at least one Lyapunov exponent is zero, automatically two of them are zero. Moreover [30], if $\mathscr{H}$ is constant then for any solution $\delta \mathbf{y}(t)$ of (3.4) one has

$$\frac{d}{dt} \langle \delta \mathbf{y}(t), \nabla \mathscr{H} \rangle = 0.$$

An important consequence of the above result is that if a solution of the variational equation is orthogonal to $\nabla \mathscr{H}$ at any time, then it will always remain orthogonal. Also, the projection of $\delta \mathbf{y}(t)$ onto such a vector is constant. Besides, the vector $\nabla \mathscr{H}$ is a Lyapunov vector associated with a zero Lyapunov exponent [30]. In fact, any given conserved quantity gives two zero Lyapunov exponents. Also, by Noether's theorem, a symmetry in the dynamics implies a zero Lyapunov exponent. A completely different behavior is associated to the projection onto the tangential direction of the flow: no answer can be given for $\langle \delta \mathbf{y}(t), \mathbf{f}(t, \mathbf{y}) \rangle = \langle \delta \mathbf{y}(t), J \nabla \mathscr{H} \rangle$. So, even the first Lyapunov vector cannot be made orthogonal to the flow at every instant.

These two special directions $J \nabla \mathcal{H}$ and $\nabla \mathcal{H}$ are usually considered as marginal. The displacement in the direction of $J \nabla \mathcal{H}$ gives just a displacement in the reference trajectory and the displacement in the direction of $\nabla \mathcal{H}$ will give rise to a transfer to a nearby orbit with a Hamiltonian value different from that of the reference one. Therefore, several strategies have been designed to separate these Lyapunov vectors from the rest [30] (in [31] a special algorithm is designed to maintain these vectors for dissipative systems). But what happens if all the Lyapunov exponents vanish as it happens for regular orbits? This is precisely the situation where the Chaos Indicators based on first order variational equations seem to have problems and it may generate spurious patterns. So, a priori, methods based on second (or higher) order variational equations (OFLI2) are more suitable when all the Lyapunov exponents vanish. The main reason of this better performance is just a probabilistic approach. As it is shown below, the appearance of spurious structures is related with the points where the initial conditions of the variational equations are aligned close to the directions with some minimal Lyapunov exponents, and in the case of the OFLI2, as we have two terms there is a low chance that our fixed initial conditions point at the same time to these special directions for the first and second order terms.

In conservative Hamiltonians with one degree of freedom the situation is quite simple: for each orbit both Lyapunov exponents vanish. The direction tangent to the flow generates a very low value of the variational Chaos Indicators because for periodic orbits the ratio $\|\mathbf{f}(t)\|/\|\mathbf{f}(t_0)\|$ has only small variations. Thus we may expect that if the initial conditions of the variational equations follow that direction only for some orbits, then we will obtain spurious patterns. The methods that use first order variational equations (MEGNO and FLI) exhibit a spurious line, as shown in Fig. 3.3. In order to have an initial vector $\delta \mathbf{y}_0 = (\delta x_0, \delta y_0)^\top$ for the variational equations tangent to the flow in the pendulum equations

$$\dot{x} = y, \quad \dot{y} = -\sin x,$$

we need

$$y_0\, \delta y_0 = -\sin(x_0)\, \delta x_0.$$

And so, for $\delta y_0 \neq 0$,

$$y_0 = -\frac{\delta x_0}{\delta y_0}\, \sin(x_0).$$

In Fig. 3.4 we show (in red online) the curve of points where the initial variation vector $(1, 1)^\top$ is tangent to the flow. This figure coincides quite well with the observed spurious patterns in Fig. 3.3.

According to the above discussion, it seems reasonable to avoid the tangent direction when we use first order variational equations. In Hamiltonian systems with just one degree of freedom there is actually a single choice to avoid the tangent: the

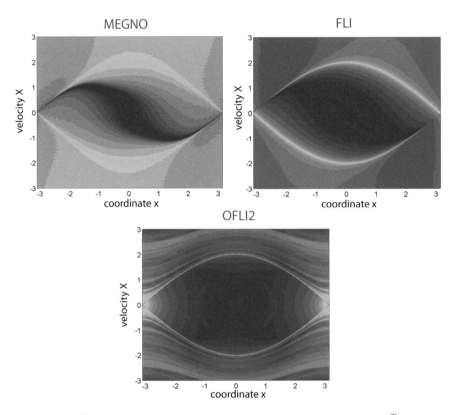

Fig. 3.3 MEGNO and FLI plots for the pendulum problem using the vector $(1, 1)^{\top}$ as initial conditions of the variational equations, and OFLI2 plot

vector orthogonal to the flow, that in this case coincides with the gradient of the Hamiltonian $\nabla \mathcal{H}$. Then the vector $\delta \mathbf{y}$ evolving in time always has an orthogonal component although a tangent one will also appear. This fact prevent us from approaching the Lyapunov vector as fast as we would like.

Let us see how this strategy works in the Hénon–Heiles Hamiltonian. We compute the FLI indicators using initial vectors $\nabla \mathcal{H} / \|\nabla \mathcal{H}\|$. The results shown in Fig. 3.5 suggest that now the FLI indicator works quite well, as OFLI2 does automatically. Therefore, as proposed in [19], a good set of initial conditions for the variational equations for any variational chaos indicator is given by $\delta \mathbf{y}_0 = \nabla \mathcal{H} / \|\nabla \mathcal{H}\|$.

To summarize this problem of spurious structures we remark that, obviously, this problem has been tackled by many other researchers using other approaches. One option is to use random initial conditions for the variational equations (as used in some articles for the SALI method [10]) and another option is to compute the chaos indicator values for two different initial orthogonal vectors for the variational equations and to use their mean value (as done for the FLI method in [32]). Both

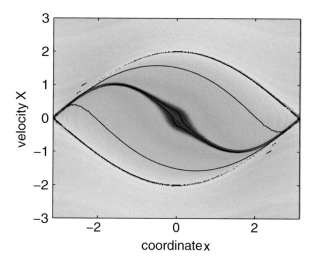

Fig. 3.4 FLI plot of Fig. 3.3, its contour plot and (in *red*) the theoretical predicted spurious pattern for the chosen initial conditions of the variational equations $(1, 1)^\top$

options provide better results than the standard application, and in most cases they eliminate the spurious structures. The only problem is that we may miss some of the information inside the regular region, and in order to obtain a result that fits well with, for instance, the skeleton of periodic orbits (with the OFLI or OFLI2 indicators), we have to use just one initial condition because we compare one value with the ones close to it. In that situation the above approaches, OFLI2 or $\delta \mathbf{y}_0 = \nabla \mathscr{H} / \|\nabla \mathscr{H}\|$ for OFLI, provide a quite suitable option.

3.2.2   Locating Periodic Orbits

A quite important problem in studying Dynamical Systems is to locate the position of the periodic orbits, what Poincaré called the "skeleton" of the system. There are several methods in literature, but a good point of the OFLI and OFLI2 methods is that they can be used to locate approximate values of periodic orbits. Note that for a continuous flow there always exists a differential rotation along any trajectory that produces an increase of the tangential component to the flow of the variational equations. Therefore the FLI and similar techniques cannot detect periodic orbits, whereas the orthogonal versions, OFLI and OFLI2, as they cancel that component when computing values for a periodic orbit their value reaches, after a transient behavior, a constant value [16]. In our case, to compute the skeleton of periodic orbits after a first scanning with the OFLI2, we have used a systematic search approach that takes advantage of the symmetries of the system (see [33] for a complete description). Another standard method is the use of modified Newton

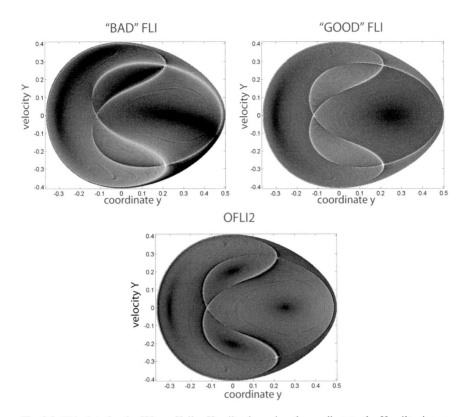

Fig. 3.5 FLI plots for the Hénon–Heiles Hamiltonian using the gradient to the Hamiltonian as initial conditions of the variational equations (*good* FLI) and another initial vector (*bad* FLI), and OFLI2 plot

methods [34], that permits to obtain the initial conditions of the periodic orbits with any desired precision.

On Fig. 3.6 we show the Poincaré sections and the OFLI2 pictures at $t_f = 300$ for the Hénon–Heiles problem for different values of the energy E on the surface $x = 0$. For $E = 1/12$ most of the orbits are regular as shows the Poincaré sections and as reflects the OFLI2 picture. Note that the OFLI2 locates the separatrices and no spurious structures are present. For $E = 1/8$ the OFLI2 gives much more information than the Poincaré sections and locates, without a selection of the orbits, the periodic orbits and the chain of islands inside the chaotic sea.

In order to study with more detail the use of the OFLI and OFLI2 indicators in the location of periodic orbits we show on the top of Fig. 3.7 the skeleton of symmetric periodic orbits for the Hénon–Heiles system up to multiplicity 12 [35, 36] on the line $x = Y = 0$. Note that in such a plot any point corresponds to the initial conditions of one orbit. The plot correspond to values of the energy below the escape region. The forbidden region is located outside the thick black line. We note

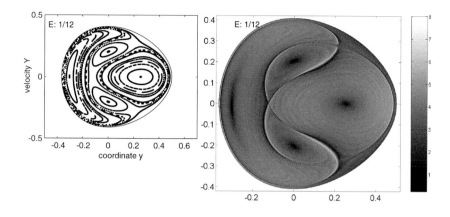

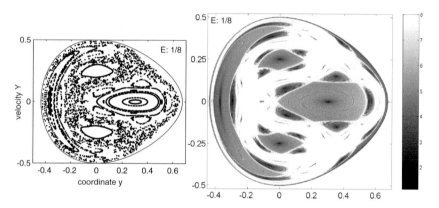

Fig. 3.6 Poincaré surfaces of section and OFLI2 plots for the Hénon–Heiles system on the section $x = 0$ with energy $E = 1/12$ and $E = 1/8$

the presence of the families of the normal modes $\Pi_{4,7,8}$ (the black lines originating at $E = 0$) that configure the behavior for large E as the other families of periodic orbits accumulate around them and define the boundaries of the exit basins [35]. On the bottom picture we present the values of the OFLI2 indicator showing how the minimum values are related with the location of the periodic orbits. The three normal modes are pointed out with solid red dots, whereas several higher order multiplicities are pointed out with red circles. Note, as said in [16], that the OFLI and OFLI2 indicators just can tell us that in the neighborhood of a minimum value there should be a periodic orbit, but to locate the exact values of the periodic orbits one has to use, as commented before, another method using these values as approximations (for instance by applying a Newton method). In any case, to locate periodic orbits is not easy as for resonant orbits close to a periodic orbit the OFLI and OFLI2 indicators behave temporarily as a constant value, and only after a transient time their value increase.

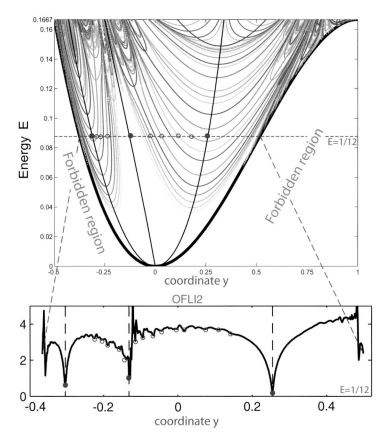

Fig. 3.7 *Top*: Skeleton of symmetric periodic orbits for the Hénon–Heiles system up to multiplicity 12 on the line $x = Y = 0$. *Bottom*: OFLI2 for the value of the energy= $1/12$ pointing out the minimum values at the location of several periodic orbits. The three *solid red dots* point the location of three normal modes

3.3 Applications: Hamiltonian Systems

Most of the chaos indicators are designed for Hamiltonian systems. In this section we show several examples of applications of the OFLI2 indicator in different studies [18, 35–41].

As it has been noted in the previous section the OFLI2 indicator permits to locate periodic orbits. Therefore, it is interesting to study the skeleton of periodic orbits of a system and to compare with the OFLI2 plot in order to complete the analysis. We present an example of this application in the Copenhagen problem [39].

The three-body problem is one of the oldest problems in dynamical systems. The Restricted Three-Body Problem (RTBP) supposes that the mass of one of the three bodies is negligible. It was considered by [42] and [43], and it can serve as an example of classical chaos. For the remaining two bodies, the case of equal masses was first investigated by Strömgren and his colleagues of the Copenhagen group. Its name is derived from the series of papers published by them starting in 1913.

Defining the distances to the respective primaries as:

$$r_1^2 = (x + \mu)^2 + y^2,$$
$$r_2^2 = (x - (1 - \mu))^2 + y^2,$$

where $\mu = m_1/(m_1 + m_2)$ with m_1 and m_2 the masses of the two bodies, the equations of motion of the Restricted Three-Body Problem are

$$\ddot{x} - 2\dot{y} = x - (1 - \mu)\frac{x + \mu}{r_1^3} - \mu\frac{x - 1 + \mu}{r_2^3},$$
$$\ddot{y} + 2\dot{x} = y - (1 - \mu)\frac{y}{r_1^3} - \mu\frac{y}{r_2^3}.$$

The only known integral of motion of this problem is the Jacobian constant

$$C = -(\dot{x}^2 + \dot{y}^2) + 2\frac{1 - \mu}{r_1} + 2\frac{\mu}{r_2} + x^2 + y^2$$

that can be used to define the effective energy as $E_J = -C/2$. Since the Copenhagen problem is the particular case of $m_1 = m_2$ we have to take the value $\mu = 1/2$. That is, it is the restricted planar three-body problem where the two big spherical and homogeneous primaries have equal masses and rotate with constant angular velocity in circular orbits around their centre of mass, while a small massless particle moves under the resultant Newtonian action of the two primaries. On the top of Fig. 3.8 we draw a simple picture of the problem with the two primaries in blue color. Note that this problem has been revisited during the last 20 years (>1990) after the discovery of many extrasolar planetary systems which are two-body systems where the two primaries have almost equal masses.

In Fig. 3.8 we show the evolution with the energy of the OFLI2 and the skeleton of symmetric periodic orbits for the Copenhagen problem [39], studying the variation in the x-axis by fixing $y = \dot{x} = 0$ and depending on the energy E_J. We note that these pictures give us a clear idea of the evolution of the system. When E_J is low the system has large forbidden regions and the motion is highly regular. Increasing E_J the system is more and more complex. The OFLI2 plot shows that the chaotic behavior appears mainly in the range $E_J \in [-1.75, 0]$. When E_J is very high the behavior is again more regular. The position of the two primaries are marked with discontinuous vertical lines. The skeleton of symmetric periodic orbits is done up to global multiplicity $m = 4$. On the figure we have used a color code for the different multiplicities. The region with a great number of periodic orbits denotes

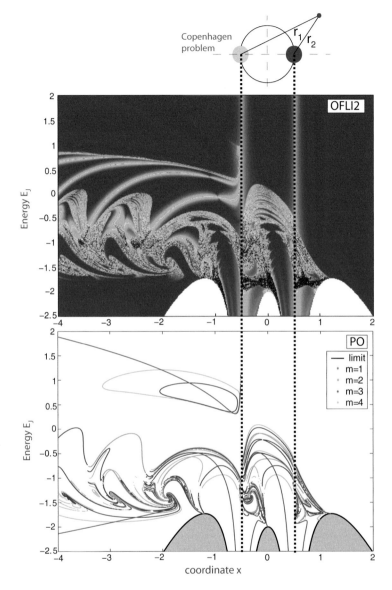

Fig. 3.8 OFLI2 plot and the skeleton of symmetric periodic orbits (PO) up to multiplicity $m =$ 4 showing the evolution with energy (coordinate x versus energy E_J plots) for the Copenhagen problem

also the regions with chaotic behavior (see the OFLI2 plot). Note that each point on the curves stands for the initial conditions of a symmetric periodic orbit and each curve is a family of periodic orbits.

Thanks to the OFLI2 indicator one is able to not only compare with the skeleton of periodic orbits, but also to join both methods of analysis giving a much more detailed study of how the system changes with a parameter. To that goal we show another example [40], with an interesting generic family of Hamiltonian systems, which is given by $\mathscr{H} = \frac{1}{2}(\dot{x}^2 + \dot{y}^2) + V(x, y)$ with the quartic potential

$$V(x, y) = \frac{1}{2}n(x^2 + y^2) + \alpha x^2 y^2 + \frac{1}{4}\beta(x^4 + y^4), \tag{3.7}$$

that was proposed by Andrle [44] for a stellar system with an axis and a plane of symmetry, and later used in many applications. This potential depends on parameters $n, \alpha, \beta \in \mathbb{R}$, and it is known to be integrable for some values of the parameters. This Hamiltonian presents the D_4 symmetry, that is, it is invariant under a rotation by $\pi/2$. It has also the time-reversal symmetry. One particular and interesting case is the dihedral potential [45] ($n = -2, \alpha = 1/4, \beta = 1$)

$$V(x, y) = \frac{1}{4}(x^2 + y^2)^2 - (x^2 + y^2) - \frac{1}{4}x^2 y^2. \tag{3.8}$$

This problem also appears when studying the Bogdanov–Takens bifurcation at the origin. This bifurcation has interest in fluid dynamics related to convection problems in a container such as in a magnetoconvention [46, 47] model with a vertical magnetic field, but it can also appear in models with a salt gradient (thermohaline convection), a Coriolis force or other stabilizing effects (see [46] for more details). Due to this stabilizing vertical gradient, the otherwise $O(2)$ symmetry of the convection problem is broken and a D_4 symmetric system (assuming a square container; other symmetries are possible in other containers) remains that depends on a parameter. The variation of this parameter allows to study the Bogdanov–Takens codimension two bifurcation on those systems. In a limit case the system is Hamiltonian [45] with the potential of Eq. (3.8).

Figure 3.9 shows the evolution of the dihedral potential as the energy grows. We have combined the skeleton of periodic orbits together with several OFLI2 plots for some values of the energy. This plot shows not only the periodic orbits, but also the KAM tori around the periodic orbits and the chaotic regions. We can see how the system evolves as the energy grows. For very low values of the energy ($E = -0.7, E = -0.5$), the system appears to be very chaotic and it is divided in two disconnected regions. When the energy increases to $E = 0.0$, the two regions touch and they are connected at $x = 0$. In the OFLI2 plots we can see several complex structures with islands due to periodic orbits (blue), separated by chaotic regions (red). If we further increase the energy, we see at $E = 2.0, E = 5.0$ and $E = 10.0$ that the two regions are now completely connected and new structures appear

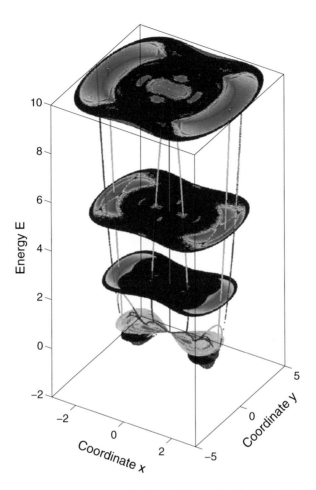

Fig. 3.9 Several OFLI2 plots on the plane (x, y) for different values $\{-0.7, -0.5, 0, 2, 5, 10\}$ of the energy together with the skeleton of symmetric periodic orbits for the dihedral squared symmetric Hamiltonian

and merge. As observed, there is a correspondence with the skeleton of periodic orbits. At $E = 5.0$ new structures appear that evolve as the energy grows, and at $E = 10.0$ those islands have merged into a bigger structure. A similar evolution happens around the edges, where some families appear, giving rise to some new islands. These plots show us how the system evolves when the energy changes, but to complement this study we have to analyze the bifurcations of the system (for more details see [40]).

3.3.1 Detecting New Dynamics: "Safe Regions"

Another interesting application of the fast chaos indicators, and in particular of OFLI2 is to detect new phenomena. The option of obtaining a large number of biparametric plots in a reasonable CPU time permits to explore large regions to find interesting behaviours. For instance, for the Hénon–Heiles system it was more or less established that just a few after the escape energy all the stable regions disappear. In [48], after a systematic study of biparametric plots it was shown that what disappear are just the stable regions due to KAM tori and several stable regions appear inside the escape region. These regions are denominated "safe regions" in [35, 48] as they permit to have bounded regular regions in large areas of fast escape dynamics.

In Fig. 3.10 we show on the top an OFLI2 plot (1000×1000 points) of the Hénon–Heiles system on the (y, E) plane. Up to the escape energy $E_e = 1/6$ most of the orbits are escape orbits. Thanks to a large number of OFLI2 plots it has been possible to locate several "safe regions" far from the escape energy. On the bottom plot we present a magnification of one of such regions with some key bifurcations. In the plot we show the OFLI2 plots of just the bounded regular region on color scale, and on the right we present schematic Poincaré sections computed from the normal form of the different bifurcations. We suppose that the bifurcation occurs at the value of the parameter $\mathscr{P}_B$ (in our case the energy E) and we write $\mathscr{P} = \mathscr{P}_B + \varepsilon$. Note that we only show one direction, from $\varepsilon < 0$ to $\varepsilon > 0$, but it could be the opposite depending on the particular bifurcation. The main family of multiplicity $m = 1$ is the one in black. The other families bifurcate from this on the points marked with a circle. The main family, and in fact all the safe region, appears with a saddle-node bifurcation (SN) and corresponds to the case where two periodic orbits are created (or destroyed), one stable and another one unstable. This is the only way of creating new families of periodic orbits, apart from the boundaries of the domain of definition of the Poincaré map. Another bifurcation that appear in this region is an isochronous pitchfork bifurcation (P). It is a symmetric pitchfork bifurcation: from a symmetric periodic orbit two new isochronous periodic orbits are created but with fewer symmetries (the symmetry in this case is on the y-axis, not D_3) and the main symmetric periodic orbit changes its stability character after the bifurcation. Besides, we plot the generic touch-and-go bifurcation, as an example of a known generic bifurcation where an unstable periodic orbit of multiplicity $m = 3$ touches [33] the center $m = 1$ periodic orbit and "bounces" while the main orbit remains stable.

Therefore, we have shown that the use of a fast chaos indicator like OFLI/OFLI2 provides a quite useful tool in the numerical study of Hamiltonian systems.

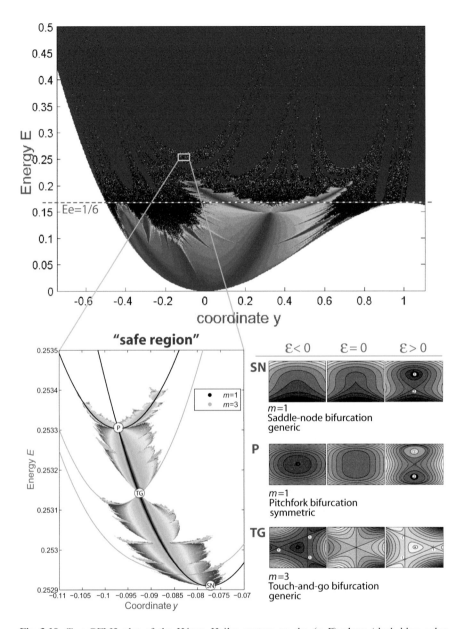

Fig. 3.10 *Top*: OFLI2 plot of the Hénon–Heiles system on the (y, E) plane (*dark blue color* denotes escape orbits). *Bottom*: magnification of a bounded region ("safe region") inside the escape region (only the OFLI2 values of the bounded regular orbits are shown). On the *left*, both OFLI2 and families of periodic orbits (the main bifurcation parameters are $\mathscr{P}_B(\text{P}) = E_P \simeq 0.25331$, $\mathscr{P}_B(\text{TG}) = E_{\text{TG}} \simeq 0.25314$ and $\mathscr{P}_B(\text{SN}) = E_{\text{SN}} \simeq 0.25292$). On the *right*, schematic Poincaré surface sections of some of the bifurcations involved in this region

3.4 Applications: Dissipative Systems

The fast chaos indicators may be also used in the study of dissipative systems, and not only on Hamiltonian dynamics. As now the dynamics is quite different it is advisable to use the fast techniques as preliminary studies to select the correct region to later perform a more detailed analysis. In our studies [49–55] we have called this combined use BPD (Biparametric Phase Diagrams), as the global procedure is designed mainly for biparametric plots, first using OFLI2 and later refined using MLE to obtain the final plots. We illustrate this procedure with two paradigmatic dissipative systems: the Lorenz and the Rössler model.

The Lorenz model [56] is one of the most classical low-dimensional chaotic problems because it is one of the first models with the presence of chaotic behavior and chaotic attractors.

The Lorenz model has attracted the attention of a large number of researchers and a great number of papers continue to appear covering the model [2, 49–51, 57–61]. The Lorenz model is a simplification of a more complicated model, presented by Saltzman [62], to describe buoyancy-driven convection patterns in the classical rectangular Rayleigh-Bénard problem applied to the thermal convection between two plates perpendicular to the direction of the Earth's gravitational force. After several simplifications, Lorenz arrived at his famous equations:

$$\frac{dx}{dt} = -\sigma x + \sigma y, \qquad \frac{dy}{dt} = -xz + rx - y, \qquad \frac{dz}{dt} = xy - bz, \qquad (3.9)$$

where t is a dimensionless time, and σ (the Prandtl number), r and b are three dimensionless control parameters.

In [49], the authors made an extensive numerical study of the Lorenz model based of the OFLI2 chaos indicator. From the numerical tests, they conjectured that the region of parameters where the Lorenz model is chaotic is bounded for fixed r. We have determined the complete parametric chaotic region for the classical Lorenz system [49, 50]. In Fig. 3.11 we show some of the findings from [49–51]. The three plots Fig. 3.11a1, a2 and a3 are done first with the OFLI2 Chaos Indicator (to locate the interesting values of parameters) and later with the Maximum Lyapunov Exponent (to provide a more standard output) using $(x(0), y(0), z(0)) = (60, 60, 10)$ as the initial conditions and fixing the third parameter (the one that is not on the biparametric plot) at the Saltzman values $(r, b, \sigma) = (28, 8/3, 10)$. The white color is associated with chaotic behavior, whereas the black color is associated with regular behavior. The biparametric plot -a1.a- is a magnification of a1, and on the top we present a MLE curve at $(b, \sigma) = (8/3, 10)$. One of the remarkable properties of Fig. 3.11 is that in the (σ, b) plane (plot -a2-) the chaotic region seems to be bounded. In [50] such a conjecture was proved, establishing that the Lorenz system (3.9) has, for any given fixed parameter value $r > 1$, a chaotic region bounded in b, and if $b \geq \epsilon > 0$ then the region is bounded in σ too (in fact, outside a bounded region every positive semiorbit of the Lorenz system converges to an equilibrium).

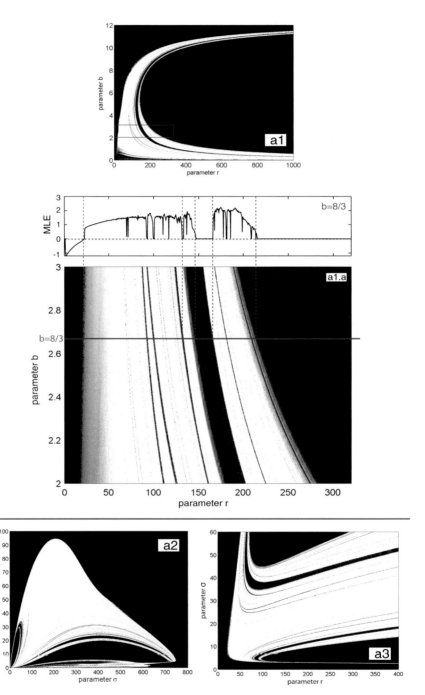

Fig. 3.11 OFLI2 and MLE (BPD) plots for the Lorenz model depending on the parameters of the system (*black color* denotes regular behavior and *white* chaotic behavior). Panel -**a1**- presents the BPD plot for the (r, b) parametric plane (the panel -**a1.a**- depicts a magnification and the *top plot* shows the MLE in the *straight line* $b = 8/3$), -**a2**- the (σ, b) plane and -**a3**- the (r, σ) plane

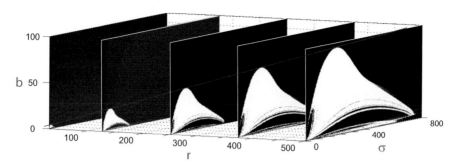

Fig. 3.12 Three-dimensional parametric evolution of the Lorenz model giving a complete study of the chaotic region

Moreover, joining several BPD (MLE and OFLI2 plots) we have been able to provide a complete three-dimensional parametric evolution of the Lorenz model giving a detailed study of the complete chaotic region [49, 50]. This is shown in Fig. 3.12.

Another low dimensional paradigmatic problem that have been frequently studied is the Rössler model [63]. The Rössler equations are given by

$$\begin{aligned}
\dot{x} &= -(y + z), \\
\dot{y} &= x + ay, \\
\dot{z} &= b + z(x - c),
\end{aligned} \tag{3.10}$$

with $a, b, c \in \mathbb{R}$, and they are assumed to be positive and dimensionless. This model is a famous prototype of a continuous dynamical system exhibiting chaotic behavior with minimum ingredients.

The Rössler system is not always dissipative as the divergence is given by $\mathrm{div}\{\mathbf{f}(x, y, z)\} = a - c + x$. Thus, in a large region of parameters (especially when a grows) and for large values of the variable x, we will have a positive divergence and so escape orbits. Therefore, apart from regular and chaotic orbits, the Rössler system also has escape orbits with transient chaos or regular behavior before escaping. This fact makes more difficult a theoretical and numerical analysis of this problem. Note that there may exist escape trajectories when the divergence of the system is negative: the initial volume eventually shrinks along the target trajectory, but the trajectory goes to infinity without bounds. But in the dissipative case escape orbits will not occur in large sets in the phase space as it happens in the positive divergence case, where this is the normal behavior.

In Fig 3.13 we present some BPD diagrams on the (c, b) plane for different values of the parameter a. The blue color is associated with regular behavior, the red color with chaotic one and white color denotes escape orbits. The plots show us a pattern structure that is repeated when the parameter c grows, that consist on interlacing bands in groups of two. Also different bands of regular motion appear inside the chaotic region [52]. When the parameter a grows, more and more of such pair of

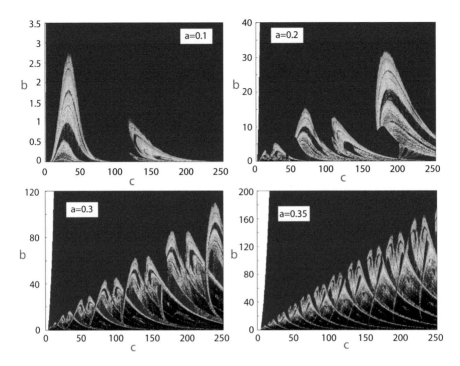

Fig. 3.13 BPD diagrams of the Rössler system on the (c, b) plane for different values of the parameter a

structures appear and the escape region grows in size, becoming the dominating behavior.

In the Rössler model, when one plots a BPD in the parametric plane (a, c) it is possible to observe several spiral structures [52, 64, 65] formed by branches of regular and chaotic orbits. These structures configure the global parametric organization of the Rössler system, and in fact of any dissipative system with Shilnikov saddle-foci. Therefore, thanks to the fast chaos indicator OFLI2 and the MLE (BPD) we can easily detect these interesting spirals. The remaining question is: how are they formed? To answer that question in [53, 54] an extensive use of BPD, combined with bifurcation analysis, has been done.

Figure 3.14 outlines the theoretical skeleton [53, 54] of the bifurcation unfolding around the spiral structure. The top picture sketches phenomenologically a caricature of the bifurcation structure of the spiral structure along with "shrimps" (also called swallows, periodic connecting windows, crossroads, ... in literature). In it, the saddle-node bifurcation curves originating from the B point (a Belyakov point) demarcate the boundaries of "shrimps" near the spiral. Indeed, the spiral can generate an infinite chain of "shrimps". A BPD zoom of the Rössler bifurcation diagram in the bottom picture depicts a few such "shrimps", S_{2j} and $S_{2j\pm1}$, which are singled out by the saddle-node curves (solid thick red curves). The

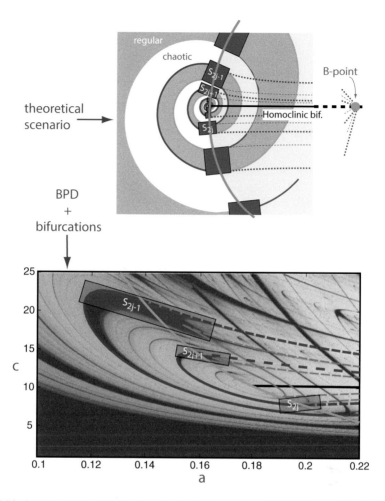

Fig. 3.14 Outline of the spiral structures in dissipative systems with Shilnikov saddle-foci (*gray* stands for regular motion, whereas *light gray* stands for chaotic motion). *Top*: Phenomenological sketch of the spiral structure formed by the "shrimps". *Bottom*: magnification of the bifurcation portrait of the spiral hub, overlaid with the homoclinic bifurcation curve (*black*) and the principal folded (*thick red*) and cusp-shaped (*thin blue*) bifurcation curves setting the boundaries for largest "shrimps" in the Rössler model

cusp-shaped saddle-node bifurcation curves (light thin blue curves) join the successive shrimps. Thus, both fold- and cusp-shaped bifurcation curves of saddle-node periodic orbits determine the local structure of the hub and the "shrimps." The latter serve as connection centers between hubs that contribute toward the formation of characteristic spiral structures in the bifurcation diagram of the system.

This generic scenario explains the formation of the spiral structures and "shrimps" in the biparameter space of a system with a Shilnikov saddle-focus [53, 54]. The skeleton of the structure is due to fold- and cusp-shaped bifurcation curves of saddle-node periodic orbits that accompany the homoclinics of the saddle-focus. These bifurcation curves distinctively shape the "shrimps" zones in the vicinity of the spiral hub. Again, a massive use of the OFLI2 indicator has permitted to study in detail such a new phenomena.

3.5 Conclusions

In the literature there are a large plethora of fast chaos indicators. Among them, the OFLI and OFLI2 (Orthogonal Fast Lyapunov Indicators) provide with some of the fastest methods, but they also permit to locate periodic orbits in Hamiltonian systems as shown in this paper. Besides, we have given a detailed analysis of the way to minimize the appearance of spurious structures in generic variational indicator methods (choosing as initial conditions of the variational equations the direction of the gradient of the Hamiltonian function). This problem is automatically avoided by the OFLI2 method. The combined used of OFLI/OFLI2 with other techniques, as skeletons of periodic orbits, bifurcation analysis and so on, gives rise to a quite powerful numerical study of Hamiltonian and dissipative dynamical systems. This fact has been shown by presenting several new phenomena detected in Hamiltonian dynamics (new insights in Copenhagen and dihedral problems, and location of "safe regions" in open systems) and dissipative systems (statement of boundness of chaotic region in Lorenz system and the location and study of spiral structures in systems with a Shilnikov saddle-focus). Finally, the efficient numerical integration of the flow and the automatic determination of the partial derivatives of the solution is performed by using the Taylor series method (TIDES software). This numerical ODE solver provides with the most efficient numerical method for high-precision requirements.

Appendix: Adaptive ODE Integrator—Taylor Series Method

A basic feature in using any chaos indicator is to obtain numerically solutions of ODE systems and variational equations. In Dynamical Systems there are also more requirements, for example, in the process of determination of periodic orbits we obviously have to integrate the differential system, normally for a short time, with very high precision, especially for highly unstable periodic orbits. Moreover, in the study of the bifurcations and stability of periodic orbits we also have to integrate the first order variational equations using as initial conditions the identity matrix as also occurs in the use of chaos indicators. To reach this goal we may, obviously, use any numerical ODE method like, for example, a Runge–Kutta method. The last few years, in the computational dynamics community [66] one of the preferred methods is the Taylor series method.

The Taylor method is one of the oldest numerical methods for solving ordinary differential equations but it is scarcely used in the numerical analysis community. Its formulation is quite simple [67]. Let us consider the initial value problem $\dot{\mathbf{y}} = \mathbf{f}(t, \mathbf{y})$. Now, the value of the solution at t_i (that is, $\mathbf{y}(t_i)$) is approximated by $\mathbf{y}_i$ from the nth degree Taylor series of $\mathbf{y}(t)$ at $t = t_i$ (the function $\mathbf{f}$ has to be a smooth function). So, denoting $h_i = t_i - t_{i-1}$,

$$\mathbf{y}(t_0) =: \mathbf{y}_0,$$

$$\mathbf{y}(t_i) \simeq \mathbf{y}_{i-1} + \mathbf{f}(t_{i-1}, \mathbf{y}_{i-1})\, h_i + \frac{1}{2!} \frac{d\mathbf{f}(t_{i-1}, \mathbf{y}_{i-1})}{dt} h_i^2 + \dots$$

$$\dots + \frac{1}{n!} \frac{d^{n-1}\mathbf{f}(t_{i-1}, \mathbf{y}_{i-1})}{dt^{n-1}} h_i^n =: \mathbf{y}_i.$$

Therefore, the problem is reduced to the determination of the Taylor coefficients $\{1/(j+1)!\, d^j\mathbf{f}/dt^j\}$. This may be done quite efficiently by means of the automatic differentiation (AD) techniques (for more details see [20]). Note that the Taylor method has several good features; one of them is that it gives directly a dense output in the form of a power series being therefore quite useful when an *event location* criteria may be used (as in the computation of Poincaré sections), it can be formulated as an interval method giving guaranteed integration methods (used, by instance, in the computer assisted proof of chaos [1] and skeletons of periodic orbits [68]), Taylor methods may manage directly high order differential equations just taking into account that the Taylor coefficients for the solution and its derivatives are evidently related, Taylor methods of degree n are also of order n and so Taylor methods of high degree give us numerical methods of high order (therefore, they are very useful for high-precision solution of ODEs, as needed, for example, in some fine studies in dynamical systems [69] and in the computation of unstable periodic orbits [34, 70]).

Just as a short look at the practical implementation of the Taylor series method we remark that in the literature there are efficient variable-stepsize variable-order (VSVO) formulations. For example, in [20, 71, 72] the variable-stepsize formulation is based on the error estimator using the last two coefficients and gives the following stepsize prediction

$$h_{i+1} = \mathtt{fac} \cdot \min \left\{ \left(\frac{\mathtt{Tol}}{\|\{\frac{1}{(n-1)!} \mathbf{f}^{(n-2)}(t_i)\|_\infty} \right)^{\frac{1}{n-1}}, \left(\frac{\mathtt{Tol}}{\|\frac{1}{n!} \mathbf{f}^{(n-1)}(t_i)\|_\infty} \right)^{\frac{1}{n}} \right\}$$

where $\mathtt{fac}$ is a *safety factor* and $\mathtt{Tol}$ the user error tolerance. A very simple order selection that only depends on the user error tolerance is given [73] by the formula $n(\mathtt{Tol}) = -\frac{1}{2} \ln \mathtt{Tol}$. See [20, 71] for a more extensive analysis and comparison with variable-stepsize variable-order formulations of the Taylor method. In Fig. 3.15 we present some comparisons on the Hénon–Heiles problem with initial conditions $(x_0, y_0, X_0, Y_0) = (0, 0.52, 0.371956090598519, 0)$ and $E = 0.157494996$ in the time interval $[0, 200]$ using the Taylor method (software TIDES [72]) and

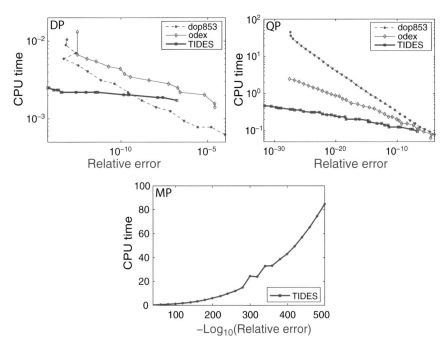

Fig. 3.15 Comparison of the CPU time in seconds vs. relative error in the numerical integration of a KAM orbit of the Hénon–Heiles problem using two well established codes, `dop853` (explicit Runge–Kutta method) and `odex` (extrapolation method), and the Taylor series method (`TIDES` code) with variable-order and variable-stepsize in double-precision (DP), quadruple-precision (QP) and multiple-precision (MP)

the well established codes `dop853` and `odex` developed by Hairer and Wanner [74]. These codes are based on an explicit Runge–Kutta of order 8(5,3) given by Dormand and Prince with stepsize control and dense output and the extrapolation method, respectively. All the methods are compared only in double and quadruple precision using the Lahey LF 95 compiler (`fortran`) because the `dop853` cannot be directly used in multiple precision. The multiple-precision tests are done using C++ and the GMP and MPFR [75] multiple precision packages. From Fig. 3.15 we note that for low precision the `dop853` code is a bit faster but when the precision demands are increased the Taylor method is by far the fastest, being for very high precision the only reliable method. Moreover, we can appreciate the different slope of the variable order method (Taylor method) and the fixed order one (`dop853`), being clear that for high precision the variable order schemes become the more competitive because they are more versatile.

For the computation of the OFLI and OFLI2 we are interested not only in the differential equations but also in the variational equations. In order to avoid their explicit generation we have devised [21] an alternative that permits us to obtain the solution of the variational equations without computing them explicitly. Therefore, we have to obtain a numerical solution of $\mathbf{y}(t)$ and $\mathscr{L}_{\delta\mathbf{y}(t_0)}\mathbf{y}(t)$, being $\mathscr{L}_{\delta\mathbf{y}(t_0)}\mathbf{y}(t)$ the Lie derivative of the solution $\mathbf{y}(t)$ with respect to the vector $\delta\mathbf{y}(t_0)$ (that is, in this case the directional derivative). Note that the partial derivatives of the solution with respect to the initial conditions are given by

$$\Pi = \left(\mathscr{L}_{\mathbf{e}_1}\mathbf{y}(t) \mid \mathscr{L}_{\mathbf{e}_2}\mathbf{y}(t) \mid \ldots \mid \mathscr{L}_{\mathbf{e}_n}\mathbf{y}(t) \right)$$

with $(\mathbf{e}_1, \mathbf{e}_2, \ldots, \mathbf{e}_n)$ the canonical base of $\mathbb{R}^n$.

The Taylor series method computes the Taylor series of the solution of the differential equation and the Taylor series of the partial derivatives of the solution

$$\delta\mathbf{y}(t_i) = \frac{\partial\mathbf{y}(t_i)}{\partial\mathbf{y}(t_0)} \cdot \delta\mathbf{y}(t_0) = \mathscr{L}_{\delta\mathbf{y}(t_0)}\mathbf{y}(t_i)$$

$$\simeq \mathscr{L}_{\delta\mathbf{y}(t_0)}\mathbf{y}(t_{i-1}) + \mathscr{L}_{\delta\mathbf{y}(t_0)}\mathbf{f}(t_{i-1})\,h_i + \frac{1}{2!}\,\mathscr{L}_{\delta\mathbf{y}(t_0)}\mathbf{f}^{(2)}(t_{i-1})\,h_i^2$$

$$+ \ldots + \frac{1}{n!}\,\mathscr{L}_{\delta\mathbf{y}(t_0)}\mathbf{f}^{(n-1)}(t_{i-1})\,h_i^n.$$

We now may compute the coefficients $1/(j+1)!\,\mathscr{L}_{\delta\mathbf{y}(t_0)}\mathbf{f}^{(j)}(t_{i-1})$ by rules of automatic differentiation of the elementary functions ($\pm$, $\times$, $/$, $\ln$, $\sin$, $\ldots$) obtained in [21]. Automatic differentiation gives a recursive procedure to obtain the numerical value of the reiterated derivatives of the elementary functions at a given point. We present here, as example, the rules for the sum, product by a constant, product, division and real power of functions (see [21] for the complete list of rules of any elementary operation):

Proposition *If* $f(t, \mathbf{y}(t)), g(t, \mathbf{y}(t)) : (t, \mathbf{y}) \in \mathbb{R}^{s+1} \mapsto \mathbb{R}$ *are functions of class* $\mathscr{C}^n$ *and given a vector* $\mathbf{v} \in \mathbb{R}^s$, *we denote*

$$f^{[j, 0]} := \frac{1}{j!}\frac{d^j f(t)}{dt^j}, \qquad f^{[j, 1]} := \frac{1}{j!}\mathscr{L}_{\mathbf{v}}f^{(j)},$$

that is, the jth Taylor coefficient of the function $f(t, \mathbf{y}(t))$ *and of its Lie derivative with respect to* $\mathbf{v}$, *respectively. Then, we have*

(i) *If* $h(t) = f(t) \pm g(t)$ *then* $h^{[n, i]} = f^{[n, i]} \pm g^{[n, i]}$.
(ii) *If* $h(t) = \alpha f(t)$ *with* $\alpha \in \mathbb{R}$ *then* $h^{[n, i]} = \alpha f^{[n, i]}$.
(iii) *If* $h(t) = f(t) \cdot g(t)$ *then*

$$h^{[n, 0]} = \sum_{j=0}^{n} f^{[n-j, 0]} \cdot g^{[j, 0]},$$

$$h^{[n, 1]} = \sum_{j=0}^{n} \left(f^{[n-j, 0]} \cdot g^{[j, 1]} + f^{[n-j, 1]} \cdot g^{[j, 0]} \right).$$

(iv) If $h(t) = f(t)/g(t)$ *then*

$$h^{[n,0]} = \frac{1}{g^{[0,0]}} \left(f^{[n,0]} - \sum_{j=0}^{n-1} h^{[j,0]} \cdot f^{[n-j,0]} \right),$$

$$h^{[n,1]} = \frac{1}{g^{[0,0]}} \left\{ f^{[n,1]} - h^{[n,0]} \cdot f^{[0,1]} \right.$$
$$\left. - \sum_{j=0}^{n-1} \left(h^{[j,0]} \cdot f^{[n-j,1]} + h^{[j,1]} \cdot f^{[n-j,0]} \right) \right\}.$$

(v) If $h(t) = f(t)^\alpha$ *with* $\alpha \in \mathbb{R}$ *and* $f^{[0,0]} \neq 0$, *then*

$$h^{[0,0]} = (f^{[0,0]}(t))^\alpha,$$

$$h^{[n,0]} = \frac{1}{nf^{[0,0]}} \sum_{j=0}^{n-1} (n\alpha - j(\alpha+1)) h^{[j,0]} \cdot f^{[n-j,0]},$$

$$h^{[0,1]} = \frac{1}{f^{[0,0]}} \alpha\, h^{[0,0]} \cdot f^{[0,1]},$$

$$h^{[n,1]} = \frac{1}{nf^{[0,0]}} \left\{ -nh^{[n,0]} \cdot f^{[0,1]} \right.$$
$$\left. + \sum_{j=0}^{n-1} (n\alpha - j(\alpha+1)) \left(h^{[j,0]} \cdot f^{[n-j,1]} + h^{[j,1]} \cdot f^{[n-j,0]} \right) \right\}.$$

The use of high-precision numerical integrators in the determination of periodic orbits is justified, for instance, by the search of highly unstable periodic orbits [34].

In Fig. 3.16 we show some comparisons for the Lorenz model (3.9) in double precision, all obtained with the code TIDES using the traditional way to compute the solution of the variational equations (VAR), that is writing them explicitly, and with the use of the extended Taylor series method (ETS) and using TIDES with this capability (using the extended Automatic Differentiation rules of Proposition 3.5). In the pictures we present computational relative error vs. CPU time diagrams in seconds. The extended Taylor series method is the fastest option with a low difference, but the most important thing is that the difference in the formulation is very high. Everyone knows how cumbersome is to write variational equations of order one, two and higher!! The picture is done for computing the complete order two and just the partial derivative $\partial^2 x/\partial x_0^2$. Note that TIDES can compute also sensitivities with respect to parameters of a system, not only with respect to initial conditions.

To end this section we remark that the use of the Taylor series method is currently helped by the free available new state-of-the-art numerical library TIDES (Taylor Integrator of Differential EquationS) that has just been developed by Profs. Abad,

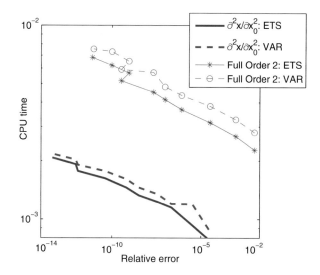

Fig. 3.16 Computational relative error in the computation of sensitivities vs. CPU time diagrams in seconds for the Lorenz model in double-precision using TIDES code using the extended Taylor series method for the solution of the variational equations (ETS) or just the standard Taylor series method with explicit formulation of the variational equations (VAR)

Barrio, Blesa and Rodríguez [72, 76]. The reader can contact the authors to obtain the software.[1]

Nowadays it is quite standard to preserve several geometric properties of the differential systems by means of "geometric integrators". This kind of methods are specially useful when we want to solve a problem with not very high precision but with a "constant" value of the energy, for instance. The problem for very long numerical integrations is that it doesn't matter how you perform the integration, finally the rounding errors of the computer will affect the integration, giving an increment of the error in the geometric object [77]. The optimal error in these quantities was studied first by Brouwer [78], who established that the error in energy grows at least as $\mathcal{O}(t^{1/2})$. This error is obtained for long integrations of careful used symplectic integrators [77] or when one is able to suppress the truncation error in any numerical integrator (and also with a careful use, of course). In other circumstances we may observe a typical linear growing $\mathcal{O}(t)$. In the case of the positions, we will have a root-mean-squared (RMS) error $\mathcal{O}(t^{3/2})$ in the best case, and a typical error $\mathcal{O}(t^2)$ (as in any non-symplectic RK code).

---

[1]TIDES: a Taylor series Integrator for Differential EquationS (GNU free software). Webpage: http://cody.unizar.es/software.html and http://sourceforge.net/projects/tidesodes/.

Now, we just show how easy is to eliminate the truncation error in the Taylor series method, and so, in TIDES. The advantage is that using the error estimator of the Taylor method, as they are based in just studying some Taylor coefficients, we may use any tolerance level. A completely different situation occurs when our error estimator is based on the substraction of two similar expressions (like in some formulations of embedded Runge–Kutta pairs where the local error estimator is given by the substraction of the solution of two methods of different order) that makes impossible to use them for tolerances lower than the rounding error due to the "catastrophic digit cancelation". So, if we fix the tolerance far below the roundoff unit of the computer we, in theory, can control the truncation error. This technique has been used previously by the group of Carles Simó [79] and by others [80–82]. We have to combine this technique with a "compensated sum" formulation [83] of the time increment as we use variable stepsize strategies (in contrast with symplectic integrators that have to use fixed stepsize implementations). So, the *truncation-free formulation* can be described as:

$$(\text{use TOL} \ll u, \text{ with } u = \text{the roundoff unit}) + (\text{"compensated sum"})$$

TIDES uses compensated sum in some stages of the method, so, if we want to preserve some geometric properties of the systems we just have to fix a low enough tolerance level. Obviously, this approach is computationally more expensive than other approaches and it is valid only if you also look for high precision numerical results.

In Fig. 3.17 we present the evolution of the error using TIDES with the truncated-free formulation. It is clear that this approach permits to achieve the optimal Brouwer's law (see Fig. 3.17), like well-programmed symplectic integrators [77], but it can be used in variable-stepsize formulations being therefore a quite flexible approach.

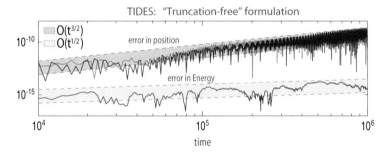

Fig. 3.17 Evolution of the error in the position and in the Energy of one orbit of the Hénon–Heiles system using a tolerance lower than the unit roundoff of the computer (*truncation-free formulation*). For the position we show the evolution of the error for the x and X variables of the Hamiltonian (3.6)

Acknowledgements The author thanks his colleagues and friends Dr. Fernando Blesa and Dr. Sergio Serrano for many interesting discussions and common work on this subject. The author thanks the referees for their help in improving the paper. The author has been supported during this research by the Spanish Research Grant MTM2012-31883, MTM2015-64095-P and by Gobierno de Aragon and Fondo Social Europeo.

References

1. Galias, Z., Zgliczyński, P.: Computer assisted proof of chaos in the Lorenz equations. Physica D **115**, 165–188 (1998)
2. Tucker, W.: A rigorous ODE solver and Smale's 14th problem. Found. Comput. Math. **2**, 53–117 (2002)
3. Skokos, C.: The Lyapunov characteristic exponents and their computation. In: Souchay, J.J., Dvorak, R. (eds.) Dynamics of Small Solar System Bodies and Exoplanets. Lecture Notes in Physics, vol. 790, pp. 63–135. Springer, Berlin, Heidelberg (2010)
4. Laskar, J.: Frequency analysis for multi-dimensional systems. Global dynamics and diffusion. Physica D **67**, 257–281 (1993)
5. Laskar, J.: Frequency analysis of a dynamical system. Celest. Mech. Dyn. Astron. **56**, 191–196 (1993)
6. Cincotta, P.M., Simó, C.: Simple tools to study global dynamics in non-axisymmetric galactic potentials - I. Astron. Astrophys. Suppl. **147**, 205–228 (2000)
7. Cincotta, P.M., Giordano, C.M., Simó, C.: Phase space structure of multi-dimensional systems by means of the mean exponential growth factor of nearby orbits. Physica D **182**, 151–178 (2003)
8. Froeschlé, C., Lega, E.: On the structure of symplectic mappings. The fast Lyapunov indicator: a very sensitivity tool. Celest. Mech. Dyn. Astron. **78**, 167–195 (2000)
9. Guzzo, M., Lega, E., Froeschlé, C.: On the numerical detection of the effective stability of chaotic motions in quasi-integrable systems. Physica D **163**, 1–25 (2002)
10. Skokos, C., Antonopoulos, C., Bountis, T.C., Vrahatis, M.N.: Detecting order and chaos in Hamiltonian systems by the SALI method. J. Phys. A **37**, 6269–6284 (2004)
11. Gottwald, G.A., Melbourne, I.: A new test for chaos in deterministic systems. Proc. R. Soc. Lond. Ser. A Math. Phys. Eng. Sci. **460**, 603–611 (2004)
12. Gottwald, G.A., Melbourne, I.: Testing for chaos in deterministic systems with noise. Physica D **212**, 100–110 (2005)
13. Barrio, R., Shilnikov, A.: Parameter-sweeping techniques for temporal dynamics of neuronal systems: case study of Hindmarsh-Rose model. J. Math. Neurosci. **1**(6), 20 (2011)
14. Barrio, R., Shilnikov, A., Shilnikov, L.: Kneadings, symbolic dynamics and painting Lorenz chaos. Int. J. Bifurcat. Chaos **22**, 1230016, 24 (2012)
15. Darriba, L.A., Maffione, N.P., Cincotta, P.M., Giordano, C.M.: Comparative study of variational chaos indicators and ODEs' numerical integrators. Int. J. Bifurcat. Chaos **22**, 1230033, 33 (2012)
16. Fouchard, M., Lega, E., Froeschlé, C., Froeschlé, C.: On the relationship between fast Lyapunov indicator and periodic orbits for continuous flows. Celest. Mech. Dyn. Astron. **83**, 205–222 (2002)
17. Barrio, R.: Sensitivity tools vs. Poincaré sections. Chaos Solitons Fractals **25**, 711–726 (2005)
18. Barrio, R.: Painting chaos: a gallery of sensitivity plots of classical problems. Int. J. Bifurcat. Chaos **16**, 2777–2798 (2006)
19. Barrio, R., Borczyk, W., Breiter, S.: Spurious structures and chaos indicators. Chaos Solitons Fractals **40**(4), 1697–1714 (2007)
20. Barrio, R., Blesa, F., Lara, M.: VSVO formulation of the Taylor method for the numerical solution of ODEs. Comput. Math. Appl. **50**, 93–111 (2005)

21. Barrio, R.: Sensitivity analysis of ODE's/DAE's using the Taylor series method. SIAM J. Sci. Comput. **27**, 1929–1947 (2006)
22. Hénon, M., Heiles, C.: The applicability of the third integral of motion: some numerical experiments. Astron. J. **69**, 73–79 (1964)
23. Geist, K., Parlitz, U., Lauterborn, W.: Comparison of different methods for computing Lyapunov exponents. Prog. Theor. Phys. **83**, 875–893 (1990)
24. Dieci, L., Van Vleck, E.S.: Lyapunov and other spectra: a survey. In: Collected Lectures on the Preservation of Stability Under Discretization (Fort Collins, CO, 2001), pp. 197–218. SIAM, Philadelphia, PA (2002)
25. Thiffeault, J.L., Boozer, A.H.: Geometrical constraints on finite-time Lyapunov exponents in two and three dimensions. Chaos **11**, 16–28 (2001)
26. Osedelec, V.I.: A multiplicative ergodic theorem. Lyapunov characteristic numbers for dynamical systems. Trans. Moscow Math. Soc. **19**, 197–231 (1968)
27. Legras, B., Vautard, R.: A guide to Liapunov vectors. In: Palmer, T. (ed.) Predictability. ECWF Seminar. vol. I, pp. 135–146. ECMWF, Reading (1996)
28. Haken, H.: At least one Lyapunov exponent vanishes if the trajectory of an attractor does not contain a fixed point. Phys. Lett. A **94**, 71–72 (1983)
29. Meyer, H.D.: Theory of the Liapunov exponents of Hamiltonian systems and a numerical study on the transition from regular to irregular classical motion. J. Chem. Phys. **84**, 3147–3161 (1986)
30. Yamaguchi, Y.Y., Iwai, T.: Geometric approach to Lyapunov analysis in Hamiltonian dynamics. Phys. Rev. E (3) **64**, 066206, 16 (2001)
31. Grond, F., Diebner, H.H., Sahle, S., Mathias, A., Fischer, S., Rossler, O.E.: A robust, locally interpretable algorithm for Lyapunov exponents. Chaos Solitons Fractals **16**, 841–852 (2003)
32. Guzzo, M., Lega, E.: The numerical detection of the Arnold web and its use for long-term diffusion studies in conservative and weakly dissipative systems. Chaos **23**, 023124 (2013)
33. Barrio, R., Blesa, F.: Systematic search of symmetric periodic orbits in 2DOF Hamiltonian systems. Chaos Solitons Fractals **41**, 560–582 (2009)
34. Abad, A., Barrio, R., Dena, A.: Computing periodic orbits with arbitrary precision. Phys. Rev. E **84**, 016701 (2011)
35. Barrio, R., Blesa, F., Serrano, S.: Fractal structures in the Hénon-Heiles Hamiltonian. Europhys. Lett. **82**, 10003 (2008)
36. Barrio, R., Blesa, F., Serrano, S.: Bifurcations and chaos in Hamiltonian systems. Int. J. Bifurcat. Chaos **20**, 1293–1319 (2010)
37. Barrio, R., Blesa, F., Serrano, S.: Qualitative analysis of the $(N+1)$-body ring problem. Chaos Solitons Fractals **36**, 1067–1088 (2008)
38. Barrio, R., Blesa, F., Elipe, A.: On the use of chaos indicators in rigid-body motion. J. Astronaut. Sci. **54**, 359–368 (2006)
39. Barrio, R., Blesa, F., Serrano, S.: Periodic, escape and chaotic orbits in the Copenhagen and the $(n+1)$-body ring problems. Commun. Nonlinear Sci. Numer. Simul. **14**, 2229–2238 (2009)
40. Blesa, F., Piasecki, S., Dena, Á., Barrio, R.: Connecting symmetric and asymmetric families of periodic orbits in squared symmetric Hamiltonians. Int. J. Mod. Phys. C **23**, 1250014, 22 (2012)
41. Blesa, F., Seoane, J.M., Barrio, R., Sanjuán, M.A.F.: To escape or not to escape, that is the question—perturbing the Hénon-Heiles Hamiltonian. Int. J. Bifurcat. Chaos **22**, 1230010, 9 (2012)
42. Euler, L.: Theoria motuum lunae (E418) Petrop. Parisin. Lond. (1772)
43. Jacobi, C.G.J.: Comp. Rend. **3**, 59–61 (1836)
44. Andrle, P.: A third integral of motion in a system with a potential of the fourth degree. Bull. Astron. Inst. Czechoslov. **17**, 169–175 (1966)
45. Armbruster, D., Guckenheimer, J., Kim, S.h.: Chaotic dynamics in systems with square symmetry. Phys. Lett. A **140**, 416–420 (1989)
46. Rucklidge, A.: Global bifurcations in the Takens-Bogdanov normal form with D4 symmetry near the O(2) limit. Phys. Lett. A **284**, 99–111 (2001)

47. Proctor, M.R.E., Weiss, N.O.: Magnetoconvection. Rep. Prog. Phys. **45**, 1317 (1982)
48. Barrio, R., Blesa, F., Serrano, S.: Bifurcations and safe regions in open Hamiltonians. New J. Phys. **11**, 053004 (2009)
49. Barrio, R., Serrano, S.: A three-parametric study of the Lorenz model. Physica D **229**, 43–51 (2007)
50. Barrio, R., Serrano, S.: Bounds for the chaotic region in the Lorenz model. Physica D **238**, 1615–1624 (2009)
51. Barrio, R., Blesa, F., Serrano, S.: Behavior patterns in multiparametric dynamical systems: Lorenz model. Int. J. Bifurcat. Chaos **22**, 1230019, 14 (2012)
52. Barrio, R., Serrano, S.: Qualitative analysis of the Rössler equations: bifurcations of limit cycles and chaotic attractors. Physica D **238**, 1087–1100 (2009)
53. Barrio, R., Blesa, F., Serrano, S., Shilnikov, A.: Global organization of spiral structures in biparameter space of dissipative systems with Shilnikov saddle-foci. Phys. Rev. E **84**, 035201 (2011)
54. Barrio, R., Blesa, F., Serrano, S.: Topological changes in periodicity hubs of dissipative systems. Phys. Rev. Lett. **108**, 214102 (2012)
55. Serrano, S., Barrio, R., Dena, A., Rodríguez, M.: Crisis curves in nonlinear business cycles. Commun. Nonlinear Sci. Numer. Simul. **17**, 788–794 (2012)
56. Lorenz, E.: Deterministic nonperiodic flow. J. Atmos. Sci. **20**, 130–141 (1963)
57. Doedel, E.J., Krauskopf, B., Osinga, H.M.: Global bifurcations of the Lorenz manifold. Nonlinearity **19**, 2947–2972 (2006)
58. Dullin, H.R., Schmidt, S., Richter, P.H., Grossmann, S.K.: Extended phase diagram of the Lorenz model. Int. J. Bifurcat. Chaos **17**, 3013–3033 (2007)
59. Guckenheimer, J., Holmes, P.: Nonlinear Oscillations, Dynamical Systems, and Bifurcations of Vector Fields. Applied Mathematical Sciences, vol. 42. Springer, New York (1990)
60. Shil'nikov, A.L., Shil'nikov, L.P., Turaev, D.V.: Normal forms and Lorenz attractors. Int. J. Bifurcat. Chaos **3**, 1123–1139 (1993)
61. Viana, M.: What's new on Lorenz strange attractors? Math. Intelligencer **22**, 6–19 (2000)
62. Saltzman, B.: Finite amplitude free convection as an initial value problem - 1. J. Atmos. Sci. **19**, 329–341 (1962)
63. Rössler, O.E.: An equation for continuous chaos. Phys. Lett. A **57**, 397–398 (1976)
64. Gaspard, P., Kapral, R., Nicolis, G.: Bifurcation phenomena near homoclinic systems: a two-parameter analysis. J. Stat. Phys. **35**, 697–727 (1984)
65. Gallas, J.A.C.: The structure of infinite periodic and chaotic hub cascades in phase diagrams of simple autonomous flows. Int. J. Bifurcat. Chaos **20**, 197–211 (2010)
66. Guckenheimer, J., Meloon, B.: Computing periodic orbits and their bifurcations with automatic differentiation. SIAM J. Sci. Comput. **22**, 951–985 (2000)
67. Corliss, G., Chang, Y.F.: Solving ordinary differential equations using Taylor series. ACM Trans. Math. Softw. **8**, 114–144 (1982)
68. Barrio, R., Rodríguez, M., Blesa, F.: Computer-assisted proof of skeletons of periodic orbits. Comput. Phys. Commun. **183**, 80–85 (2012)
69. Simó, C.: Dynamical systems, numerical experiments and super-computing. Mem. Real Acad. Cienc. Artes Barcelona **61**, 3–36 (2003)
70. Dena, Á., Rodríguez, M., Serrano, S., Barrio, R.: High-precision continuation of periodic orbits. Abstr. Appl. Anal. 12 (2012). Art. ID 716024
71. Barrio, R.: Performance of the Taylor series method for ODEs/DAEs. Appl. Math. Comput. **163**, 525–545 (2005)
72. Abad, A., Barrio, R., Blesa, F., Rodríguez, M.: Algorithm 924: TIDES, a Taylor series integrator for differential equationS. ACM Trans. Math. Softw. **39**, 5:1–5:28 (2012)
73. Jorba, À., Zou, M.: A software package for the numerical integration of ODEs by means of high-order Taylor methods. Exp. Math. **14**, 99–117 (2005)
74. Hairer, E., Nørsett, S.P., Wanner, G.: Solving Ordinary Differential Equations. I. Springer Series in Computational Mathematics, vol. 8. Springer, Berlin (1993)

75. Fousse, L., Hanrot, G., Lefèvre, V., Pélissier, P., Zimmermann, P.: MPFR: a multiple-precision binary floating-point library with correct rounding. ACM Trans. Math. Softw. **33**, 13:1–13:15 (2007)
76. Abad, A., Barrio, R., Blesa, F., Rodríguez, M.: TIDES tutorial: integrating ODEs by using the Taylor series method. Monografías de la Academia de Ciencias de la Universidad de Zaragoza **36**, 1–116 (2011)
77. Hairer, E., McLachlan, R.I., Razakarivony, A.: Achieving Brouwer's law with implicit Runge-Kutta methods. BIT **48**, 231–243 (2008)
78. Brouwer, D.: On the accumulation of errors in numerical integration. Astron. J. **30**, 149–153 (1937)
79. Simó, C.: Global dynamics and fast indicators. In: Global Analysis of Dynamical Systems, pp. 373–389. Institute of Physics, Bristol (2001)
80. Sharp, P.W.: N-body simulations: the performance of some integrators. ACM Trans. Math. Softw. **32**, 375–395 (2006)
81. Grazier, K.R., Newman, W.I., Hyman, J.M., Sharp, P.W.: Long simulations of the Solar System: Brouwer's law and chaos. ANZIAM J. **46**, C1086–C1103 (2004/05)
82. Grazier, K.R., Newman, W.I., Hyman, J.M., Sharp, P.W., Goldstein, D.J.: Achieving Brouwer's law with high-order Störmer multistep methods. ANZIAM J. **46**, C786–C804 (2004/05)
83. Higham, N.J.: Accuracy and Stability of Numerical Algorithms, 2nd edn. SIAM, Philadelphia, PA (2002)

Chapter 4
Theory and Applications of the Mean Exponential Growth Factor of Nearby Orbits (MEGNO) Method

Pablo M. Cincotta and Claudia M. Giordano

Abstract In this chapter we discuss in a pedagogical way and from the very beginning the *Mean Exponential Growth factor of Nearby Orbits* (MEGNO) method, that has proven, in the last ten years, to be efficient to investigate both regular and chaotic components of phase space of a Hamiltonian system. It is a fast indicator that provides a clear picture of the resonance structure, the location of stable and unstable periodic orbits as well as a measure of hyperbolicity in chaotic domains which coincides with that given by the maximum Lyapunov characteristic exponent but in a shorter evolution time. Applications of the MEGNO to simple discrete and continuous dynamical systems are discussed and an overview of the stability studies present in the literature encompassing quite different dynamical systems is provided.

4.1 Introduction

One of the most challenging aspects of dynamical systems, particularly of those that present a divided phase space, is the understanding of global properties in phase space. Unfortunately, for instance, global instabilities of near-integrable multidimensional Hamiltonian systems are far from being well understood, so in this chapter we should focus on local features, that is, the dynamical behavior in a small domain around a given point of the phase space of the system.

An example of the study of the local dynamics in "every" point of phase space concerns the so-called *chaos detection tools*. This implies the characterization of the dynamical flow around a given initial condition, that is for instance, how two orbits starting very close to each other evolve with time t. A well known result is that for ordered or regular motion, the separation between these two initially

P.M. Cincotta (✉) • C.M. Giordano
Grupo de Caos en Sistemas Hamiltonianos, Facultad de Ciencias Astronómicas y Geofísicas,
Universidad Nacional de La Plata and Instituto de Astrofísica de La Plata (CONICET), La Plata,
Argentina
e-mail: pmc@fcaglp.unlp.edu.ar; giordano@fcaglp.unlp.edu.ar

© Springer-Verlag Berlin Heidelberg 2016
Ch. Skokos et al. (eds.), *Chaos Detection and Predictability*, Lecture Notes
in Physics 915, DOI 10.1007/978-3-662-48410-4_4

nearby orbits grows linearly with time (or in some particular cases at some power of t); while in those domains where the motion is unstable, chaotic, this separation grows exponentially with t. The rate of this exponential divergence, defined as the limit when $t \rightarrow \infty$, is given by the so-called maximum Lyapunov Characteristic Exponent (mLCE). Therefore if we know how to compute efficiently this separation for large times we can obtain a picture of the local dynamics at any given point of phase space. Indeed, in case of regular motion the mLCE vanishes and it has a positive value for chaotic motion (and for unstable periodic orbits).

Another way to characterize the local dynamics is through a spectral analysis. In fact, regular motion proceeds on invariant tori with a constant frequency vector while, when the dynamics is chaotic, the frequencies are no longer local integrals of motion but change with time. Therefore if we managed to develop an accurate technique to measure the frequency of the motion, we could be able to separate the dynamics in regular and chaotic. Moreover, in the regular regime it would be possible to compute the full set of local integrals of motion (that is, the components of the frequency vector).

Since the eighties and mid of the nineties two well known techniques have been available in the literature, an algorithm to compute the mLCE, see for instance [5], and the so called Frequency Map Analysis (FMA [43]). The first one obviously provides the rate of divergence of nearby orbits while the second one is a very precise method to obtain the frequencies of the motion. Both approaches were widely used in many physical and astronomical applications; in particular the FMA was the natural technique to investigate the dynamics of planets and, by means of this tool it was shown that the Solar System as a whole dynamical system is not stable and in fact it is chaotic or marginal unstable [42, 44, 45].

Actually, the mLCE and the FMA (besides the well known Poincaré surface of section for systems with two degrees of freedom), were popular chaos detection tools in dynamical astronomy at those times.

However, computers were not fast enough to deal with large samples of orbits and quite long integration times. For a set of $M \gtrsim 10^6$ orbits, $\sim 12 \times M > 10^7$ nonlinear coupled differential equations should be numerically integrated over long time intervals and with high accuracy in order to get numerical values of the mLCEs close to the expected theoretical ones. For instance, for regular motion the theoretical mLCE $\rightarrow 0$ when $t \rightarrow \infty$ as $\sim \ln t/t$, so for an evolution time $t \sim 10^4$, a null mLCE numerically means $\sim 10^{-3}$. Thus, it was not possible to distinguish a regular orbit from a chaotic one with a mLCE $\sim 10^{-3}$. Therefore, much larger evolution times would be necessary to discriminate the nature of the motion by the numerical asymptotic value. Thus it becomes clear that 20 years ago, this was a severe restriction to derive precise values of the mLCE.

Since in the end of the nineties several *fast dynamical indicators* appeared in the literature, some of the most popular ones in dynamical astronomy are largely discussed in the present volume. All of them rest on the same theoretical arguments behind the mLCE, by following the evolution of the flow in a small neighborhood of a given initial condition. Besides, a few new techniques, based on spectral analysis have also been developed which are in fact, slight variations of the FMA. At the end of this chapter we will briefly refer to several of such chaos detection tools.

4.1.1 The MEGNO: Brief History

The MEGNO belongs to the class of the so-called fast dynamical indicators. It is, in fact, a byproduct of a former fast indicator, the *Conditional Entropy of Nearby Orbits*, first proposed in [57] and improved in [10] and [11].

The MEGNO was announced in [11], but neither a description of the method nor a name was provided. In [12], the MEGNO was introduced, but that work was not devoted exclusively to the MEGNO, but to discuss analytical and numerical methods for describing global dynamics in non-axisymmetric galactic potentials in both regimes, regular and chaotic. The MEGNO was addressed there just as an additional and simple tool, and its name (MEGNO) was proposed, following the strong suggestion of one of the reviewers of the paper. The MEGNO was introduced as an efficient way to derive accurately the mLCE. Indeed, in the Introduction of that paper, the authors wrote ... *Alternative techniques were proposed to separate ordered and stochastic motion, to classify orbits in families, to describe the global structure of phase space, but not to get the LCN in shorter times. In Sect. 3 we shall resume this point together with some comparisons with the new technique here presented (MEGNO).... This new tool has proven to be useful for studying global dynamics and succeeds in revealing the hyperbolic structure of phase-space, the source of chaotic motion. The MEGNO provides a measure of chaos that is proportional to the LCN, so that it allows to derive the actual LCN but in realistic physical times...*

It was in [13] that the MEGNO was discussed in detail and a generalization of the original method was presented with applications to both multidimensional Hamiltonian flows and maps.

This chapter is organized as follows. In Sect. 4.2 we address the theory of the MEGNO in a simple fashion, without any intention to enunciate theorems and their concomitant proofs: just several numerical examples would serve to show the expected theoretical behavior of this dynamical indicator. In Sect. 4.3 we present some applications to Hamiltonian flows and symplectic maps. We also yield the results of an exhaustive comparative study of different indicators of chaos in Sect. 4.4. We discuss further applications of the MEGNO that can be found along the literature, from realistic planetary models to bifurcation analysis, in Sect. 4.5. A thorough discussion is provided in the last section.

4.2 The Mean Exponential Growth Factor of Nearby Orbits (MEGNO)

Herein we address the MEGNO's theory following the original presentation given in [13] but in a more pedagogical way.

To that aim, let us consider the phase space state vector

$$\mathbf{x} = (\mathbf{p}, \mathbf{q}) \in B \subset \mathbb{R}^{2N}, \tag{4.1}$$

and introduce the vector field, also defined in B

$$\mathbf{v}(\mathbf{x}) = \left(-\frac{\partial H}{\partial \mathbf{q}}, \frac{\partial H}{\partial \mathbf{p}} \right), \tag{4.2}$$

where $H(\mathbf{p}, \mathbf{q})$ refers to an N-dimensional Hamiltonian, assumed to be autonomous just for the sake of simplicity. The formulation given below however is completely independent of the system being Hamiltonian as well as of the phase space coordinates adopted to express the state vector $\mathbf{x}$. In case of a Hamiltonian system, since the motion in general takes place on a compact energy surface $M_h = \{\mathbf{x} = (\mathbf{p}, \mathbf{q}) \in B : H(\mathbf{p}, \mathbf{q}) = h\}$, thus $\mathbf{x} \in B' \subseteq M_h$, where $\dim(M_h) = 2N - 1$.

Therefore, the equations of motion in B' have the simple form

$$\dot{\mathbf{x}} = \mathbf{v}(\mathbf{x}). \tag{4.3}$$

Let $\varphi(t)$ denote a given solution of the flow (4.3), for a given initial condition $\mathbf{x}_0$,

$$\varphi(t) = \left\{ \mathbf{x}(t; \mathbf{x}_0), \mathbf{x}_0 \in B' \right\}. \tag{4.4}$$

For any such an orbit φ the mLCE, $\sigma(\varphi)$, is defined as

$$\sigma(\varphi) = \lim_{t \to \infty} \sigma_1(\varphi(t)), \qquad \sigma_1(\varphi(t)) = \frac{1}{t} \ln \frac{\|\delta(\varphi(t))\|}{\|\delta_0\|}, \tag{4.5}$$

where $\delta(\varphi(t))$ and δ_0 are "infinitesimal displacements" from φ at times t and 0, respectively, and $\|\cdot\|$ denotes the usual Euclidean norm.[1]

In fact, $\delta(\varphi(t))$ is the time evolution of the difference $\varphi'(t) - \varphi(t)$, being $\varphi'(t)$ a nearby orbit to $\varphi(t)$ whose initial condition is $\mathbf{x}_0' = \mathbf{x}_0 + \delta\mathbf{x}_0$, for $\|\delta\mathbf{x}_0\|$ small enough. The evolution of $\varphi'(t) - \varphi(t)$ after linearizing the flow around $\varphi(t)$ is then computed. Therefore we are evaluating the flow (4.3) and its first variation on a single orbit, instead of computing the evolution of $\varphi'(t)$ and $\varphi(t)$ and performing their difference. This is the very same way in which the algorithm to compute the

---

[1] Let us note that any other norm could be used all the same.

mLCE was developed. Though it would be possible to integrate the flow to get φ' and φ starting at $\mathbf{x}_0'$ and $\mathbf{x}_0$ respectively, when performing the difference $\varphi'(t) - \varphi(t)$, both in M_h, then after a large but finite time t, the separation between the two orbits would reach, in a chaotic domain, an upper bound $\|\varphi'(t) - \varphi(t)\| \leq d$, where d is the maximum size of the accessible region in M_h. The limiting case when $\|\varphi'(t) - \varphi(t)\| = d$ corresponds to the completely ergodic case, in which any orbit, and also the difference of nearby ones, could fill densely the energy surface M_h.

In any case, the computation of $\varphi'(t) - \varphi(t)$ would provide the right physical insight about the nature of the dynamics in a small neighborhood of $\mathbf{x}_0$, but computationally this is not the best option, since the mLCE measures the divergence of $\delta(t) = \|\varphi'(t) - \varphi(t)\|$ when $t \to \infty$ and $\delta_0 = \|\varphi'(0) - \varphi(0)\| \to 0$.

It is well known that σ provides relevant information about the flow in a small domain around φ. Indeed, recasting (4.5) in the form

$$\sigma(\varphi) = \lim_{t \to \infty} \frac{1}{t} \int_0^t \frac{\dot{\delta}(\varphi(s))}{\delta(\varphi(s))} \, ds = \overline{\left(\frac{\dot{\delta}}{\delta} \right)}, \tag{4.6}$$

where $\delta \equiv \|\boldsymbol{\delta}\|$ is the Euclidean norm, $\dot{\delta} \equiv d\delta/dt = \dot{\boldsymbol{\delta}} \cdot \boldsymbol{\delta}/\|\boldsymbol{\delta}\|$, and $\overline{(\cdot)}$ denotes time-average, thus it is explicit that the mLCE measures the "mean exponential rate of divergence of nearby orbits".

Thus defined, the so-called tangent vector $\boldsymbol{\delta}$ satisfies the first variational equation of the flow (4.3):

$$\dot{\boldsymbol{\delta}} = \Lambda(\varphi(t))\boldsymbol{\delta}, \tag{4.7}$$

where $\Lambda(\varphi(t)) \equiv D_{\mathbf{x}}\mathbf{v}(\varphi(t))$ is the Jacobian matrix of the vector field $\mathbf{v}$ evaluated on $\varphi(t)$.

Let us now introduce a slightly different sensitive function on the orbit $\varphi(t)$ which is closely related to the integral in (4.6); the *Mean Exponential Growth factor of Nearby Orbits* (MEGNO), $Y(\varphi(t))$, through

$$Y(\varphi(t)) = \frac{2}{t} \int_0^t \frac{\dot{\delta}(\varphi(s))}{\delta(\varphi(s))} s \, ds. \tag{4.8}$$

Recall that in case of an exponential increase of δ, as it occurs for an unstable periodic orbit or a chaotic one, $\delta(\varphi(t)) = \delta_0 \exp(\sigma t)$, $\sigma > 0$, $Y(\varphi(t))$ can be considered as a weighted variant of the integral in (4.6). Indeed, instead of the instantaneous rate of growth, σ, we average the logarithm of the growth factor, $\ln(\delta(\varphi(t))/\delta_0) = \sigma t$. Further variants will be considered in Sect. 4.2.2 where the generalization of the MEGNO is addressed.

In what follows we consider some, though quite special, very representative solutions of (4.7) in order to show how $Y(\varphi(t))$ serves to provide clear indication on the character of the motion in each case.

Thus, let us first consider any orbit $\varphi_q(t)$ on a N-dimensional irrational torus in a non-isochronous or nonlinear system. Therefore we can locally define action-angle variables $(\mathbf{I}, \theta)$ such that $\theta(t) = \omega(\mathbf{I})t + \theta_0$, $\mathbf{I} \equiv \mathbf{I}_0$, being $\mathbf{I}_0$ a constant and, for any set of generalized coordinates $(\mathbf{p}, \mathbf{q})$ the solution of (4.3) can be expanded in Fourier series in θ with coefficients that depend on $\mathbf{I}$. Therefore for any such quasiperiodic orbit, φ_q, the solution of (4.7) in generalized coordinates has the form

$$\delta\left(\varphi_q(t)\right) \approx \delta_0\left(1 + w_q(t) + t\left(\lambda_q + u_q(t)\right)\right), \tag{4.9}$$

where $\lambda_q > 0$ is the absolute value of the linear rate of divergence around φ_q, $w_q(t)$ and $u_q(t)$ are oscillating functions (in general quasiperiodic and with zero average) of bounded amplitude, that satisfy $|u_q(t)| \leq b_q < \lambda_q$, for some positive constant b_q.[2] The quantity λ_q is a measure of the lack of isochronicity around the orbit and it is related to the absolute value of the maximum eigenvalue of the nonlinearity matrix

$$\frac{\partial \omega_i}{\partial I_j} = \frac{\partial^2 H}{\partial I_i \partial I_j}.$$

Recall that for a linear or quasi-linear system, such as the harmonic oscillator, $\lambda = 0$ for all φ. Indeed, the linear divergence of two nearby quasiperiodic orbits reflects the fact that they move on nearby N-dimensional tori. Since we assume that ω depends on $\mathbf{I}$, two nearby tori have a small different action vector, say $\mathbf{I}$ and $\mathbf{I} + \delta\mathbf{I}$, and thus $\omega(\mathbf{I} + \delta\mathbf{I}) = \omega(\mathbf{I}) + \delta\omega$. However if $\det\left(\partial\omega_i/\partial I_j\right) = 0$, the system behaves as a linear one and no divergence between two nearby orbits is expected.

From (4.8) and (4.9), keeping in mind that $|u_q|$ is bounded by b_q, it is straightforward to see that $Y(\varphi_q(t))$ oscillates around 2 with bounded amplitude, verifying that

$$|Y\left(\varphi_q(t)\right) - 2| \leq 4\ln\frac{\lambda_q + b_q}{\lambda_q - b_q} \approx 8\frac{b_q}{\lambda_q}, \qquad t \to \infty, \tag{4.10}$$

where the last approximation holds if $b_q \ll \lambda_q$. The time evolution of $Y(\varphi_q(t))$ is given by

$$Y\left(\varphi_q(t)\right) \approx 2 - \frac{2\ln(1 + \lambda_q t)}{\lambda_q t} + O\left(\varphi_q(t)\right), \tag{4.11}$$

where O denotes an oscillating term (with zero average) due to the quasiperiodic character of both $w_q(t)$ and $u_q(t)$. Though

$$\lim_{t \to \infty} Y\left(\varphi_q(t)\right) \tag{4.12}$$

---

[2] Anyway (4.9) could be empirically derived by numerical means.

does not exist due to the oscillatory term $O\left(\varphi_q(t)\right)$ in (4.11), introducing the time-average

$$\overline{Y}(\varphi_q(t)) \equiv \frac{1}{t} \int_0^t Y(\varphi_q(s))\mathrm{d}s, \qquad (4.13)$$

it can readily be seen from (4.10), (4.11) and (4.13) that

$$\overline{Y}\left(\varphi_q\right) \equiv \lim_{t\to\infty} \overline{Y}\left(\varphi_q(t)\right) = 2. \qquad (4.14)$$

Therefore, for quasiperiodic motion, $\overline{Y}(\varphi)$ converges to a constant value, which is independent of $\varphi_q(t)$.

The above results still hold in case of a regular orbit $\varphi(t)$ that is not purely stable quasiperiodic. Let us restrict ourselves to 2-dimensional (2D) Hamiltonian systems, though the arguments given below could be straightforwardly extended to higher dimensions and let $\varphi(t)$ be close to a stable periodic orbit, $\varphi_s(t)$. Since $O(\varphi(t))$ in (4.11) involves nearly periodic terms, and both λ and b/λ are small, it follows from (4.10) and (4.11) that $Y(\varphi(t))$ oscillates around 2 with a small amplitude and that $\overline{Y}(\varphi(t))$ converges to 2 slower the smaller is λ. When $\varphi(t) \to \varphi_s(t)$, both $u(t)$, $\lambda \to 0$, and $\overline{Y} \to 0$ as $t \to \infty$. In this limiting case, the oscillations of $Y(\varphi(t))$ about 0 are due to the presence of the term $w(t)$ in (4.9).

Meanwhile, whenever $\varphi(t)$ is close to an unstable periodic orbit, $\varphi_u(t)$, $Y(\varphi(t))$ behaves in a different fashion since in such a case, the motion in any small neighborhood of $\varphi_u(t)$, U, is mainly determined by its associated stable and unstable manifolds. For a sufficiently large motion time, $\varphi(t)$ will pass close to $\varphi_u(t)$ several times. Suppose that between two successive close approaches with $\varphi_u(t)$, $\varphi(t)$ spends a time Δ_1 within U and a time Δ_2 outside U. During the interval Δ_1, $\delta(\varphi(t)) \approx \delta(\varphi_u(t)) \approx \delta_0 \exp(\sigma t)$ with $\sigma > 0$, while, during Δ_2, $\delta(\varphi(t))$ approximately obeys (4.9). The "interaction time" between $\varphi(t)$ and $\varphi_u(t)$, Δ_1, is larger the closer the orbits are to each other. Thus, $Y(\varphi(t))$ should exhibit quasiperiodic oscillations modulated by periodic pulses, of period $\sim \Delta_2$, width $\sim \Delta_1$ and similar amplitude. Analogous considerations apply to $\overline{Y}(\varphi(t))$ but, due to the averaging, the amplitude of the pulses should decrease as $\sim 1/t$. In general, $\overline{Y}(\varphi(t))$ will approach 2 from above and, after a total evolution time t, $\overline{Y}(\varphi(t))$ will be larger the smaller is the distance $\|\varphi(t) - \varphi_u(t)\|$. In the limit, when $\varphi(t) \to \varphi_u(t)$, $\Delta_1 \to t$ and $\delta(\varphi(t))$ grows exponentially with time, so that $\overline{Y}(\varphi(t)) \gg 2$ (see the forthcoming Eq. (4.17) and the so-called "right-stop" criterion discussed in Sect. 4.2.4 that applies for maps).

In case of an irregular orbit, $\varphi_i(t)$, within any chaotic component, the solution of (4.7), besides oscillation terms which are irrelevant in this case, is

$$\delta(\varphi_i(t)) \approx \delta_0 \mathrm{e}^{\sigma_i t}, \qquad (4.15)$$

σ_i being the $\varphi_i(t)$'s mLCE. Thus,

$$Y(\varphi_i(t)) \approx \sigma_i t + \tilde{O}(\varphi_i(t)), \tag{4.16}$$

with $\tilde{O}$ some oscillating term of bounded amplitude which is in general neither periodic nor quasiperiodic, but it has zero average.[3] Note that in a chaotic domain the orbits proceed on a D-dimensional manifold where $N < D < 2N - 1$. In these domains, tori are in general destroyed and the dynamics is said to be hyperbolic since a chaotic orbit could be thought as a slight distortion of an unstable P-periodic orbit with $P \gg 1$.

On averaging (4.16) over a large time interval, we obtain

$$\overline{Y}(\varphi_i(t)) \approx \frac{\sigma_i}{2} t, \quad t \to \infty. \tag{4.17}$$

Therefore, for a chaotic orbit, $Y(\varphi_i(t))$ and $\overline{Y}(\varphi_i(t))$ grow linearly with time, at a rate equal to the mLCE of the orbit or one half of it, respectively (see below). Only when the phase space has an hyperbolic structure, does Y grow with time. Otherwise, it saturates to a constant value, even in the degenerated cases in which δ grows with some power of t, say n, and therefore $\overline{Y} \to 2n$ as $t \to \infty$.

The MEGNO's temporal evolution allows for being summed up as a single expression valid for any kind of motion, which is not the case for σ_1 or any other chaos indicator. In fact, the asymptotic behavior of $\overline{Y}(\varphi(t))$ may be written in the fashion

$$\overline{Y}(\varphi(t)) \approx a_\varphi t + b_\varphi \tag{4.18}$$

where $a_\varphi = \sigma_\varphi/2$ and $b_\varphi \approx 0$ for chaotic motion, while $a_\varphi = 0$ and $b_\varphi \approx 2$ for stable quasiperiodic motion. Departures from the value $b_\varphi \approx 2$ indicate that φ is close to some periodic orbit, being $b_\varphi \lesssim 2$ and $b_\varphi \gtrsim 2$ for stable or near-unstable periodic orbits, respectively.

Notice that $\hat{\sigma}_1 \equiv Y(\varphi(t))/t$ verifies

$$\hat{\sigma}_1(\varphi_q(t)) \approx \frac{2}{t}, \qquad \hat{\sigma}_1(\varphi_i(t)) \approx \sigma_i, \qquad t \to \infty, \tag{4.19}$$

which show that, for regular motion $\hat{\sigma}_1$ converges to 0 faster than σ_1, which it does as $\ln t/t$, while for chaotic motion both magnitudes approach the positive mLCE at a similar rate.

As it turns out from (4.18) and perhaps the key point of the MEGNO method (but not widespread used) is that, since for chaotic motion $\overline{Y}$ grows linearly with time with a rate $\sigma/2$, a very accurate estimate of the mLCE can be obtained in rather

---

[3]Since the motion is bounded in phase space, any orbit $\varphi(t)$ should be an oscillating function of time of bounded amplitude, despite if it is regular or chaotic. For unstable or chaotic orbits the main secular growth is given by the exponential term and therefore it is always possible to separate it from a purely oscillating term with zero average.

short times by means of a linear least squares fit on $\overline{Y}(\varphi(t))$. The main feature of this procedure is that it takes advantage of all the dynamical information contained in $\overline{Y}(\varphi(t))$ regarding the whole interval $(t_0, t), t \gg t_0$ and on the fact that $\overline{Y}$ has a smooth behavior. Since for purely quasiperiodic orbits $\overline{Y}(\varphi(t))$ approaches the constant value 2 quite faster than for nearly stable and near-unstable periodic orbits, the mLCE derived from a linear least squares fit of the MEGNO would also yield information on elliptic and hyperbolic points as well.

4.2.1 Comparison of Theoretical and Numerical Results

In order to illustrate the predicted MEGNO's behavior, we regard the well known 2D Hénon–Heiles model [35],

$$H(p_x, p_y, x, y) = \frac{1}{2}(p_x^2 + p_y^2) + \frac{1}{2}(x^2 + y^2) + x^2 y - \frac{y^3}{3}, \qquad (4.20)$$

where x, y, p_x, $p_y \in \mathbb{R}$. This Hamiltonian was proposed in the sixties to investigate the existence of the so-called third integral of motion in the Galaxy. We consider the energy level $h = 0.118$. The phase space at this energy level displays at least two main unconnected chaotic domains having different mLCE's as shown by the Poincaré surfaces of section presented in Fig. 4.1 (see, for instance, [11]).

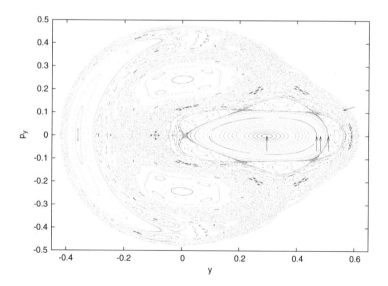

Fig. 4.1 (y, p_y)-Surfaces of section for the Hénon–Heiles Hamiltonian for $h = 0.118, x = 0$, $p_x > 0$. The *arrows* indicate the location of the five initial conditions, *from left to right* (sp), (up), (qp), (c1), (c2). See text

We picked up initial conditions for five representative orbits from the surface $x = 0$: one close to the stable 1-periodic orbit at $(y, p_y) = (0.295456, 0)$ (sp); another one looking like stable quasiperiodic at $(0.483, 0)$ (qp); a third one at $(0.46912, 0)$ also quasiperiodic but close to an unstable 4-periodic orbit (up); and two irregular orbits, one in the stochastic layer surrounding a 5-periodic island chain (or at a $5 : m$ resonance for $m \in \mathbb{Z}_0$) (c1) at $(0.509, 0)$, and the other one lying in a large chaotic sea (c2) at $(0.56, 0.112)$.

We computed Y and $\overline{Y}$ by means of (4.8) and (4.13) respectively; note that the renormalization of δ, if necessary, proceeds naturally from (4.8). Along this work all the numerical integrations were carried out by recourse to a Runge–Kutta 7/8th order integrator (the Dopri8 routine, see [58] and [34]), the accuracy in the conservation of the energy in this case being $\sim 10^{-13}$. The initial tangent vector δ is chosen at random and with unit norm.[4]

In Fig. 4.2 we show that both Y and $\overline{Y}$ evolve with time as predicted. Indeed, in Fig. 4.2a we observe that, for the stable quasiperiodic orbit (qp), Y oscillates around the value 2 with an amplitude $\lesssim 1$, while $\overline{Y}$ shows a very fast convergence to the actual average (see below).

Figure 4.2b displays the typical behavior of a trajectory close to an unstable periodic orbit. While the (up) orbit is "far away" from the hyperbolic point, both Y and $\overline{Y}$ evolve as in the quasiperiodic case. However, when this quasiperiodic orbit passes close to the unstable one, the mutual interaction causes the oscillations of Y to exhibit a strong modulation, which is damped in $\overline{Y}$ as t increases. Thus, after the first close approach at $t \sim 2000$, $\overline{Y} > 2$ (mainly due to the cumulative effect on the average) but, for t large enough, it asymptotically approaches 2.

Also for the irregular orbits (c1) and (c2) we compute the time-evolution of Y and $\overline{Y}$. The results are given in Fig. 4.2c, where both Y and $2\overline{Y}$ are plotted together to show that, as follows from (4.16) and (4.17), both quantities have the same time-rate. Since the trajectories belong to unconnected chaotic domains, the time-rate (i.e. the mLCE) is different for the two orbits.

In Fig. 4.2d, the temporal evolution of $\overline{Y}$ for all the three regular orbits are compared. For the stable quasiperiodic orbit (qp), $\overline{Y}$ reaches 2 much faster than for the orbit (sp), which is close to a stable periodic one. In fact, $\overline{Y}(\varphi_{sp}) \lesssim 2$ over the full time interval. The time evolution of both, $\overline{Y}(\varphi_{sp})$ and $\overline{Y}(\varphi_{qp})$, fit very well (4.11), on neglecting oscillations and being $\lambda_{sp} < \lambda_{qp}$. We note again just for $\overline{Y}$, that the orbits (qp) and (up) evolve in a rather similar way, as long as the interaction between (up) and its nearby unstable periodic orbit is weak. Therefore, a least squares fit on $\overline{Y}$ could distinguish clearly quasiperiodic orbits from stable and unstable periodic orbits.

In order to show that $\hat{\sigma}_1 \to \text{mLCE}$ when $t \to \infty$, in Fig. 4.2e we display its time evolution together with that of σ_1 for three of the orbits, namely, (sp), (c1) and

---

[4]One should verify that the tangent vector has a non-vanishing component normal to the flow, particularly in the regular component, in order to ensure the linear divergence of nearby orbits (see [16]).

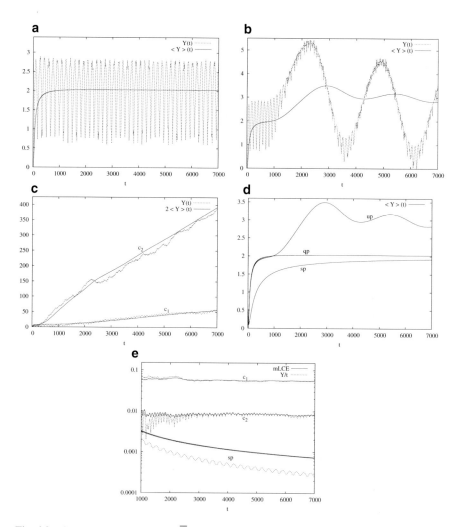

Fig. 4.2 Time evolution of Y and $\overline{Y}$ ($< Y >$ in the figure) for the orbits: (**a**) (qp) stable quasiperiodic; (**b**) (up) quasiperiodic but close to an unstable 4-periodic orbit; (**c**) (c1) and (c2) irregular, embedded in two different chaotic domains. (**d**) $\overline{Y}$ ($< Y >$ in the figure) for three regular orbits: (sp) close to a stable periodic orbit, (qp), and (up); (**e**) time evolution of $\hat{\sigma}_1$ (Y/t in the figure) and the mLCE, σ_1 for (sp), (c1) and (c2) computed using the algorithm given in [5]

(c2). We observe that for the chaotic orbits, both magnitudes converge to the same positive mLCE at the same rate. For the regular orbit (sp) instead, we note that $\hat{\sigma}_1$ decreases faster than σ_1, the expected final values (see (4.19) and discussion below), 0.0013 and 0.00028 respectively, being the latter close to the computed one.

In the case of chaotic motion, both Y and $\overline{Y}$ evolve almost linearly with time over the whole time interval, as shown in Fig. 4.2c. The deviations from the linear trend, for instance in (c2), are presumably caused by stickiness. Indeed, during those

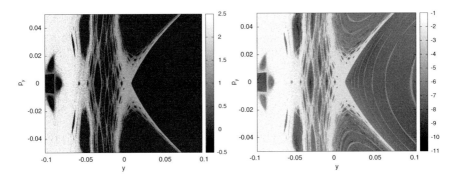

Fig. 4.3 (y, p_y)-Surface of section of the Hénon–Heiles Hamiltonian for $h = 0.118, x = 0, p_x > 0$. The *left panel* corresponds to MEGNO contour plot in logarithmic scale ($\log 2 \approx 0.3$) and in the *right panel* the mLCE, also in logarithmic scale, derived from a linear least squares fit on $\overline{Y}$. See text for details

time intervals, Δt_s, in which Y is almost flat, the orbit remains close to some small stability domain embedded in the chaotic sea. In this particular example, stickiness does not significantly reduce the linear trend but, whenever it is strong, it does influence the mean time-rate of both Y and $\overline{Y}$ and consequently, the derived mLCE. However, the same effect would be present in the numerical computation of the mLCE, since the stickiness phenomena affects the evolution of $\delta(t)$ and therefore if $\delta(t)$ does not increase exponentially within Δt_s, the evolution of $\sigma_1(t)$ would decrease with time as $\sim \ln t / t$ while $t \in \Delta t_s$.

Finally, in Fig. 4.3 we present a small domain of the (y, p_y)-surface of section of the Hénon–Heiles Hamiltonian for $h = 0.118$, $x = 0$, $p_x > 0$ given in Fig. 4.1, the contour plots providing, in logarithmic scale, the MEGNO and the mLCE computed by a least squares fit on the time evolution of $\overline{Y}$ over the time interval (t_0, t) with $t = 10^4$, $t_0 = 2 \times 10^3$. A given $t_0 > 0$ is adopted in order to avoid the initial transient; thus, the least squares fit is performed over the 80 % of the full time interval. From these two plots we observe that the MEGNO provides a clear picture of the dynamics but the accurate value of the mLCE obtained following this alternative procedure furnishes more information than the MEGNO itself. Indeed, both plots show up the very same information in the chaotic domain, however, the MEGNO does not separate clearly the thin unstable domain inside the stability island as the mLCE computed by a least squares fit does. Note that using a simple least squares fit on the time evolution on $\overline{Y}$ over the 80 % of the whole time interval, we reach values of the mLCE for regular motion of the order of 10^{-10} or lower considering motion times $t \sim 10^4$, when the expected lower value of the mLCE by recourse to the classical algorithm would be $\sim 10^{-3}$. This is, in our opinion, one of the main results provided by the MEGNO: its very accurate determination of the positive and null mLCE, for chaotic and regular motion respectively.

Further details on the MEGNO's performance when applied to the study of the dynamics of 2D Hamiltonians, as well as other advantages of deriving the mLCE from a least squares fit on $\overline{Y}$ are given in [12].

4.2.2   Generalization of the MEGNO

Let us generalize the MEGNO by introducing the exponents (m, n) such that

$$Y_{m,n}\left(\varphi(t)\right) = (m+1)\, t^n \int_0^t \frac{\dot{\delta}(\varphi(s))}{\delta(\varphi(s))}\,(s)^m\,ds, \qquad (4.21)$$

now defining

$$\overline{Y}_{m,n}\left(\varphi(t)\right) = \frac{1}{t^{m+n+1}} \int_0^t Y_{m,n}\left(\varphi_q(s)\right)\,ds, \qquad (4.22)$$

and analyze whether any benefit would turn out when taking values for the exponents (m, n), $m \geq 0$ other than the natural choice $(1, -1)$ which yielded (4.8) and (4.13). Note that in the limit when $t \to \infty$, $Y_{0,-1} \to \sigma$ as defined in (4.6).

The time evolution of $Y_{m,n}$ for regular, quasiperiodic motion, is given by

$$Y_{m,n}\left(\varphi_q(t)\right) \approx (m+1)\left(\sum_{k=0}^{m-1} \frac{(-1)^k t^{m+n-k}}{(m-k)\lambda_q^k}\right)$$

$$+(m+1)\left((-1)^m \frac{t^n \ln(1 + \lambda_q t)}{\lambda_q^m}\right) + O\left(\varphi_q(t)\right), \quad (4.23)$$

which naturally reduces to (4.11) for $(m, n) = (1, -1)$. This expression is obtained by replacing the value of $\delta\left(\varphi_q(t)\right)$ given by (4.9) in (4.21). Notice that for t large enough we get

$$\frac{Y_{m,n}\left(\varphi_q(t)\right)}{t^{m+n}} \approx \frac{(m+1)}{m}, \qquad (4.24)$$

so the ratio $Y_{m,n}/t^{m+n}$ saturates to a constant as $t \to \infty$.

Moreover, from both (4.22) and (4.23) it follows that

$$\overline{Y}_{m,n}\left(\varphi_q(t)\right) \approx \frac{(m+1)}{m\,(m+n+1)}, \qquad t \to \infty, \qquad (4.25)$$

which is also a fixed constant not depending on the orbit.

For an irregular orbit, φ_i, with mLCE σ_i, we have

$$\frac{Y_{m,n}\left(\varphi_i(t)\right)}{t^{m+n}} \approx \sigma_i\, t + \tilde{O}\left(\varphi_i(t)\right), \tag{4.26}$$

while, on considering a sufficiently large time, we obtain

$$\overline{Y}_{m,n}\left(\varphi_i(t)\right) \approx \frac{\sigma_i\, t}{(m+n+2)}. \tag{4.27}$$

For a chaotic orbit then, both $Y_{m,n}/t^{m+n}$ and $\overline{Y}_{m,n}$, thus defined, grow linearly with time, at a rate that is proportional to the mLCE of the orbit.

Therefore, the asymptotic behavior of $\overline{Y}_{m,n}$ can be recast as

$$\overline{Y}_{m,n}\left(\varphi(t)\right) \approx a_\varphi t + b_\varphi, \tag{4.28}$$

where now $a_\varphi = \sigma_i/(m+n+2)$ and $b_\varphi \approx 0$ for irregular, chaotic motion, while $a_\varphi = 0$ and $b_\varphi \approx (m+1)/m(m+n+1)$ for stable, quasiperiodic motion. As it turns out from (4.28), the mLCE can also be recovered by a simple linear least squares fit on $\overline{Y}_{m,n}\left(\varphi(t)\right)$.

Notice that $\hat{\sigma}_{1,m,n} = Y_{m,n}/t^{m+n+1}$ satisfies

$$\hat{\sigma}_{1,m,n}(\varphi_q(t)) \approx \frac{(m+1)}{m\,t}, \qquad \hat{\sigma}_{1,m,n}(\varphi_i(t)) \approx \sigma_i, \qquad t \to \infty, \tag{4.29}$$

so that, for regular motion, $\hat{\sigma}_{1,m,n}$ also converges to 0 faster than $\sigma_1 \sim \ln t/t$, while for chaotic motion, both magnitudes approach the positive mLCE at a similar rate.

An exhaustive comparison of the generalized MEGNO's performance for different exponents (m, n) revealed that, besides the natural choice $(1, -1)$, the values $(2, 0)$ serve to distinguish regular from chaotic behavior in a quite efficient manner (see below).

Just for the sake of illustration, let us turn back to the 2D Hénon–Heiles example given in Sect. 4.2.1. For the same three regular orbits labeled as (sp), (qp) and (up), we computed both $Y_{m,n}$ and $\overline{Y}_{m,n}$, by means of (4.21) and (4.22) respectively, for three different choices of (m, n), namely, $(1, -1)$, $(2, 0)$ and $(3, 1)$.

In Fig. 4.4 we show that for regular motion, $\overline{Y}_{m,n}$ evolves with time as predicted by (4.25). Indeed, the temporal evolution of $\overline{Y}_{m,n}$ for the three regular orbits is seen to tend to the asymptotic values 2, $1/2$ and $4/15$, when the exponents are $(1, -1)$, $(2, 0)$ and $(3, 1)$, respectively. We note that, for the stable quasiperiodic orbit (qp), $\overline{Y}_{m,n}$ converges to the value given in (4.25), a faster convergence being observed the larger is m. Also for the orbit close to a stable periodic one (sp), does $\overline{Y}_{m,n}$ reach the constant value (4.25) faster as a larger exponent m is considered. Notice however that for $m = 2$ much smaller oscillations around the asymptotic value (4.25) are observed in the case of the trajectory close to an unstable periodic orbit (up). Let us note that the exponent n is dummy in the present discussion.

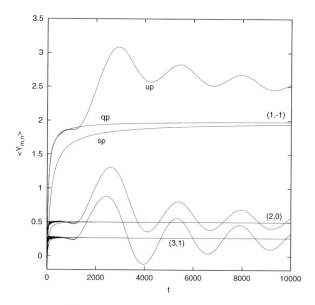

Fig. 4.4 Time evolution of $\overline{Y}_{m,n}$ ($< Y >_{m,n}$ in the plot) for the regular orbits (sp), (qp) and (up) in the Hénon–Heiles model, for different values of the exponents (m, n). Figure taken from [13]

From this comparison we conclude that the choice of exponents $(2, 0)$ allows for clearly separating regular and chaotic regime even in rather short evolution times. Furthermore, if we use as a dynamical indicator the quantity $4\overline{Y}_{2,0}$, we see that for regular orbits it tends to 2, as $\overline{Y}_{1,-1}$ does, while for orbits with exponential instability it tends to $\sigma_i t$. Then, either a linear fit or simply $4\overline{Y}_{2,0}(\varphi_i(t))/t$ provides an estimate of the mLCE. However, the choice $(1, -1)$ for the exponents offers the additional benefit of more clearly identifying stable and unstable periodic motion as well. Anyway, though all the eventual advantages of the generalized MEGNO showed above, the use of the classical MEGNO, $\overline{Y}_{1,-1}$, is widespread. Therefore, we will show the results for $\overline{Y}_{2,0}$ when dealing with discrete applications and discuss below an interesting connection between the classical MEGNO and another well known chaos indicator.

4.2.3 The Connection Between the MEGNO and the FLI

The standard MEGNO, defined adopting the value of the exponents $(1, -1)$, exhibits an intrinsic relation with the classical Fast Lyapunov Indicator (FLI) [53], as we will see in the sequel. For that sake we recall that in [18] the authors define the FLI for a given solution of the flow (4.3), $\varphi(t)$, in terms of the norm of the tangent vector

$\delta \equiv \|\delta\|$ as

$$\text{FLI}(\varphi(t)) = \ln \delta(\varphi(t)), \qquad (4.30)$$

expression that has been used to obtain analytical results in both [18] and [33].

Thus, the time average of the FLI in the interval $(0, t)$ is given by

$$\overline{\text{FLI}}(\varphi(t)) = \frac{1}{t} \int_0^t \ln \delta(\varphi(s)) ds. \qquad (4.31)$$

The MEGNO is twice a time weighted average of the relative divergence of orbits as it can be seen from (4.8). In order to show the relation between the MEGNO and the FLI, let us rewrite (4.8) in the fashion:

$$Y(\varphi(t)) = \frac{2}{t} \int_0^t \frac{d}{ds} \left(\ln \delta(\varphi(s)) \right) ds. \qquad (4.32)$$

After a simple manipulation we obtain

$$Y(\varphi(t)) = 2 \left\{ \ln \delta(\varphi(t)) - \frac{1}{t} \int_0^t \ln \delta(\varphi(s)) ds \right\}, \qquad (4.33)$$

where the value $\delta(0) = 1$ has been taken. From (4.30), (4.31) and (4.33) we conclude that the MEGNO is twice the difference between the FLI and its time average over the interval $(0, t)$,

$$Y(\varphi(t)) = 2 \left\{ \text{FLI}(\varphi(t)) - \overline{\text{FLI}}(\varphi(t)) \right\}. \qquad (4.34)$$

This result serves to understand two facts that have been recently mentioned in the literature. One point is that the MEGNO criterion takes advantage of the dynamical information of the evolution of the tangent vector along the complete orbit, as stated in [70] and [37]. Equation (4.34) tells us exactly in which way it encompasses this information: at every time the MEGNO subtracts from the FLI its average value.

The other point worth discussing, which is explicitly mentioned in [7] and [3], is the reason by which the MEGNO gives account of the degree of chaoticity of an orbit in an absolute scale while the FLI just gives relative values; i.e. in the case of regular orbits the MEGNO tends asymptotically towards a constant value (2), while the FLI behaves logarithmically, not allowing to count with a time independent criterion to establish the threshold that separates chaotic from regular motion.

Just to illustrate this situation let us consider the case of an ideal KAM regular orbit. Therefore the norm of the tangent vectors behaves as (4.9) and besides oscillations $\delta(\varphi_q(t)) \approx 1 + \lambda t$ ($\lambda > 0$ and $\delta_0 = 1$). In this case it is

$$\text{FLI}(\varphi_q(t)) \approx \ln(1 + \lambda t) \qquad (4.35)$$

and

$$\overline{\text{FLI}}((\varphi_q(t))) \approx \ln(1 + \lambda t) + \frac{\ln(1 + \lambda t)}{\lambda t} - 1. \tag{4.36}$$

Therefore, on regarding (4.34) there results

$$Y(\varphi_q(t)) = 2 \left\{ 1 - \frac{\ln(1 + \lambda t)}{\lambda t} \right\}, \tag{4.37}$$

and we rediscover the already mentioned asymptotic limit of the MEGNO for regular orbits.

On the other hand, in the case of an ideal chaotic orbit, with $\delta(\varphi_i(t)) \approx e^{\sigma t}$ (being σ the mLCE), the MEGNO-FLI relation allows to prove that both indicators behave similarly, that is linearly with time with a slope equal to σ.

In order to show the MEGNO-FLI relation we consider again the Hénon–Heiles model for the same energy level, $h = 0.118$, and two orbits one quasiperiodic at $y = 0.2, p_y = 0$ inside the largest island, and a chaotic one at $y = -0.18, p_y = 0$ in the chaotic sea. Just to eliminate oscillations, we compute $\overline{Y}(\varphi(t))$ and the average of $\{\text{FLI}(\varphi(t)) - \overline{\text{FLI}}(\varphi(t))\}$ for these two orbits. The results presented in Fig. 4.5 show an excellent agreement between both magnitudes.

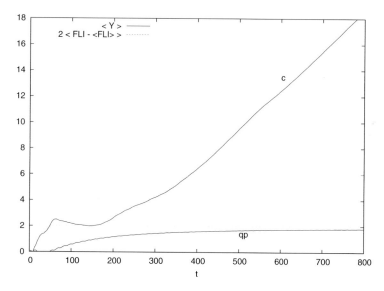

Fig. 4.5 Illustration of the relation (4.34)—in average—for a quasiperiodic orbit (qp) and a chaotic (c) one for the Hénon–Heiles system at $h = 0.118$. Note that both curves are indistinguishable

Therefore in view of the close relation between the MEGNO and the FLI, any improvement concerning the FLI, as for instance the alternative version of the FLI, the so-called Orthogonal Fast Lyapunov Indicator (OFLI)—see [16] and the corresponding chapter in this volume—, applies naturally to improve the MEGNO itself.

4.2.4 The MEGNO for Maps

In this subsection we briefly show how the MEGNO should be implemented to discrete dynamical systems. For dealing with maps, this numerical tool is defined essentially as before, but summing over the iterates of the map instead of integrating with respect to t, and taking the differential map in place of the first variational equations.

For a given initial point P_0, iterates under a given map T are computed yielding points $P_k = T^k(P_0)$. An initial random and unitary tangent vector $\mathbf{v}_0$, is transported under the differential map DT, to obtain vectors $\mathbf{v}_k = DT^k(P_0)\mathbf{v}_0$. Then, after N iterates, the (generalized) MEGNO is computed by means of

$$Y_{m,n}(N) = (m+1) N^n \sum_{k=1}^{N} \ln\left(\frac{\|\mathbf{v}_k\|}{\|\mathbf{v}_{k-1}\|}\right) k^m, \tag{4.38}$$

and

$$\overline{Y}_{m,n}(N) = \frac{1}{N^{m+n+1}} \sum_{k=1}^{N} Y_{m,n}(k). \tag{4.39}$$

We have considered different values for the exponents m and n. Again, it turned out that the larger m, the faster $\overline{Y}_{m,n}$ converges to a constant value for regular motion, but, for m rather large, small oscillations show up. However, the bumpy late evolution of $\overline{Y}_{m,n}$ (which is also present in the continuous case, as Fig. 4.4 shows, in the case of (up) orbits) is diminished if the iteration is stopped when the distance between the initial and final points is minimum ("right-stop" condition). On returning close to the initial point, the effect of the periodic or quasiperiodic oscillations added to a regular behavior is minimized. This sort of refinement in regards to the stop time in the case of maps has proven rather efficient in smoothing such oscillations.

The choice $(2,0)$ for the exponents, together with the "right-stop" condition, have shown to provide a fairly good fast dynamical indicator for maps. A minor additional modification is also convenient with the choice $(m,n) = (2,0)$. Let us define the parameter

$$\hat{Y}_{2,0}(N) = \frac{4\overline{Y}_{2,0}(N) - 2}{N}, \tag{4.40}$$

which when $N \to \infty$, $\hat{Y}_{2,0} \to 0^-$ for orbits lying on tori, while $\hat{Y}_{2,0} \to \sigma_i$ in the case of chaotic orbits that lie in a higher dimensional domain. So, negative values (close to 0) of $\hat{Y}_{2,0}(N)$ arise for regular orbits (provided N is taken not too small), while small positive values would identify mild chaos.

4.3   Applications

4.3.1   A System of Continuous Time: The Arnold Model

Let us consider the well known classical Arnold Hamiltonian [1], which is the paradigmatic model that leads to the so-called (and perhaps controversial) Arnold diffusion. We will address this simple but very representative nonlinear model because, in our opinion, it has not been discussed in a plain manner for non-mathematical readers yet. In fact, though Sect. 7 in [8] is devoted to present Arnold diffusion in an heuristic way by recourse to this model, unfortunately that section of the outstanding review by B. Chirikov seems not to be widespread in the nonlinear community. The Arnold model is also well discussed in the lectures of Giorgilli [24], though in a more mathematical fashion.

The Arnold Hamiltonian has the form

$$H(I_1, I_2, \theta_1, \theta_2, t; \varepsilon, \mu) = \frac{1}{2}(I_1^2 + I_2^2) + \varepsilon(\cos\theta_1 - 1)(1 + \mu B(\theta_2, t))$$

$$B(\theta_2, t) = \sin\theta_2 + \cos t, \tag{4.41}$$

with $I_1, I_2 \in \mathbb{R}$, $\theta_1, \theta_2, t \in \mathbb{S}^1$; where μ should be exponentially small with respect to ε, so that $\varepsilon\mu \ll \varepsilon \ll 1$ (just in Arnold formulation, however see below).

For $\varepsilon = 0$ we have two integrals of motion, namely I_1 and I_2 which determine the invariant tori supporting the quasiperiodic motion with frequencies $\omega_1 = I_1$, $\omega_2 = I_2$. Therefore we have a very simple dynamical system consisting of two uncoupled free rotators, so that, $\theta_1(t) = I_1 t + \theta_1^0$, $\theta_2(t) = I_2 t + \theta_2^0$.

For $\varepsilon \neq 0$, $\mu = 0$ we still have two integrals,

$$H_1(I_1, \theta_1; \varepsilon) = \frac{1}{2}I_1^2 + \varepsilon(\cos\theta_1 - 1), \quad I_2, \tag{4.42}$$

and the unperturbed Hamiltonian could be written as

$$H_0(I_1, I_2, \theta_1; \varepsilon) = H_1(I_1, \theta_1; \varepsilon) + \frac{1}{2}I_2^2. \tag{4.43}$$

Notice that H_1 is the pendulum model for the resonance $\omega_1 = 0$; $H_1 \equiv h_1 = -2\varepsilon$ corresponds to the exact resonance or stable equilibrium point at $(I_1, \theta_1) = (0, \pi)$

while $h_1 = 0$ to the *separatrix* and thus $(I_1, \theta_1) = (0, 0)$ is the unstable point or *whiskered torus*.[5]

The associated frequencies are now $\omega_1 = \omega_p(h_1, \varepsilon)$, $\omega_2 = I_2$, where $\omega_p(h_1, \varepsilon)$ is the pendulum frequency,

$$\omega_p(h_1, \varepsilon) = \frac{\pi \omega_0(\varepsilon)}{2K\left(k_{h_1}\right)}, \qquad -2\varepsilon \leq h_1 < 0,$$

(4.44)

$$\omega_p(h_1, \varepsilon) = \frac{\pi \omega_r(h_1, \varepsilon)}{2K\left(k_{h_1}^{-1}\right)}, \qquad h_1 > 0;$$

where $k_{h_1}^2 = (h_1 + 2\varepsilon)/2\varepsilon$, $\omega_0(\varepsilon) \equiv \sqrt{\varepsilon}$ is the small oscillation frequency, $\omega_r(h_1, \varepsilon) = \omega_0(\varepsilon)k_{h_1}$ is the half-rotation frequency and $K(\kappa)$ is the complete elliptical integral of the first kind. For rotations, the second in (4.44) provides the half-rotation frequency, in order to avoid the jump of a factor 2 between the frequency at both sides of the separatrix. Therefore in the oscillation regime $\omega_p(h_1, \varepsilon) \leq \omega_0(\varepsilon)$ and close to the separatrix for both oscillations and rotations, $\omega_p(|h_1| \ll 1, \varepsilon) \equiv \omega_{sx}(h_1, \varepsilon)$ takes the asymptotic form

$$\omega_{sx}(h_1, \varepsilon) = \frac{\pi \omega_0(\varepsilon)}{\ln\left(\frac{32\varepsilon}{|h_1|}\right)}, \qquad \omega_{sx}(h_1, \varepsilon) \to 0 \quad \text{as} \quad |h_1| \to 0.$$

(4.45)

In the rotation regime, for h_1 large enough $2\omega_p(h_1, \varepsilon) \approx \sqrt{2h_1} \approx I_1$. Figure 4.6 shows the dependence of ω_p on h_1, for $\varepsilon = 0.15$.

The resonance $\omega_1 = 0$ has a half-width $(\Delta I_1)^r = 2\sqrt{\varepsilon}$ in action space, so the variation of I_1 is bounded by $|\Delta I_1| \leq 2\sqrt{\varepsilon}$ while I_2 remains constant. Therefore in (I_1, I_2) plane, $\omega_1 \to \omega_{sx}(h_1, \varepsilon) \to 0$ when $I_1 \to 2\sqrt{\varepsilon}$.

For $\varepsilon \neq 0, \mu \neq 0$ the original system (4.41) can be written as

$$H(I_1, I_2, \theta_1, \theta_2, t; \varepsilon, \mu) = H_0(I_1, I_2, \theta_1; \varepsilon) + \mu V(\theta_1, \theta_2, t; \varepsilon),$$ (4.46)

$$\mu V(\theta_1, \theta_2, t; \varepsilon) = \varepsilon\mu(\sin\theta_2 + \cos t)(\cos\theta_1 - 1),$$

where H_0 is given by (4.43) and $\theta_2(t) = \omega_2 t + \theta_2^0$. Therefore the full Hamiltonian is a simple pendulum and a free rotator coupled by $V(\theta_1, \theta_2, t; \varepsilon)$.

Since the perturbation depends on θ_2 and t, it affects the phase oscillations at the resonance $\omega_1 = 0$ and leads to the formation of the stochastic layer around its separatrix. Moreover, due to the dependence of V on θ_2, the perturbation changes not only I_1 but I_2 as well, and then motion along the stochastic layer should proceed. Due to the stochasticity of the motion inside the layer, the variation of I_2 should be

---

[5]The whiskered torus is a generalization of a saddle equilibrium point and it is defined as the connected intersection of the stable and unstable manifolds or, in Arnold language, arriving and departing whiskers, W^- and W^+ respectively (see [1, 24] for further details).

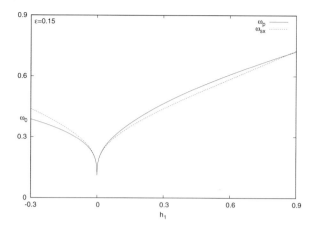

Fig. 4.6 Frequency of the pendulum H_1 given by (4.44), ω_p, and the approximation (4.45), ω_{sx}, for $\varepsilon = 0.15$. The separatrix corresponds to $h_1 = 0$ and the small oscillation frequency, $\omega_0 = \sqrt{\varepsilon} \approx 0.39$. Within the oscillation domain $h_1 \leq 0$ and $\omega(h_1) \leq \omega_0$, while for h_1 large enough, $2\omega \approx \sqrt{2h_1}$

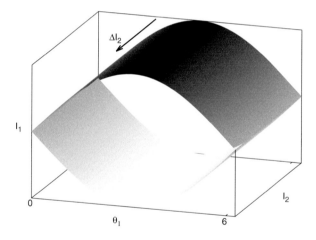

Fig. 4.7 Sketch of diffusion along I_2. The *arrow* indicates that ΔI_2 lies on the stochastic layer of the resonance $\omega_1 = 0$

also stochastic, giving rise to diffusion in I_2, as sketched in Fig. 4.7. In consequence, as I_2 would change unboundedly, a gross instability could set up. This is the way in which Arnold diffusion is described in an heuristic way in [8]. However, in this model, since the perturbation V vanishes at $I_1 = 0, \theta_1 = 0$, it is possible to build up a *transition chain* [1, 24] such that if ω_2 is irrational, then all tori defined by

$I_1 = 0, I_2 = \omega_2 > 0$ are *transition tori*,[6] and when $t \to \infty$, $|I_2(t) - I_2(0)| = \mathcal{O}(1)$, independently of ε and also of μ. Therefore a "large variation" of I_2 could take place. Let us state that by "large variation" we mean that I_2 could vary over a finite domain, which does not imply that it can be proved that I_2 changes without any bound. In fact, this is an open subject of research from a theoretical point of view. Therefore, any demonstration that diffusion might spread along the *resonance web* is quite far to be obtained, as pointed out in [46] and [9].

In the full Hamiltonian (4.46) however, $\omega_1 = 0$ is just one of the six first order resonances. Indeed, multiplying the different harmonics and using trigonometric relationships in $\mu V(\theta_1, \theta_2, t; \varepsilon)$ we obtain the following primary resonances at order ε and $\varepsilon\mu$:

$$\omega_1 = 0, \quad \omega_2 = 0, \quad \omega_1 \pm \omega_2 = 0 \quad \omega_1 \pm 1 = 0, \qquad (4.47)$$

which are depicted in Fig. 4.8 in frequency space, illustrating their respective widths. In (4.47) but in the action or energy space, we should use either the approximations

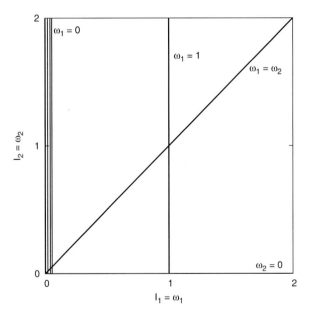

Fig. 4.8 Primary resonances in Arnold model (4.46) in the domain $I_1, I_2 \geq 0$ and considering $(\omega_1, \omega_2) = (I_1, I_2)$. The resonance $\omega_1 = 0$ has an amplitude $V_{10} = \varepsilon$ while for the rest, $V_{mn} = \varepsilon\mu \ll V_{10}$

---

[6]Roughly, a transition torus is a whiskered torus satisfying that points belonging to its arriving whisker W^-, intersect any manifold which is transverse to its departing whisker W^+. Therefore a transition chain is a set of k transition tori satisfying that W_l^+ of the l-transition torus intersects transversally W_{l+1}^- of the $(l+1)$-transition torus.

$\omega_1 \approx I_1$ in case $I_1 \gg 2\sqrt{\varepsilon}$, while $\omega_1 = \omega_p(h_1, \varepsilon)$ for $I_1 < 2\sqrt{\varepsilon}\,(h_1 < 0)$ or $\omega_1 = 2\omega_p(h_1, \varepsilon)$ in case $I_1 \gtrsim 2\sqrt{\varepsilon}\,(h_1 > 0)$.

For $I_1 \gg 2\sqrt{\varepsilon}$ the resonance "lines" intersect at seven fixed different points namely, $(I_1, I_2) = (0,0), (0, \pm 1), (\pm 1, \pm 1)$.[7] Hence, as pointed out by Chirikov [8], the diffusion would spread over all this resonance set. Notice however, that for $\varepsilon\mu \ll \varepsilon \ll 1$ the diffusion rate should be negligible along all resonances except for $\omega_1 = 0$, since this resonance is the one that has the main strength, its amplitude being ε, while all the remaining resonances have amplitudes $\varepsilon\mu \ll \varepsilon$. Indeed, it can be shown (see for instance [8] and [9]) that the diffusion rate depends exponentially on $-1/\sqrt{V_{mn}}$, where V_{mn} stands for the amplitude of the above considered resonances.

Considering the fully perturbed motion, besides the ones given in (4.47), the full set of resonances is an integer linear combination of the form

$$m_1\omega_1 + m_2\omega_2 + m_3 = 0, \qquad m_1, m_2, m_3 \in \mathbb{Z}, \qquad (4.48)$$

where again, $\omega_1 \approx I_1$ or $\omega_p(h_1, \varepsilon)$ depending on the value of $I_1/2\sqrt{\varepsilon}$. Therefore, the true picture of the Arnold web in action space[8] should be much more complex than the one presented in Fig. 4.8, since in that case it is assumed that $\varepsilon \ll \varepsilon\mu \ll 1$ and away from the origin it holds that $I_1 \gg 2\sqrt{\varepsilon}$ so that $\omega_1 = 2\omega_p \sim I_1$. In this case we expect vertical resonances for $m_2 = 0$, horizontal ones for $m_1 = 0$ and an infinite but countable set of curves for $m_1, m_2 \neq 0$ (see below).

For the sake of illustration, we present first the result of a numerical experiment adopting $\varepsilon = 0.05$ and $\mu = 0.0001$, such that the condition $\varepsilon\mu \ll \varepsilon \ll 1$ is fulfilled. Figure 4.9 shows the actual resonances while plotting just the MEGNO values larger than 2.05 for 10^6 initial conditions in the I_1, I_2 space with $\theta_1 = \pi, \theta_2 = t = 0$ after a total motion time 10^4. This plot should be compared to Fig. 4.8 where the main resonances, $\omega_1 = 0, \omega_2 = 0, \omega_1 = \omega_2$, and $\omega_1 \approx 1$ are clearly distinguished.

The expected width of the main resonance $\omega_1 = 0$, $(\Delta I_1)^r \approx 0.45$ is fully consistent with the computed one, and regarding the rest of the resonances, their width should be rather small, close to 4×10^{-3}, and thus they show up approximately as a single curve. Some other resonances do not appear as lines, while the one at $\omega_1 = 1$ do not arise exactly at $I_1 = 1$. Indeed, if we take the resonance condition given by (4.48) for $m_2 \neq 0$, we can rewrite it using the right value for ω_1,

$$\omega_2 = -\frac{m_1}{m_2}\omega_p(h_1, \varepsilon) - \frac{m_3}{m_2}, \qquad (4.49)$$

---

[7]Note that for $I_1 \sim 2\sqrt{\varepsilon}$, $\omega_1 = 2\omega_p(h_1, \varepsilon)$ and the resonances should not intersect in the same set of points, since for instance the resonance $\omega_1 = \omega_2$ leads to a curve in the (I_1, I_2) plane that changes with ε.

[8]The web of all resonances such as (4.48) for all $m_1, m_2, m_3 \in \mathbb{Z}$.

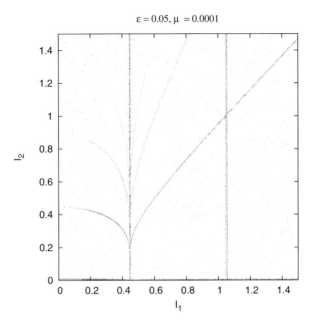

Fig. 4.9 Actual resonances in the Arnold model according to a MEGNO mapping on the (I_1, I_2)-plane for $\theta_1 = \pi, \theta_2 = t = 0$ and $\varepsilon\mu \ll \mu \ll \epsilon$. Region in *black* corresponds to chaotic domains, while those in *white* correspond to periodic or quasiperiodic motion (see text)

and several of the observed curves follow the very same pattern of $\omega_p(h_1, \varepsilon)$ given in Fig. 4.6.[9] In order to compare both figures recall that in Fig. 4.9, $\theta_1 = \pi$ so h_1 and I_1 are simply related by $I_1^2 = 2h_1 + 4\varepsilon$.

Many other resonances are obtained by means of the MEGNO for two sets of larger values of ε and μ, the results being displayed in Fig. 4.10. These are somewhat closer to a more realistic case since in a generic Hamiltonian, it is not possible to reduce the "perturbation" in such a way that it becomes exponentially small with respect to the integrable part. The assumption $\varepsilon \gtrsim \mu$ represents a typical situation in a system involving an integrable Hamiltonian plus a perturbation, which in fact is an artificial separation in a real problem (see for instance [66]).

In Fig. 4.10 we use the (h_1, I_2)-plane to display the resonances just to simplify the comparison of the pattern shown by high order resonances with the plot in Fig. 4.6. Several resonances of the form (4.49) can be observed, namely those of very low order, like $\omega_1 = 0$ of width 2ε (measured in h_1) where the separatrix appears at $h_1 = 0$. Many other high order ones show up exhibiting a similar pattern as that of $\omega_p(h_1, \varepsilon)$. Close to the separatrix all resonances accumulate at $(h_1, I_1) = (0, 0)$ following the very same behavior as ω_p.

---

[9]See next page for the estimation of the right position in the (I_1, I_2)-plane of the $\omega_1 = 1$ resonance.

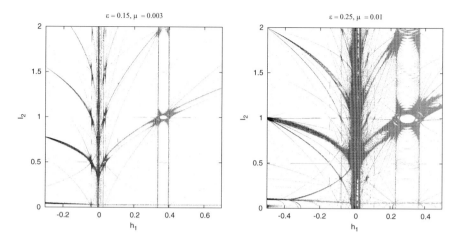

Fig. 4.10 True pictures through a MEGNO map on the (h_1, I_2)-plane for 10^6 initial values of (h_1, I_2) for a total motion time 10^4. *White* corresponds to regions of regular motion where $\overline{Y} < 2.05$, while those in *black* correspond to chaotic motion ($\overline{Y} \geq 2.05$). Section at $\theta_1 = \pi, \theta_2 = t = 0$ for $h_1 \geq -2\varepsilon, I_2 = \omega_2 > 0$

Let us take for instance the MEGNO's plot for $\varepsilon = 0.15$. The resonance $\omega_1 = 0$ should have a half-width $2\varepsilon = 0.3$, which is fully consistent with the computed one, and the separatrix appears at $h_1 = 0$ as expected. For the resonance $\omega_1 = 1$, the approximate value of $I_1^r = 1$ in fact corresponds to $h_1^r = 0.2$. However if we use the approximation (4.45) for the resonance condition $2\omega_p(h_1, \varepsilon) = 1$, it leads to $h_1^r \approx 0.4$ ($I_1^r \approx 1.2$) very close to the computed one. The obtained picture for $\varepsilon = 0.25$ shows a similar structure but, as expected, resonances are wider and many other high order resonances appear, particularly in the region close to the separatrix.

In both MEGNO contour plots the center of any resonance "channel" corresponds to 2D elliptic tori while the borders (the stochastic layer or homoclinic tangle) to 2D hyperbolic tori. At the intersection of two or more resonances a periodic orbit appears, which could be stable or unstable. In general, the intersection of two elliptic 2D tori leads to a stable periodic orbit and to a small domain of stable motion. From Fig. 4.10 we see that the MEGNO plots reveal the stability character of all the periodic orbits as well as a clear picture of the dynamics on the whole domain. However, from these plots nothing could be inferred concerning diffusion in action space, since as we have already pointed out, the MEGNO, as most chaos indicators, only provides information about the local dynamics of the Hamiltonian flow. Therefore we only have at hand just the behavior of the flow in any rather small open domain of every selected point in the grid. Nothing could be said about if it is possible that a chaotic orbit could explore a finite domain.

4.3.2   Models of Discrete Time

4.3.2.1   The Rational Shifted Standard Map

Let us consider the so-called Rational Shifted Standard Map (RSSM—see [13] for some additional details). This is a 2D area-preserving discrete dynamical system given by

$$y' = y + \varepsilon f(x), \qquad\qquad x' = x + \varepsilon y', \qquad\qquad (4.50)$$

with $x \in [0, 2\pi)$, $y \in [0, 2\pi/\varepsilon)$, and where

$$f(x) = \frac{\sin (x + \varphi)}{1 - \mu \cos x} - \Delta, \qquad \Delta = \frac{\mu \sin \varphi}{\sqrt{1 - \mu^2} + 1 - \mu^2}. \qquad (4.51)$$

Notice that (4.50) and (4.51) define some sort of Standard Map (SM) modified in order to have a no longer symmetric nor entire function f. Indeed, symmetry is lost through the introduction of the phase φ, while the insertion of the denominator, with the parameter $\mu \in [0, 1)$, breaks the entire character of f. The quantity Δ is fixed so that f has zero average, in order the RSSM be area-preserving.

After rescaling the y-variable by means of $y \to \varepsilon y$ such that both $x, y \in [0, 2\pi)$, the RSSM reads

$$y' = y + \varepsilon^2 f(x), \qquad\qquad x' = x + y', \qquad\qquad (4.52)$$

and adopts an even closer form to that of the SM.

On expanding

$$\frac{1}{1 - \mu \cos x} = 1 + \mu \cos x + \mu^2 \cos^2 x + \mu^3 \cos^3 x + \dots, \qquad (4.53)$$

and adopting $\varphi = 0$ in order to emphasize the comparison with the SM, after taking into account some trivial trigonometric identities, there results

$$f(x) = \frac{\sin x}{1 - \mu \cos x} = \left(1 + \frac{\mu^2}{4} \right) \sin x + \frac{\mu}{2} \sin 2x + \frac{\mu^2}{4} \sin 3x + \dots. \qquad (4.54)$$

To analyze the effect of changing φ, we perform the shift $x \to x + \varphi$ after which the equation for x in (4.52) remains invariant. On fixing $\varphi = \pi$

$$f(x) = \frac{\sin x}{1 + \mu \cos x} = \left(1 + \frac{\mu^2}{4} \right) \sin x - \frac{\mu}{2} \sin 2x + \frac{\mu^2}{4} \sin 3x + \dots, \qquad (4.55)$$

the map for $x, y \in [0, 2\pi)$ is seen to have a different dependence on the parameters than in the case in which $\varphi = 0$. Therefore a strong dependence of the dynamics on

φ is expected. Herein we will consider the two limiting values, $\varphi = 0, \pi$, in order to reduce the number of free parameters and to clearly show the differences with the SM.

Thus, from (4.54) and (4.55) it becomes clear that the RSSM shows up all the harmonics, instead of the solely term in $\sin x$ present in the SM. Furthermore, the resonances' width depends not only on ε^2, as it is the case in the SM, but on μ as well, and the resonance structure of both maps is similar when $\mu \to 0$. In the RSSM, for $\mu \neq 0$, all resonances (like $y/2\pi = 0, 1/3, 1/2, 2/3$) appear at order ε^2 and at different orders in μ, while in the SM, for instance the semi-integer resonance as $y = 1/2$ appears at ε^4 and those at $y = 1/3, 2/3$ show up at order ε^6, so as μ increases the resonances' interaction in the RSSM is stronger than in the SM.

The potential function for $f \equiv -V'$ is

$$V(x) = \pm \frac{1}{\mu} \ln \left\{ 1 - \mu \cos x \right\}, \qquad \mu \neq 0. \tag{4.56}$$

Expanding $V(x)$ in powers of μ and using the 2π-periodic δ in its Fourier form, the potential $U(x)$ of the corresponding Hamiltonian has the form

$$U(x) = \frac{\varepsilon^2}{4\pi^2} \left\{ \left(1 + \frac{\mu^2}{4} \right) \sum_{n=-\infty}^{\infty} \cos(x + nt) + \right.$$

$$\left. + \frac{\mu}{4} \sum_{n=-\infty}^{\infty} \cos(2x + nt) + \frac{\mu^2}{12} \sum_{n=-\infty}^{\infty} \cos(3x + nt) + \dots \right\}, \tag{4.57}$$

while the kinetic energy is given by $\hat{y}^2/2$, being $\hat{y} = y/2\pi$. Thus we can easily see how resonances appear at different orders in ε and μ.

The MEGNO has been applied to (4.52) in an equispaced grid of 1000×1000 pixels in the domain $(x/2\pi, y/2\pi) \in [0, 1) \times [0, 1)$, to obtain $\hat{Y}_{2,0}(N)$ for $N = 11{,}000$ (see (4.40) and discussion below). The results for $\varphi = 0$ and $\varphi = \pi$ are presented in Fig. 4.11, for $\varepsilon = 0.8$ and two different values of μ. There the pixels corresponding to initial conditions of regular behavior are plotted in white and those of chaotic behavior in black.[10] While for $\varphi = 0$ the regular regime prevails (plots on the left), the dynamics for $\varphi = \pi$ displays several chaotic domains (plots on the right) surrounding stochastic layers of resonances or as it seems, a connected chaotic open domain, but rotational invariant curves (joining the vertical boundaries) still exist. The variation of φ from 0 to π has a quite notorious effect on the dynamics as already mentioned in the theoretical discussion. The figures on the top corresponds to $\mu = 0.1$ while those on the bottom to $\mu = 0.2$. We can notice that increasing the value of μ changes the stability of the periodic orbit at $(0.5, 0.5)$ in the case

---

[10]We take slightly different threshold values in the figures just to display the global behavior, since for $\varphi = 0$ the map is mostly regular while for $\varphi = \pi$ it is strongly chaotic.

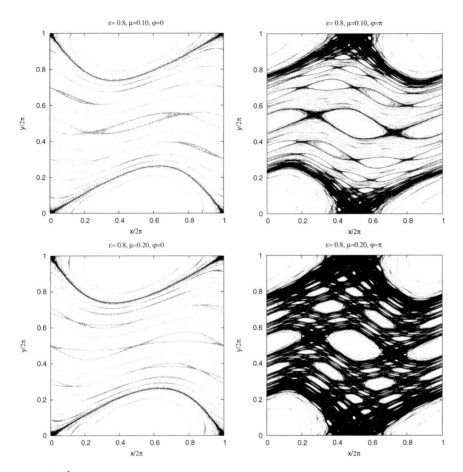

Fig. 4.11 $\hat{Y}_{2,0}$-levels for the RSSM corresponding to $\varepsilon = 0.8$, for $\varphi = 0$ (on the *left*) and $\varphi = \pi$ (on the *right*). The figures on the *top* correspond to $\mu = 0.1$ and those on the *bottom* to $\mu = 0.2$ Regions of regular behavior are depicted in *white* and those of chaotic behavior in *black*. The threshold values are 2×10^{-4} for $\varphi = 0$ and 2×10^{-2} for $\varphi = \pi$

of $\varphi = 0$. Meanwhile, for $\varphi = \pi$ the chaotic regime increases as larger values of μ are adopted. Notice that the MEGNO also succeeds in unveiling the high order resonance structure of this map.

4.3.2.2 The Coupled Rational Shifted Standard Map

Let us now turn to the Coupled Rational Shifted Standard Map (CRSSM), consisting of two coupled RSSM, defined by

$$y_1' = y_1 + \varepsilon_1 f_1(x_1) + \gamma_+ f_3(x_1 + x_2) + \gamma_- f_3(x_1 - x_2),$$
$$y_2' = y_2 + \varepsilon_2 f_2(x_2) + \gamma_+ f_3(x_1 + x_2) - \gamma_- f_3(x_1 - x_2),$$

$$x_1' = x_1 + \varepsilon_1 y_1',$$
$$x_2' = x_2 + \varepsilon_2 y_2', \tag{4.58}$$

with $x_i \in [0, 2\pi)$, $y_i \in [0, 2\pi/\varepsilon_i)$, $i = 1, 2$, and where

$$f_i(x) = \frac{\sin(x + \varphi_i)}{1 - \mu_i \cos x} - \Delta_i, \qquad \Delta_i = \frac{\mu_i \sin \varphi_i}{\sqrt{1 - \mu_i^2} + 1 - \mu_i^2}, \quad i = 1, 3,$$

$$\tag{4.59}$$

with $\mu_i \in [0, 1)$ and again Δ_i fixed so that the f_i have zero average. Notice that two coupling terms in $(x_1 + x_2)$ and $(x_1 - x_2)$ have been added, γ_+ and γ_- being the coupling parameters. This map provides a more realistic representation of nonlinear resonance interactions than two coupled Standard Maps, so its dynamics would well serve as an improved simple model for many dynamical scenarios.

Again as in the RSSM, rescaling the y-variables, the CRSSM can be recast as

$$y_1' = y_1 + \varepsilon_1^2 f_1(x_1) + \varepsilon_1 \gamma_+ f_3(x_1 + x_2) + \varepsilon_1 \gamma_- f_3(x_1 - x_2),$$
$$y_2' = y_2 + \varepsilon_2^2 f_2(x_2) + \varepsilon_2 \gamma_+ f_3(x_1 + x_2) - \varepsilon_2 \gamma_- f_3(x_1 - x_2),$$
$$x_1' = x_1 + y_1',$$
$$x_2' = x_2 + y_2', \tag{4.60}$$

where $(x_i, y_i) \in [0, 2\pi) \times [0, 2\pi)$.

The full set of primary resonances is determined by

$$k_1 y_1 + k_2 y_2 + 2\pi k_3 = 0, \qquad k_1, k_2, k_3 \in \mathbb{Z}. \tag{4.61}$$

Therefore, in the action plane, horizontal resonances correspond to the uncoupled (x_2, y_2) map and appear for $k_1 = 0$, the vertical ones correspond to the uncoupled (x_1, y_1) map obtained by setting $k_2 = 0$, while the coupling resonances given by $y_2 = -(k_1 y_1 + 2\pi k_3)/k_2$ are dense (but countable) in the (y_1, y_2)-space.

The MEGNO has been computed for an equispaced grid of 1000×1000 pixels in the domain $(y_1/2\pi, y_2/2\pi) \in [0, 1) \times [0, 1)$. The initial values for the remaining variables are $x_1 = 0, x_2 = 0$. The "right-stop" condition described in Sect. 4.2.4 has been applied so that for each initial condition the iteration is stopped after N iterates, with $10000 < N < 11000$, when the distance between the N-th iteration of the map and the initial condition is minimum.

A difference should be remarked with the action space of Arnold model discussed in the previous section. Indeed, by making the cross product $(x_1, y_1) \times (x_2, y_2)$ at $x_1 = x_2 = 0$, depending on the adopted value of φ and considering Fig. 4.11, we should expect that in the (y_1, y_2)-plane, only do the hyperbolic 2D tori (or to be precise, the homoclinic tangle) show up for $\varphi = 0$ while for $\varphi = \pi$, the picture should be similar to that of Fig. 4.10, since both, the elliptic and hyperbolic 2D tori would be present, as well as the nearly resonant 3D tori that are trapped in resonances.

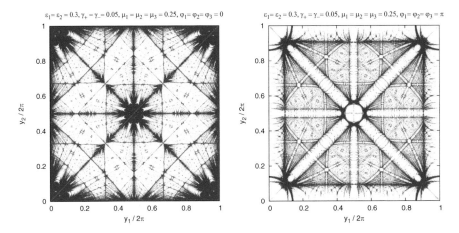

Fig. 4.12 $\hat{Y}_{2,0}$-levels for the CRSSM for different values of the parameters φ_i, 0 in the plot on the *left* and π in that on the *right*. The contour plots correspond to $\hat{Y}_{2,0}$ binned in three intervals; pixels corresponding to initial conditions of regular behavior are plotted in *white*, and those of chaotic behavior in *black* (*red*). With *gray* (*green*) we identify mild chaotic or even quasiperiodic motion

In order to illustrate the efficiency of the MEGNO to display the full dynamics of this 4D map, we show the results for $\Delta_i = 0$, i.e, $\varphi_i = 0$ and $\varphi_i = \pi$, $i = 1, 2, 3$, $\varepsilon_1 = \varepsilon_2 = 0.3$, $\mu_1 = \mu_2 = \mu_3 = 0.25$, and $\gamma_+ = \gamma_- = 0.05$, given in Fig. 4.12. The contour-like plots exhibit the obtained values for $\log(\hat{Y}_{2,0})$ given by (4.40) scaled in order to range from $-5 \leq \log(\hat{Y}_{2,0}) < -3$, to $\log(\hat{Y}_{2,0}) \geq -3$. Recall that $\hat{Y}_{2,0} \to 0^-$ for quasiperiodic motion while $\hat{Y}_{2,0} > 0$ indicates chaotic dynamics. The initial conditions corresponding to regular orbits have been depicted in white, while those in black (red) are chaotic. The orbits holding intermediate values of $\hat{Y}_{2,0}$ are plotted in gray (green) and considered as, possibly, quasiperiodic or mildly chaotic.

Though the Arnold web could be obtained also by means of other chaos indicators, let us mention that since the MEGNO and its generalized version have a clear threshold value, both of them allow for separating regular and chaotic orbits, providing for the latter a measure of the mLCE. Therefore, instead of performing an automatic contour plot it is possible to select the MEGNO ranges to be depicted. This has several benefits when we are interested in separating regular motion and chaotic motion with different degrees of hyperbolicity.

The resonances defined by (4.61) can be clearly distinguished. The wider ones are the integer resonances of the uncoupled maps, however for $\varphi = 0$ almost all resonances have a rather small width due to the fact that in such a case we see only the hyperbolic part, except for a few high order resonances which show up as narrow "channels". The opposite picture corresponds to $\varphi = \pi$, where most resonances reveal their 2D elliptic and hyperbolic tori. Note that the periodic orbit at each resonance intersection is, as expected, unstable for $\varphi = 0$, while it is stable

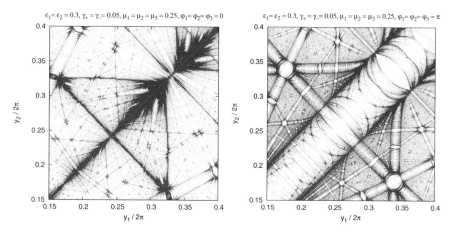

Fig. 4.13 Zoom of Fig. 4.12 in the domain $0.15 \leq y_1, y_2 \leq 0.4$

for $\varphi = \pi$. The complement of the set of 2D elliptic tori, 2D hyperbolic tori and periodic orbits, corresponds to 3D tori, where the motion is quasiperiodic.

In Fig. 4.13 we present a zoom of Fig. 4.12 corresponding to the region $0.15 \leq y_1, y_2 \leq 0.4$. In these $\hat{Y}_{2,0}$-contour plots we can distinguish many resonances of very high order as well as the dynamics in the resonance crossings. We can also notice how the stable and unstable manifolds bend to lead to either a regular or a chaotic domain. These manifolds are very important since they are the objects able to carry the motion arriving along one of the resonances either to the "other parts" of the resonance or to a different resonance. Besides, from the numerical results provided by the MEGNO, we could infer the true effect of the intersection of resonances of different order.

4.4   Comparison of Different Chaos Indicators

Some comparisons between particular indicators in the framework of specific studies were carried out, as those given for instance in [3, 47] and [37]. However, no systematic comparisons of the performance of several chaos indicators had been accomplished up to our comparative studies given in [48] and [49], which we briefly describe in the forthcoming subsections.

4.4.1   Comparative Studies for a Hamiltonian Flow

As already mentioned, the standard MEGNO has become a widespread technique for the study of Hamiltonian systems, particularly in the field of dynamical astronomy and astrodynamics, then, a comparison with other dynamical indicators was in order. Therefore, in [49] a rather complex nonlinear system was addressed that reproduces many characteristics of real elliptical galaxies, namely, the self-consistent model introduced in [56]. Such a model was used as the scenario for a comprehensive comparison between the MEGNO and the mLCE, and even with the FLI. A detailed numerical and statistical study of a sample of orbits in the triaxial galactic system showed that the MEGNO is a suitable fast indicator to separate regular from chaotic motion and that it is particularly useful to investigate the nature of orbits that have a small but positive mLCE.

A rather good correlation was obtained between the MEGNO and the mLCE values for short, moderate and large integration times when considering just chaotic orbits, while the MEGNO provided much better results for regular motion. The FLI also looked like a reliable fast indicator, but since it has no reference value for regular motion, it might be useful to explore the phase space rather than to investigate the nature of a given orbit, unless of course the time evolution of such indicator was followed.

In [50] the same self-consistent triaxial stellar dynamical model was studied for different energy levels by means of some selected variational indicators and spectral analysis methods. Therein, the comparison of several variational indicators on different scenarios was addressed. Indeed, the Average Power Law Exponent (APLE) [47] and the MEGNO's estimation of the mLCE by a least squares fit of its time evolution were compared. The spectral analysis method selected for that investigation was the Frequency Modified Fourier Transform (FMFT) [63], which is just a slight variation of the FMA. Besides, a comparative study of the APLE, the FLI, the Orthogonal Fast Lyapunov Indicator (OFLI) [16] and the estimation of the mLCE obtained from the MEGNO's slope yielded as a result that the latter could be an appropriate alternative to the MEGNO when studying large samples of initial conditions. In fact, it succeeded in separating the chaotic and the regular components and in identifying the different levels of hyperbolicity (or exponential rate of divergence of nearby orbits) as well. Further, it turned out to be more reliable than the FMFT while describing chaotic domains.

4.4.2   Comparative Studies for Maps

In [48] the efficiency of several variational indicators of chaos when applied to mappings was compared. We considered the mLCE, the MEGNO, the Smaller Alignment Index (SALI) [67], its generalized version, the Generalized Alignment Index (GALI) [69], the FLI [19], the Dynamical Spectra of stretching numbers (SSN) [71] and the corresponding Spectral Distance (D) and the Relative Lyapunov Indicator (RLI) [62], which is based on the evolution of the difference between two close orbits. As a result of several experiments presented therein concerning two different 4D mappings, namely, a variant of Froeschlé's symplectic mapping [14, 17, 67, 68] and a system comprising two coupled Standard Maps, it was shown that a package composed of the FLI and the RLI (when a global analysis of the phase portrait is pursued) and of the MEGNO and the SALI (if the objective is the analysis of individual orbits) turned out to be the best choices to yield a good description of the dynamics of the systems under study.

4.5   Further Applications of the MEGNO

In recent years the MEGNO has been widely used mainly in the field of dynamics of multi-planet extrasolar systems to address stability and habitability studies as well as in the solar system, galactic dynamics, astrobiology and chemistry. Herein, we include some references that would serve as illustration of this issue. We refer the reader to the original papers for details regarding the concomitant physical problems.

For extrasolar dynamical studies see for instance [20, 26–32, 52, 55, 61]. For research concerning Solar System dynamics we refer for example to [22] and [37], where the MEGNO technique is applied to the investigation of the dynamics of Jovian irregular satellites, to [60] which is devoted to the resonant structure of Jupiter's Trojan asteroids, its long-term stability and diffusion or to [21] where the evection resonance is considered. Interesting results in astrobiology are obtained while studying the dynamical habitability of exoplanetary systems (see [15, 36, 54]). Further applications of the MEGNO can be found in the study of space debris motion as in [38, 39, 70] among others, and of the chaotic motion of geosynchronous satellites as in [6, 40, 41].

As far as galactic studies are concerned, we can refer for instance to [7, 51, 72].

The use of this chaos indicator in rigid-body motion can be found in [2], and in the realm of chemistry in the analysis of intramolecular dynamics [65], or while revisiting the problem of driven coupled Morse oscillators [64]. Finally, bifurcations and chaos in different scenarios are studied by means of the MEGNO for instance in [4, 23, 25, 59], among many others.

4.6   Discussion

In this review we have described a rather simple technique, the *Mean Exponential Growth factor of Nearby Orbits* (MEGNO), which succeeds in providing detailed indications on the dynamics of continuous dynamical systems and maps. The intrinsic connection of this technique with the FLI and the mLCE is also presented.

The MEGNO furnishes an efficient algorithm that allows not only to clearly identify regular and irregular motion as well as stable and unstable periodic orbits, but also to obtain a quite good estimate of the mLCE in comparatively very short evolution times, for both ordered and chaotic components of phase space. This is a particular feature of this indicator that is not shared with many other techniques. In fact, we could deem that the derivation of the mLCE by a least squares fit of the time evolution of the MEGNO is an alternative algorithm to get the time-scale for exponential divergence of nearby orbits but in rather short times in comparison with the classical approach.

Thus, by the application of this single tool it is possible to grasp the dynamics of the system over the whole phase space, and this procedure is a first attempt to get dynamical information about the motion using the whole orbit.

Moreover, there exists profuse numerical evidence of the MEGNO being a fast indicator capable of unveiling the hyperbolic structure of the phase space, as well as yielding a clear picture of the resonance structure in any dimensional systems. Besides, the MEGNO is shown to provide the actual size of a resonance of very high order as well as to reveal its internal structure.

Let us mention that the application of this technique to many different dynamical systems along the literature shows that it could be useful to investigate stability domains in exoplanetary models, chemical dynamics, space debris as well as to discuss purely theoretical features like bifurcation analysis.

Finally, regarding which is the more suitable chaos detection tool (based on the evolution of the tangent vector) we claim from our experience and in view of the nowadays available computational resources that, it is just a matter of the gained expertise on the adopted technique. However, let us say that the MEGNO is the one with a theoretical threshold value that allows to clearly separate regular from chaotic motion as well as it provides an accurate estimate of the mLCE by means of a very simple algorithm.

Therefore, a combination of any such indicator together with an accurate spectral technique, like the FMA, would be the best option to display the full dynamics of nonlinear systems which in general present a divided phase space.

Acknowledgements We should state clear that also Prof. C. Simó worked very hard developing this technique. Further, we are grateful to him for his illuminating discussions, comments and suggestions that helped us to write this chapter in a comprehensive way and any mistake is nothing but our own responsibility.

We also deeply appreciate the recommendations of the two thoughtful reviewers of this effort that serve to improve the final version.

This work was supported by grants from CONICET, UNLP and *Instituto de Astrofísica de La Plata*.

References

1. Arnold, V.I.: Soviet Math. Dokl. **5**, 581–585 (1964)
2. Barrio, R., Blesa, F., Elipe, A.: J. Astronaut. Sci. **54**(3–4), 359–368 (2006)
3. Barrio, R., Borczyk, W., Breiter, S.: Chaos Solitons Fractals **40**, 1697–1714 (2009)
4. Barrio, R., Blesa, F., Serrano, S.: Int. J. Bifurcat. Chaos **20**(5), 1293–1319 (2010)
5. Benettin, G., Galgani, L., Giorgilli, A., Strelcyn, J.M.: Meccanica **15**(1) Part I: Theory, 9–20; Part II: Numerical applications, 21–30 (1980)
6. Breiter, S., Wytrzyszczak, I., Melendo, B.: Adv. Space Res. **35**(7), 1313–1317 (2006)
7. Breiter, S., Fouchard, M., Ratajczak, R.: Mon. Not. R. Astron. Soc. **383**(1), 200–208 (2008)

8. Chirikov, B.V.: Phys. Rep. **52**, 263–379 (1979)
9. Cincotta, P.M.: New Astron. Rev. **46**, 13–39 (2002)
10. Cincotta, P.M., Simó, C.: Celest. Mech. Dyn. Astron. **73**(1–4), 195–209 (1999)
11. Cincotta, P.M., Simó, C.: In: Gurzadyan, V.G., Ruffini, R. (eds.) The Chaotic Universe. Advanced Series in Astrophysics and Cosmology, vol. 10, pp. 247–258. World Scientific (2000). ISBN: 978-981-279-362-1
12. Cincotta, P.M., Simó, C.: Astron. Astrophys. Suppl. **147**, 205–228 (2000)
13. Cincotta, P.M., Giordano, C.M., Simó, C.: Physica D **182**, 11–178 (2003)
14. Contopoulos, G., Giorgilli, A.: Meccanica **23**, 19–28 (1988)
15. Dvorak, R., Pilat-Lohinger, E., Bois, E., Schwarz, R., Funk, B., Beichman, C., Danchi, W., et al.: Astrobiology **10**(1), 33–43 (2010)
16. Fouchard, M., Lega, E., Froeschlé, Ch., Froeschlé, Cl.: Celest. Mech. Dyn. Astron. **83**, 205–222 (2002)
17. Froeschlé, C.: Astron. Astrophys. **16**, 172–189 (1972)
18. Froeschlé, C., Lega, E.: Celest. Mech. Dyn. Astron. **78**, 167–195 (2000)
19. Froeschlé, C., Gonczi, R., Lega, E.: Planet. Space Sci. **45**, 881–886 (1997)
20. Frouard, J., Compére, A.: Icarus **220**(1), 149–161 (2012)
21. Frouard, J., Fouchard, M., Vienne, A.: Astron. Astrophys. **515**(7) (2010). doi: 10.1051/0004-6361/200913048
22. Frouard, J., Vienne, A., Fouchard, M.: Astron. Astrophys. **532** (2011). doi: 10.1051/0004-6361/201015873
23. Gelfreich, V., Simó, C., Vieiro, A.: Physica D **243**(1), 92–110 (2013)
24. Giorgilli, A.: In: Benest, D., Froeschlé, C. (eds.) Les Methodes Modernes de la Mecanique Celeste, pp. 249–284. Frontières (1990). ISBN: 2-8633209-2
25. Gonchenko, S.V., Simó, C., Vieiro, A.: Nonlinearity **26**(3), 621–678 (2013)
26. Goździewski, K., Migaszewski, C.: Mon. Not. R. Astron. Soc. Lett. **397**(1), L16–L20 (2009)
27. Goździewski, K., Konacki, M., Maciejewski, A.J.: Astrophys. J. Lett. **622**(2I), 1136–1148 (2005)
28. Goździewski, K., Konacki, M., Wolszczan, A.: Astrophys. J. Lett. **619**(2I), 1084–1097 (2005)
29. Goździewski, K., Breiter, S., Borczyk, W.: Mon. Not. R. Astron. Soc. **383**(3), 989–999 (2006)
30. Goździewski, K., Konacki, M., Maciejewski, A.J.: Astrophys. J. Lett. **645**(1I), 688–703 (2006)
31. Goździewski, K., Migaszewski, C., Musielinski, A.: In: Proceedings of the International Astronomical Union 3, Symposium 249, pp. 447–460 (2007)
32. Goździewski, K., Slonina, M., Migaszewski, C., Rozenkiewicz, A.: Mon. Not. R. Astron. Soc. **430**(1), 533–545 (2013)
33. Guzzo, M., Lega, E., Froeschlé, C.: Physica D **163**, 1–25 (2002)
34. Hairer, E., Nørsett, S., Wanner, G.: Solving Ordinary Differential Equations I: Nonstiff Problems. Springer, New York (1987)
35. Hénon, M., Heiles, C.: Astron. J. **69**, 73 (1964)
36. Hinse, T.C., Michelsen, R., Jorgensen, U.G., Goździewski, K., Mikkola, S.: Astron. Astrophys. **488**(3), 1133–1147 (2008)
37. Hinse, T.C., Christou, A.A., Alvarellos, J.L.A., Goździewski, K.: Mon. Not. R. Astron. Soc. **404**(2), 837–857 (2010)
38. Hubaux, C., Lemâitre, A., Delsate, N., Carletti, T.: Adv. Space Res. **49**(10), 1472–1486 (2012)
39. Hubaux, Ch., Libert, A.S., Delsate, N., Carletti, T.: Adv. Space Res. **51**(1), 25–38 (2013)
40. Kuznetsov, E.D., Kaiser, G.T.: Cosmic Res. **45**(4), 359–367 (2007)
41. Kuznetsov, E.D., Kudryavtsev, A.O.: Russ. Phys. J. **52**(8), 841–849 (2009)
42. Laskar, J.: Icarus **88**, 266–291 (1990)
43. Laskar, J.: Physica D **67**, 257–281 (1993)
44. Laskar, J.: Astron. Astrophys. **287**(1), L9–L12 (1994)
45. Laskar, J., Froeschlé, C., Celletti, A.: Physica D **56**(2–3), 253–269 (1992)
46. Lochak, P.: In: Simó, C. (ed.) Hamiltonian Systems with Three or More Degrees of Freedom. NATO ASI Series, pp. 168–183. Kluwer Academic Publishers, The Netherlands (1999)
47. Lukes-Gerakopoulos, G., Voglis, N., Efthymiopoulos, C.: Physica A **387**, 1907–1925 (2008)

48. Maffione, N.P., Giordano, C.M., Cincotta, P.M.: Celest. Mech. Dyn. Astron. **111**, 285–307 (2011)
49. Maffione, N.P., Giordano, C.M., Cincotta, P.M.: Int. J. Non Linear Mech. **46**, 23–34 (2011)
50. Maffione, N.P., Darriba, L.A., Cincotta, P.M., Giordano, C.M.: Mon. Not. R. Astron. Soc. **429**(3), 2700–2717 (2013)
51. Manos, T., Bountis, T., Skokos, C.: J. Phys. A Math. Theor. **46**(25), 254017 (2013)
52. Martí, J.G., Giuppone, C.A., Beaugé, C.: Mon. Not. R. Astron. Soc. **433**(2), 928–934 (2013)
53. Mestre, M., Cincotta, P.M., Giordano, C.M.: Mon. Not. R. Astron. Soc. Lett. **404**, L100–L103 (2011)
54. Migaszewski, C., Goździewski, K., Hinse, T.C.: Mon. Not. R. Astron. Soc. **395**(3), 1204–1212 (2008)
55. Migaszewski, C., Slonina, M., Goździewski, K.: Mon. Not. R. Astron. Soc. **427**(1), 770–789 (2012)
56. Muzzio, J.C., Carpintero, D.D., Wachlin, F.C.: Celest. Mech. Dyn. Astron. **91**(1–2), 173 (2005)
57. Núñez, J.A., Cincotta, P.M., Wachlin, F.C.: Celest. Mech. Dyn. Astron. **54**(1–2), 43–53 (1996)
58. Prince, P., Dormand, J.: J. Comput. Appl. Math. **35**, 67 (1981)
59. Puig, J., Simó, C.: Regul. Chaotic Dyn. **16**(1), 61–78 (2011)
60. Robutel, P., Gabern, F.: Mon. Not. R. Astron. Soc. **372**(4), 1463–1482 (2006)
61. Saito, M.M., Tanikawa, K., Orlov, V.V.: Celest. Mech. Dyn. Astron. **116**(1), 1–10 (2013)
62. Sándor, Z., Érdi, B., Efthymiopoulos, C.: Celest. Mech. Dyn. Astron. **78**, 113–123 (2000)
63. Sidlichovský, M., Nesvorný, D.: Celest. Mech. Dyn. Astron. **65**, 137–148 (1997)
64. Sethi, A., Keshavamurthy, S.: Mol. Phys. **110**(9–10), 717–727 (2012)
65. Shchekinova, E., Chandre, C., Lan, Y., Uzer, T.: J. Chem. Phys. **121**(8), 3471–3477 (2004)
66. Simó, C.: In: Broer, H.W., Krauskopf, B., Vegter, G. (eds.) Global Analysis of Dynamical Systems, pp. 373–390. IOP Publishing, Bristol (2001)
67. Skokos, Ch.: J. Phys. A Math. Gen. **34**, 10029–10043 (2001)
68. Skokos, Ch., Contopoulos, G., Polymilis, C.: Celest. Mech. Dyn. Astron. **65**, 223–251 (1997)
69. Skokos, Ch., Bountis, T., Antonopoulus, Ch.: Physica D **231**, 30–54 (2007)
70. Valk, S., Delsate, N., Lemaitre, A., Carletti, T.: Adv. Space Res. **43**(10), 1509–1526 (2009)
71. Voglis, N., Contopoulos, G., Efthymiopoulos, C.: Celest. Mech. Dyn. Astron. **73**, 211–220 (1999)
72. Zorzi, A.F., Muzzio, J.C.: Mon. Not. R. Astron. Soc. **423**(2), 1955–1963 (2012)

Chapter 5
The Smaller (SALI) and the Generalized (GALI) Alignment Indices: Efficient Methods of Chaos Detection

Charalampos (Haris) Skokos and Thanos Manos

Abstract We provide a concise presentation of the Smaller (SALI) and the Generalized Alignment Index (GALI) methods of chaos detection. These are efficient chaos indicators based on the evolution of two or more, initially distinct, deviation vectors from the studied orbit. After explaining the motivation behind the introduction of these indices, we sum up the behaviors they exhibit for regular and chaotic motion, as well as for stable and unstable periodic orbits, focusing mainly on finite-dimensional conservative systems: autonomous Hamiltonian models and symplectic maps. We emphasize the advantages of these methods in studying the global dynamics of a system, as well as their ability to identify regular motion on low dimensional tori. Finally we discuss several applications of these indices to problems originating from different scientific fields like celestial mechanics, galactic dynamics, accelerator physics and condensed matter physics.

The original version of this chapter was revised.
An erratum to this chapter can be found at DOI 10.1007/978-3-662-48410-4_9

Ch. (Haris) Skokos (✉)
Department of Mathematics and Applied Mathematics, University of Cape Town, Rondebosch 7701, South Africa
e-mail: haris.skokos@uct.ac.za

T. Manos
Center for Applied Mathematics and Theoretical Physics (CAMTP), University of Maribor, Krekova 2, SI-2000 Maribor, Slovenia

School of Applied Sciences, University of Nova Gorica - Vipavska 11c, SI-5270 Ajdovščina, Slovenia

Institute of Neuroscience and Medicine Neuromodulation (INM-7), Research Center Jülich, 52425 Jülich, Germany
e-mail: t.manos@fz-juelich.de

© Springer-Verlag Berlin Heidelberg 2016
Ch. Skokos et al. (eds.), *Chaos Detection and Predictability*, Lecture Notes in Physics 915, DOI 10.1007/978-3-662-48410-4_5

5.1 Introduction and Basic Concepts

A fundamental aspect in studies of dynamical systems is the identification of chaotic behavior, both locally, i.e. in the neighborhood of individual orbits, and globally, i.e. for large samples of initial conditions. The most commonly used method to characterize chaos is the computation of the maximum Lyapunov exponent (mLE) λ_1. In general, Lyapunov exponents (LEs) are asymptotic measures characterizing the average rate of growth or shrinking of small perturbations to orbits of dynamical systems. They were introduced by Lyapunov [55] and they were applied to characterize chaotic motion by Oseledec in [70], where the Multiplicative Ergodic Theorem (which provided the theoretical basis for the numerical computation of the LEs) was stated and proved. For a recent review of the theory and the numerical evaluation of LEs the reader is referred to [81]. The numerical evaluation of the mLE was achieved in the late 1970s [11, 32, 69] and allowed the discrimination between regular and chaotic motion. This evaluation is performed through the time evolution of an infinitesimal perturbation of the orbit's initial condition, which is described by a deviation vector from the orbit itself. The evolution of the deviation vector is governed by the so-called *variational equations* [32].

In practice, λ_1 is evaluated as the limit for $t \to \infty$ of the *finite time maximum Lyapunov exponent*

$$\Lambda_1(t) = \frac{1}{t} \ln \frac{\|\mathbf{w}(t)\|}{\|\mathbf{w}(0)\|}, \tag{5.1}$$

where t denotes the time and $\|\mathbf{w}(0)\|$, $\|\mathbf{w}(t)\|$ are the Euclidean norms[1] of the deviation vector $\mathbf{w}$ at times $t = 0$ and $t > 0$ respectively. Thus

$$\lambda_1 = \lim_{t \to \infty} \Lambda_1(t). \tag{5.2}$$

The computation of the mLE was extensively used for studying chaos and it is still implemented nowadays for this purpose. Nevertheless, one of its major practical disadvantages is the slow convergence of the finite time Lyapunov exponent (5.1) to its limit value (5.2). Since $\Lambda_1(t)$ is influenced by the whole evolution of the deviation vector, the time needed for it to converge to λ_1 is not known a priori, and in many cases it may become extremely long. This delay can result in CPU-time expensive computations, especially when the study of many orbits is required for the global investigation of a system. In order to overcome this problem several other fast chaos detection techniques have been developed over the years; some of which are presented in this volume.

Throughout this chapter we consider finite-dimensional conservative dynamical systems and in particular, autonomous Hamiltonian models and symplectic maps (except from Sect. 5.4.3 where a time dependent Hamiltonian system is studied).

---

[1] We note that the value of λ_1 is independent of the used norm.

In these systems regular motion occurs on the surface of a torus in the system's phase space and is characterized by $\lambda_1 = 0$. Any deviation vector $\mathbf{w}(0)$ from a regular orbit eventually falls on the tangent space of this torus and its norm will approximately grow linearly in time, i.e. eventually becoming proportional to t, $\|\mathbf{w}(t)\| \propto t$. Consequently, $\Lambda_1(t) \propto \ln t/t$, which practically means that $\Lambda_1(t)$ tends asymptotically to zero following the power law t^{-1} because the values of $\ln t$ change much slower than t as time grows (see for example [11, 26] and Sect. 5.3 of [81]). On the other hand, in the case of chaotic orbits the use of any initial deviation vector in (5.1) and (5.2) practically leads to the computation of the mLE $\lambda_1 > 0$ because this vector eventually is stretched towards the direction associated to the mLE, assuming of course that $\lambda_1 > \lambda_2$, with λ_2 being the second largest LE. We note here that, from the first numerical attempts to evaluate the mLE [11, 32] it became apparent that a random choice of the initial deviation vector $\mathbf{w}(0)$ leads with probability one to the computation of λ_1. This means that, the choice of $\mathbf{w}(0)$ does not affect the limiting value of $\Lambda_1(t)$, but only the initial phases of its evolution. This behavior introduces some difficulties when we want to evaluate the whole spectrum of LEs of chaotic orbits because any set of initially distinct deviation vectors eventually end up to vectors aligned along the direction defined by the mLE. It is worth-noting that even in cases where we could theoretically know the initial choice of deviation vectors which would lead to the evaluation of LEs other than the maximum one, the unavoidable numerical errors in the computational procedure will lead again to the computation of the mLE [15]. This problem was bypassed by the development of a procedure based on repeated orthonormalizations of the evolved deviation vectors [10, 12–15, 78, 95].

Although the eventual coincidence of distinct initial deviation vectors for chaotic orbits with $\lambda_1 > \lambda_2$ was well-known from the early 1980s, this property was not directly used to identify chaos for about two decades until the introduction of the Smaller Alignment Index (SALI) method in [79]. In the 1990s some indirect consequences of the fact that two initially distinct deviation vectors eventually coincide for chaotic motion, while they will have different directions on the tangent space of the torus for regular ones, were used to determine the nature of orbits, but not the fact itself. In particular, in [91] the spectra of what was named the '*stretching number*', i.e. the quantity

$$\alpha = \frac{\ln\left(\frac{\|\mathbf{w}(t+\Delta t)\|}{\|\mathbf{w}(t)\|}\right)}{\Delta t}, \tag{5.3}$$

where Δt is a small time step, were considered. The main outcome of that paper was that '*the spectra for two different initial deviations are the same for chaotic orbits, but different for ordered orbits*', as was stated in the abstract of Voglis et al. [91]. This feature was later quantified in [92] by the introduction of a quantity measuring the 'difference' of two spectra, the so-called 'spectral distance'. In [92] it was shown that this quantity attains constant, positive values for regular orbits, while it becomes zero for chaotic ones. It is worth noting that in [91] it was explained that the observed behavior of the two spectra was due to the fact that the deviation vectors eventually coincide for chaotic orbits, producing the same sequences of stretching numbers, while they remain different for regular ones resulting in different spectra of stretching numbers. Nevertheless, instead of directly checking the matching (or not) of the two deviation vectors the method developed in [91, 92] requires unnecessary, additional computations as it goes through the construction of the two spectra and the evaluation of their 'distance'. Naturally, this procedure is influenced by the whole time evolution of the deviation vectors, which in turn results in the delay of the matching of the two spectra with respect to the matching of the two deviation vectors.

Apparently, the direct determination of the possible coincidence (or not) of the deviation vectors is a much faster and more efficient approach to reveal the regular or chaotic nature of orbits than the evaluation of the spectral distance, as it requires less computations (see [79] for a comparison between the two approaches). This observation led to the introduction in [79] of the SALI method which actually checks the possible coincidence of deviation vectors, while the later introduced Generalized Alignment Index (GALI) [84] extends this criterion to more deviation vectors. As we see in Sect. 5.3 this extension allows the correct characterization of chaotic orbits also in the case where the spectrum of the LEs is degenerate and the second, or even more, largest LEs are equal to λ_1.

In order to illustrate the behaviors of both the SALI and the GALI methods for regular and chaotic motion we use in this chapter some simple models of Hamiltonian systems and symplectic maps.

In particular, as a two degrees of freedom (2D) Hamiltonian model we consider the well-known Hénon-Heiles system [43], described by the Hamiltonian

$$H_2 = \frac{1}{2}(p_1^2 + p_2^2) + \frac{1}{2}(q_1^2 + q_2^2) + q_1^2 q_2 - \frac{1}{3}q_2^3. \tag{5.4}$$

We also consider the 3D Hamiltonian system

$$H_3 = \sum_{i=1}^{3} \frac{\omega_i}{2}(q_i^2 + p_i^2) + q_1^2 q_2 + q_1^2 q_3, \tag{5.5}$$

initially studied in [15, 32]. Note that ω_i in (5.5) are some constant coefficients. As a model of higher dimensions we use the ND Hamiltonian

$$H_N = \frac{1}{2}\sum_{i=1}^{N} p_i^2 + \sum_{i=0}^{N} \left[\frac{1}{2}(q_{i+1} - q_i)^2 + \frac{1}{4}\beta(q_{i+1} - q_i)^4 \right], \tag{5.6}$$

which describes a chain of N particles with quadratic and quartic nearest neighbor interactions, known as the Fermi–Pasta–Ulam β model (FPU-β) [36], where $q_0 = q_{N+1} = 0$. In all the above-mentioned ND Hamiltonian models, $q_i, p_i, i = 1, 2, \ldots N$ are respectively the generalized coordinates and the conjugate momenta defining the $2N$-dimensional ($2N$d) phase space of the system.

As a symplectic map model we consider in our presentation the $2M$-dimensional ($2M$d) system of coupled standard maps studied in [51]

$$x_j' = x_j + y_j'$$

$$y_j' = y_j + \frac{K_j}{2\pi} \sin\left(2\pi x_j\right) - \frac{\gamma}{2\pi} \left\{\sin\left[2\pi\left(x_{j+1} - x_j\right)\right] + \sin\left[2\pi\left(x_{j-1} - x_j\right)\right]\right\},$$

$$\tag{5.7}$$

where $j = 1, 2, \ldots, M$ is the index of each standard map, K_j and γ are the model's parameters and the prime ($'$) denotes the new values of the variables after one iteration of the map. We note that each variable is given modulo 1, i.e. $0 \le x_j < 1$, $0 \le y_j < 1$ and also that the conventions $x_0 = x_M$ and $x_{M+1} = x_1$ hold.

In order to make this chapter more focused and easier to read we decided not to present any analytical proofs for the various mathematical statements given in the text; we prefer to direct the reader to the publications where these proofs can be found. Nevertheless, we want to emphasize here that all the laws describing the behavior of the SALI and the GALI have been obtained theoretically and they are not numerical estimations or fits to numerical data. Indeed, these laws succeed to accurately reproduce the evolution of the indices in actual numerical simulations, some of which are presented in the following sections.

The chapter is organized as follows. In Sect. 5.2 the SALI method is presented and the behavior of the index for regular and chaotic orbits is discussed. Section 5.3 is devoted to the GALI method. After explaining the motivation that led to the introduction of the GALI, the definition of the index is given and its practical computation is discussed in Sect. 5.3.1. Then, in Sect. 5.3.2 the behavior of the index for regular and chaotic motion is presented and several example orbits of Hamiltonian systems and symplectic maps of various dimensions are used to illustrate these behaviors. The ability of the GALI to identify motion on low dimensional tori is presented in Sect. 5.3.3, while Sect. 5.3.4 is devoted to the behavior of the index for stable and unstable periodic orbits. In Sect. 5.4 several applications of the SALI and the GALI methods are presented. In particular, in Sect. 5.4.1 we explain how the SALI and the GALI can be used for understanding the global dynamics of a system, while specific applications of the indices to various dynamical models are briefly discussed in Sect. 5.4.2. The particular case of time dependent Hamiltonians is considered in Sect. 5.4.3. Finally, in Sect. 5.5 we summarize the advantages of the SALI and the GALI methods and briefly discuss some recent comparative studies of different chaos indicators.

5.2   The Smaller Alignment Index (SALI)

The idea behind the SALI's introduction was the need for a simple, easily computed quantity which could clearly identify the possible alignment of two multidimensional vectors. As has been already explained, it was well-known that any two deviation vectors from a chaotic orbit with $\lambda_1 > \lambda_2$ are stretched towards the direction defined by the mLE, eventually becoming aligned having the same or opposite directions. Thus, it would be quite helpful to devise a quantity which could clearly indicate this alignment.

Since we are only interested in the direction of the two deviation vectors and not in their actual size, we can normalize them before checking their alignment. This process also eliminates the problem of potential numerical overflow due to vectors' growth in size, which appears especially in the case of chaotic orbits. So in practice, we let the two deviation vectors evolve under the system's dynamics (according to the variational equations for Hamiltonian models, or the so-called tangent map for symplectic maps) normalizing them after a fixed number of evolution steps to a predefined norm value. For simplicity in our presentation we consider the usual Euclidean norm (denoted by $\| \cdot \|$) and renormalize the evolved vectors to unity.

In the case of chaotic orbits this procedure is schematically shown in Fig. 5.1 where the two initially distinct unit deviation vectors[2] $\hat{\mathbf{w}}_1(0)$, $\hat{\mathbf{w}}_2(0)$ converge to the same direction. We emphasize that Fig. 5.1 is just a schematic representation on the plane of the real deviation vectors which are objects evolving in multidimensional spaces. Since the mLE $\lambda_1 > 0$ denotes the mean exponential rate of each vector's stretching, they are elongated at some later time $t > 0$,[3] becoming $\mathbf{w}_1(t)$, $\mathbf{w}_2(t)$,

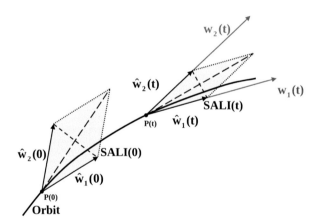

Fig. 5.1 Schematic representation of the evolution of two deviation vectors and of the corresponding SALI for a chaotic orbit. Two initially distinct unit deviation vectors $\hat{\mathbf{w}}_1(0)$, $\hat{\mathbf{w}}_2(0)$ from point $P(0)$ of a chaotic orbit become $\mathbf{w}_1(t)$, $\mathbf{w}_2(t)$ after some time $t > 0$ when the orbit reaches point $P(t)$, with $\hat{\mathbf{w}}_1(t)$, $\hat{\mathbf{w}}_2(t)$ being the unit vectors along these directions. The length of the shortest diagonals of the *grey-shaded parallelograms* defined by $\hat{\mathbf{w}}_1(0)$, $\hat{\mathbf{w}}_2(0)$ and $\hat{\mathbf{w}}_1(t)$, $\hat{\mathbf{w}}_2(t)$ are the values of the SALI(0) and the SALI(t) respectively

---

[2]We note that throughout this chapter we use the hat symbol (^) to denote a unit vector.

[3]For Hamiltonian systems the time is a continuous variable, while for maps it is a discrete one counting the map's iterations.

while the corresponding unit vectors are $\hat{\mathbf{w}}_1(t)$, $\hat{\mathbf{w}}_2(t)$. Then the diagonals of the parallelograms defined by $\hat{\mathbf{w}}_1(t)$, $\hat{\mathbf{w}}_2(t)$, both for $t = 0$ and $t > 0$, depict the sum and the difference of the two unit vectors.

In the particular case shown in Fig. 5.1 the two unit vectors tend to align by becoming equal. This means that $\|\hat{\mathbf{w}}_1(t) - \hat{\mathbf{w}}_2(t)\| \to 0$ and $\|\hat{\mathbf{w}}_1(t) + \hat{\mathbf{w}}_2(t)\| \to 2$. Of course the dynamics could have led the vectors to become opposite. In that case we get $\|\hat{\mathbf{w}}_1(t) - \hat{\mathbf{w}}_2(t)\| \to 2$ and $\|\hat{\mathbf{w}}_1(t) + \hat{\mathbf{w}}_2(t)\| \to 0$. Since we are not interested in the particular orientation of the deviation vectors, i.e. whether they become equal or opposite to each other, when we check their possible alignment, a rather natural choice is to define the minimum of norms $\|\hat{\mathbf{w}}_1(t) + \hat{\mathbf{w}}_2(t)\|$, $\|\hat{\mathbf{w}}_1(t) - \hat{\mathbf{w}}_2(t)\|$ as an indicator of the vectors' alignment. This is the reason of the appellation, as well as of the definition of the SALI in [79] as

$$\text{SALI}(t) = \min \left\{ \|\hat{\mathbf{w}}_1(t) + \hat{\mathbf{w}}_2(t)\|, \|\hat{\mathbf{w}}_1(t) - \hat{\mathbf{w}}_2(t)\| \right\}, \tag{5.8}$$

with $\hat{\mathbf{w}}_i(t) = \frac{\mathbf{w}_i(t)}{\|\mathbf{w}_i(t)\|}$, $i = 1, 2$ being unit vectors.

Naturally, in order for the SALI to be efficiently used as a chaos indicator it should exhibit distinct behaviors for chaotic and regular orbits. As explained before the SALI becomes zero for chaotic orbits. On the other hand, in the case of regular orbits deviation vectors fall on the tangent space of the torus on which the motion occurs, having in general different directions as there is no reason for them to be aligned [82, 91]. This behavior is shown schematically in Fig. 5.2. Thus, in this case the index should be always different from zero. In practice, the values of the SALI exhibit bounded fluctuations around some constant, positive number.

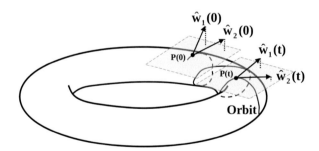

Fig. 5.2 Schematic representation of the evolution of two deviation vectors for a regular orbit. The motion takes place on a torus. We consider two initially distinct unit deviation vectors $\hat{\mathbf{w}}_1(0)$, $\hat{\mathbf{w}}_2(0)$ from point $P(0)$, which are not necessarily on the tangent space of the torus (this space is depicted as a *shaded parallelogram* passing through $P(0)$). As time evolves the deviation vectors tend to fall on the torus' tangent space and the corresponding unit vectors $\hat{\mathbf{w}}_1(t)$, $\hat{\mathbf{w}}_2(t)$ at time $t > 0$ are 'closer' to the current tangent space (i.e. the *grey-shaded parallelogram* passing through $P(t)$), as the shortening of the perpendicular to the tangent spaces *dotted lines* from the edges of the deviation vectors indicate. Since there is no reason for the alignment of the two deviation vectors, the SALI will not become zero

Thus, in order to compute the SALI we follow the evolution of two initially distinct, random, unit deviation vectors $\hat{\mathbf{w}}_1(0)$, $\hat{\mathbf{w}}_2(0)$. Choosing these vectors to be also orthogonal sets the initial SALI to its highest possible value (SALI$(0) = \sqrt{2}$) and ensures that they are considerably different from each other, which has proved to be a very good computational practice. Then, every $t = \tau$ time units we normalize the evolved vectors $\mathbf{w}_1(i\tau)$, $\mathbf{w}_2(i\tau)$, $i = 1, 2, \ldots$, to $\hat{\mathbf{w}}_1(i\tau)$, $\hat{\mathbf{w}}_2(i\tau)$ and evaluate the SALI$(i\tau)$ from (5.8). This algorithm is described in pseudo-code in Table 5.1 of the Appendix. A MAPLE code for this algorithm, developed specifically for the Hénon-Heiles system (5.4) can be found in Chap. 5 of [20].

The completely different behaviors of the SALI for regular and chaotic orbits are clearly seen in Fig. 5.3,[4] where some representative results are shown for the 2D Hamiltonian system (5.4) and the 6d symplectic map

$$
\begin{aligned}
x_1' &= x_1 + y_1' \\
y_1' &= y_1 + \frac{K}{2\pi} \sin(2\pi x_1) - \frac{\gamma}{2\pi} \{\sin[2\pi(x_2 - x_1)] + \sin[2\pi(x_3 - x_1)]\} \\
x_2' &= x_2 + y_2' \\
y_2' &= y_2 + \frac{K}{2\pi} \sin(2\pi x_2) - \frac{\gamma}{2\pi} \{\sin[2\pi(x_3 - x_2)] + \sin[2\pi(x_1 - x_2)]\} \\
x_3' &= x_3 + y_3' \\
y_3' &= y_3 + \frac{K}{2\pi} \sin(2\pi x_3) - \frac{\gamma}{2\pi} \{\sin[2\pi(x_1 - x_3)] + \sin[2\pi(x_2 - x_3)]\} ,
\end{aligned}
\tag{5.9}
$$

obtained by considering $M = 3$ coupled standard maps with $K_1 = K_2 = K_3 = K$ in (5.7). From the results of Fig. 5.3 we see that for both systems the SALI of regular orbits (black, solid curves) remains practically constant and positive, i.e.

$$
\text{SALI} \propto \text{constant}. \tag{5.10}
$$

On the other hand, the SALI of chaotic orbits (black, dashed curve in Fig. 5.3a and grey, solid curve in Fig. 5.3b) exhibits a fast decrease to zero after an initial transient time interval, reaching very small values around the computer's accuracy (10^{-16}). Actually, it was shown in [83] that the SALI tends to zero exponentially fast in such cases, following the law

$$
\text{SALI}(t) \propto \exp[-(\lambda_1 - \lambda_2)t], \tag{5.11}
$$

where λ_1, λ_2 ($\lambda_1 \geq \lambda_2$) are the first (i.e. the mLE) and the second largest LEs respectively. As an example demonstrating the validity of this exponential-decay law we plot in Fig. 5.4 the evolution of the SALI (solid curve) of the chaotic orbit of

---

[4]We note that throughout this chapter the logarithm to base 10 is denoted by log.

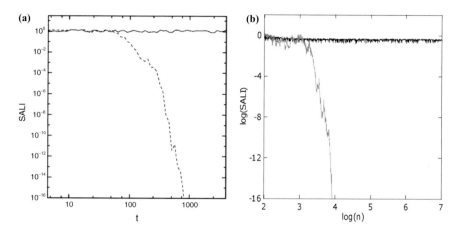

Fig. 5.3 The time evolution of the SALI for a regular and a chaotic orbit of (**a**) the 2D Hamiltonian system (5.4) for $H_2 = 0.125$ (after [83]) and (**b**) the 6d map (5.9) for $K = 3$ and $\gamma = 0.1$ (after [79]). In (**a**) the time t is continuous, while in (**b**) it is discrete and counts the map's iterations n. The initial conditions of the orbits are: (**a**) $q_1 = 0$, $q_2 = 0.1$, $p_1 = 0.49058$, $p_2 = 0$ (regular orbit; *solid curve*) and $q_1 = 0$, $q_2 = -0.25$, $p_1 = 0.42081$, $p_2 = 0$ (chaotic orbit; *dashed curve*), and (**b**) $x_1 = 0.55$, $y_1 = 0.05$, $x_2 = 0.55$, $y_2 = 0.01$, $x_3 = 0.55$, $y_3 = 0$ (regular orbit; *black curve*) and $x_1 = 0.55$, $y_1 = 0.05$, $x_2 = 0.55$, $y_2 = 0.21$, $x_3 = 0.55$, $y_3 = 0$ (chaotic orbit; *grey curve*)

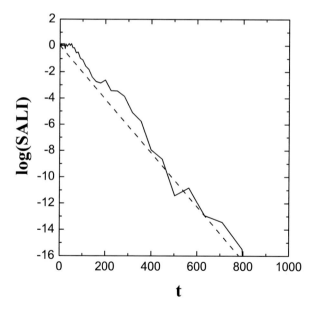

Fig. 5.4 The evolution of the SALI (*solid curve*) for the chaotic orbit of Fig. 5.3a as a function of time t. The *dashed line* corresponds to a function proportional to $\exp(-\lambda_1 t)$ for $\lambda_1 = 0.047$. Note that the t-axis is linear (after [83])

Fig. 5.3a using a linear horizontal axis for time t. Since for 2D Hamiltonian systems $\lambda_2 = 0$, (5.11) becomes

$$\text{SALI}(t) \propto \exp\left(-\lambda_1 t\right), \qquad (5.12)$$

For this particular orbit the mLE was found to be $\lambda_1 \approx 0.047$ in [83]. From Fig. 5.4 we see that (5.12) with $\lambda_1 = 0.047$ (dashed line) reproduces correctly the evolution of the SALI.[5]

Thus, the completely different behavior of the SALI for regular (5.10) and chaotic (5.11) orbits permits the clear and efficient distinction between the two cases. In [79, 83] a comparison of the SALI's performance with respect to other chaos detection techniques was presented and the efficiency of the index was discussed. A main advantage of the SALI method is its ability to detect chaotic motion faster than other techniques which depend on the whole time evolution of deviation vectors, like the mLE and the spectral distance, because the SALI is determined by the *current* state of these vectors and is not influenced by their evolution history. Hence, the moment the two vectors are close enough to each other the SALI becomes practically zero and guarantees the chaotic nature of the orbit beyond any doubt. In addition, the evaluation of the SALI is simpler and more straightforward with respect to other methods that require more complicated computations. Such aspects were discussed in [83] where a comparison of the index with the Relative Lyapunov Indicator (RLI) [77] and the so-called '0–1' test [39] was presented. Another crucial characteristic of the SALI is that it attains values in a given interval, namely $\text{SALI}(t) \in [0, \sqrt{2}]$, which does not change in time as is for example the case for the Fast Lyapunov Indicator (FLI) [37]. Thus, setting a realistic threshold value below which the SALI is considered to be practically zero (and the corresponding orbit is characterized as chaotic), allows the fast and accurate discrimination between regular and chaotic motion. Due to all these features the SALI became a reliable and widely used chaos indicator as its numerous applications to a variety of dynamical systems over the years prove. Some of these applications are discussed in Sect. 5.4.

5.3   The Generalized Alignment Index (GALI)

A fundamental difference between the SALI and other, commonly applied chaos indicators, is that it uses information from the evolution of two deviation vectors instead of just one. A consequence of this feature is the appearance of the two

---

[5]We note that here, as well as in several, forthcoming figures in this chapter, the evaluation of the LEs is done only for confirming the theoretical predictions for the time evolution of the SALI (Eq. (5.12) in the current case) and later on of the GALIs, and it is not needed for the computation of the SALI and the GALIs.

largest LEs in (5.11). After performing this first leap from using only one deviation vector, the question of going even further arises naturally. To formulate this in other words: why should we stop in using only two deviation vectors? Can we extend the definition of the SALI to include more deviation vectors? Assuming that this extension is possible, what will we gain from it? Will the use of more than two deviation vectors lead to the introduction of a new chaoticity index which will permit the acquisition of a deeper understanding of the system's dynamics, exhibiting at the same time a better numerical performance than the SALI? For instance, from (5.11) we realize that in the case of a chaotic orbit with $\lambda_1 \approx \lambda_2$ the convergence of the SALI to zero will be extremely slow. As a result long integrations would be required in order for the index to distinguish this orbit from a regular one for which the SALI remains practically constant. Although the existence of such chaotic orbits is not very probable the drawback of the SALI remains. An alternative way to state this problem is the following: can we construct a new index whose behavior in the case of chaotic orbits will depend on more LEs than the two largest ones so that it can overcome the discrimination problem for $\lambda_1 \approx \lambda_2$?

Indeed, such an index can be constructed. The key point to its development is the observation that the SALI is closely related to the area of the parallelogram defined by the two deviation vectors.[6] From the schematic representation of the deviation vectors' evolution in Fig. 5.1 we see that when the SALI vanishes one of the diagonals of the parallelogram also vanishes, and consequently its area becomes zero. The area A_2 of a usual 2d parallelogram is equal to the norm of the exterior product of its two sides $\mathbf{v}_1$, $\mathbf{v}_2$, and also equal to the half of the product of its diagonals' lengths

$$A_2 = \|\mathbf{v}_1 \times \mathbf{v}_2\| = \frac{\|\mathbf{v}_1 + \mathbf{v}_2\| \cdot \|\mathbf{v}_1 - \mathbf{v}_2\|}{2}. \tag{5.13}$$

In a similar way, the area A of the parallelogram of Fig. 5.1 is given by the generalization of the exterior product of vectors to higher dimensions, i.e. the so-called wedge product denoted by $(\wedge)$,[7] so that

$$A = \|\hat{\mathbf{w}}_1 \wedge \hat{\mathbf{w}}_2\| = \frac{\|\hat{\mathbf{w}}_1 + \hat{\mathbf{w}}_2\| \cdot \|\hat{\mathbf{w}}_1 - \hat{\mathbf{w}}_2\|}{2}. \tag{5.14}$$

Note the analogy of this equation to (5.13).[8]

Based on the fact that the SALI is related to the area of the parallelogram defined by two unit deviation vectors, the extension of the index to include more vectors

---

[6]Note that this parallelogram is not the usual 2d parallelogram on the plane because its sides (the deviation vectors) are not 2d vectors.

[7]For a brief introduction to the notion of the wedge product the reader is referred to the Appendix A of [84] and the Appendix of [81].

[8]A proof of the second equality of (5.14) can be found in the Appendix B of [84].

is straightforward: the new quantity is defined as the volume of the parallelepiped formed by more than two deviation vectors. This volume is computed as the norm of the wedge product of these vectors. These arguments led to the introduction in [84] of the Generalized Alignment Index of order k (GALI$_k$) as

$$\mathrm{GALI}_k(t) = \|\hat{\mathbf{w}}_1(t) \wedge \hat{\mathbf{w}}_2(t) \wedge \ldots \wedge \hat{\mathbf{w}}_k(t)\|, \qquad (5.15)$$

where $\hat{\mathbf{w}}_i$ are unit vectors as in (5.8). In this definition the number of used deviation vectors should not exceed the dimension of the system's phase space, because in this case the k vectors will become linearly dependent and the corresponding volume will be by definition zero, as is for example the area defined by two vectors having the same direction. Thus, for an ND Hamiltonian system with $N \geq 2$ or a $2N$d symplectic map with $N \geq 1$, we consider only GALIs with $2 \leq k \leq 2N$.

By its definition the GALI$_k$ is a quantity clearly indicating the linear dependence (GALI$_k = 0$) or independence (GALI$_k > 0$) of k deviation vectors. The SALI has the same discriminating ability as SALI $= 0$ indicates that the two vectors are aligned, i.e. they are linearly dependent, while SALI > 0 implies that the vectors are not aligned, which means that they are linearly independent. Actually, the connection between the two indices can be quantified explicitly. Indeed, it was proved in the Appendix B of [84] that

$$\mathrm{GALI}_2 = \mathrm{SALI} \cdot \frac{\max\{\|\hat{\mathbf{w}}_1(t) + \hat{\mathbf{w}}_2(t)\|, \|\hat{\mathbf{w}}_1(t) - \hat{\mathbf{w}}_2(t)\|\}}{2}. \qquad (5.16)$$

Since the $\max\{\|\hat{\mathbf{w}}_1(t) + \hat{\mathbf{w}}_2(t)\|, \|\hat{\mathbf{w}}_1(t) - \hat{\mathbf{w}}_2(t)\|\}$ is a number in the interval $[\sqrt{2}, 2]$ we conclude that

$$\mathrm{GALI}_2 \propto \mathrm{SALI}, \qquad (5.17)$$

which means that the GALI$_2$ is practically equivalent to the SALI. This is another evidence that the GALI definition (5.15) is a natural extension of the SALI for more than two deviation vectors.

5.3.1   Computation of the GALI

Let us discuss now how one can actually calculate the value of the GALI$_k$ for an ND Hamiltonian system ($N \geq 2$) or a $2N$d symplectic map ($N \geq 1$). For this purpose we consider the $k \times 2N$ matrix

$$\mathbf{A}(t) = \begin{bmatrix} w_{11}(t) & w_{12}(t) & \cdots & w_{1\,2N}(t) \\ w_{21}(t) & w_{22}(t) & \cdots & w_{2\,2N}(t) \\ \vdots & \vdots & & \vdots \\ w_{k1}(t) & w_{k2}(t) & \cdots & w_{k\,2N}(t) \end{bmatrix} \qquad (5.18)$$

having as rows the $2N$ coordinates of the k unit deviation vectors $\hat{\mathbf{w}}_i(t)$ with respect to the usual orthonormal basis $\hat{\mathbf{e}}_1 = (1, 0, 0, \ldots, 0)$, $\hat{\mathbf{e}}_2 = (0, 1, 0, \ldots, 0)$, \ldots, $\hat{\mathbf{e}}_{2N} = (0, 0, 0, \ldots, 1)$. We note that the elements of $\mathbf{A}(t)$ satisfy the condition $\sum_{j=1}^{2N} w_{ij}^2(t) = 1$ for $i = 1, 2, \ldots, k$ as each deviation vector has unit norm.

We can now follow two routes for evaluating the $\text{GALI}_k(t)$. According to the first one we compute the GALI_k by evaluating the norm of the wedge product of k vectors as

$$\text{GALI}_k(t) = \left\{ \sum_{1 \le i_1 < i_2 < \cdots < i_k \le 2N} \left(\det \begin{bmatrix} w_{1i_1}(t) & w_{1i_2}(t) & \cdots & w_{1i_k}(t) \\ w_{2i_1}(t) & w_{2i_2}(t) & \cdots & w_{2i_k}(t) \\ \vdots & \vdots & & \vdots \\ w_{ki_1}(t) & w_{ki_2}(t) & \cdots & w_{ki_k}(t) \end{bmatrix} \right)^2 \right\}^{1/2},$$

(5.19)

where the sum is performed over all the possible combinations of k indices out of $2N$ (a proof of this equation can be found in [84]). In practice this means that in our calculation we consider all the $k \times k$ determinants of $\mathbf{A}(t)$. Equation (5.19) is particularly useful for the theoretical description of the GALI's behavior (actually expressions (5.22) and (5.23) below were obtained by using this equation), but not very efficient from a practical point of view. The reason is that the number of determinants appearing in (5.19) can increase enormously when N grows, leading to unfeasible numerical computations.

A simpler, straightforward and computationally more efficient approach to evaluate the GALI_k was developed in [85], where it was proved that the index is equal to the product of the singular values z_i, $i = 1, 2, \ldots, k$ of $\mathbf{A}^T(t)$ (the transpose of matrix $\mathbf{A}(t)$), i.e.

$$\text{GALI}_k(t) = \prod_{i=1}^{k} z_i(t).$$

(5.20)

We note that the singular values of $\mathbf{A}^T(t)$ are obtained by performing the Singular Value Decomposition (SVD) procedure to $\mathbf{A}^T(t)$. According to the SVD method (see for instance Sect. 2.6 of [74]) the $2N \times k$ matrix $\mathbf{A}^T$ is written as the product of a $2N \times k$ column-orthogonal matrix $\mathbf{U}$ ($\mathbf{U}^T \cdot \mathbf{U} = \mathbf{I}_k$, with $\mathbf{I}_k$ being the $k \times k$ unit matrix), a $k \times k$ diagonal matrix $\mathbf{Z}$ having as elements the positive or zero singular values z_i, $i = 1, \ldots, k$, and the transpose of a $k \times k$ orthogonal matrix $\mathbf{V}$ ($\mathbf{V}^T \cdot \mathbf{V} = \mathbf{I}_k$), i.e.

$$\mathbf{A}^T = \mathbf{U} \cdot \mathbf{Z} \cdot \mathbf{V}^T.$$

(5.21)

In practice, in order to compute the GALI of order k we follow the evolution of k initially distinct, random, orthonormal deviation vectors $\hat{\mathbf{w}}_1(0)$, $\hat{\mathbf{w}}_2(0)$, \ldots, $\hat{\mathbf{w}}_k(0)$. Similarly to the computation of the SALI, choosing orthonormal vectors ensures that all of them are sufficiently far from linear dependence and gives to the GALI_k

its largest possible initial value $\text{GALI}_k = 1$. Afterwards, every $t = \tau$ time units we normalize the evolved vectors $\mathbf{w}_1(i\tau)$, $\mathbf{w}_2(i\tau)$, ..., $\mathbf{w}_k(i\tau)$, $i = 1, 2, \ldots$, to $\hat{\mathbf{w}}_1(i\tau)$, $\hat{\mathbf{w}}_2(i\tau)$, ..., $\hat{\mathbf{w}}_k(i\tau)$ and set them as rows of a matrix $\mathbf{A}(i\tau)$ (5.18). Then, according to (5.20) the $\text{GALI}_k(i\tau)$ is computed as the product of the singular values of matrix $\mathbf{A}^T(i\tau)$. This algorithm is described in pseudo-code in Table 5.2 of the Appendix. A MAPLE code computing all the possible GALIs (i.e. GALI_2, GALI_3 and GALI_4) for the 2D Hamiltonian (5.4) can be found in Chap. 5 of [20].

5.3.2   Behavior of the GALI for Chaotic and Regular Orbits

After defining the new index and explaining a practical way to evaluate it, let us discuss its ability to discriminate between chaotic and regular motion. As we have already mentioned, in the case of a chaotic orbit all deviation vectors eventually become aligned to the direction defined by the largest LE. Thus, they become linearly dependent and consequently the volume they define vanishes, meaning that the GALI_k, $2 \leq k \leq 2N$, will become zero. Actually, in [84] it was shown analytically that in this case the $\text{GALI}_k(t)$ decreases to zero exponentially fast with an exponent which depends on the k largest LEs as

$$\text{GALI}_k(t) \propto \exp\left\{-\left[(\lambda_1 - \lambda_2) + (\lambda_1 - \lambda_3) + \cdots + (\lambda_1 - \lambda_k)\right]t\right\}. \qquad (5.22)$$

Note that for $k = 2$ we get the exponential law (5.11) in agreement with the equivalence between the GALI_2 and the SALI (5.17).

Let us now consider the case of regular motion in a ND Hamiltonian system or a $2N$d symplectic map with $N \geq 2$. In general, this motion occurs on an Nd torus in the system's $2N$d phase space. As we discussed in Sect. 5.2, in this case any deviation vector eventually falls on the Nd tangent space of the torus (Fig. 5.2). Consequently, the k initially distinct, linearly independent deviation vectors we follow in order to compute the evolution of the GALI_k eventually fall on the Nd tangent space of the torus, without necessarily having the same directions. Thus, if we do not consider more deviation vectors than the dimension of the tangent space ($k \leq N$) we end up with k linearly independent vectors on the torus' tangent space and consequently the volume of the parallelepiped they define (i.e. the GALI_k) will be different from zero. As we see later on, numerical simulations show that in this case the GALI_k exhibits small fluctuations around some positive value. If, on the other hand, we consider more deviation vectors than the dimension of the tangent space ($N < k \leq 2N$) the deviation vectors eventually become linearly dependent, as we end up with more vectors in the torus' tangent space than the space's dimension. Thus, the volume that these vectors define will vanish and the GALI_k will become zero. Specifically, in [84] it was shown analytically that in this case the GALI_k tends to zero following a power law whose exponent depends on the torus dimension and on the number k

of deviation vectors considered, i.e. $GALI_k \propto t^{-2(k-N)}$. In summary the behavior of the $GALI_k$ for regular orbits is

$$GALI_k(t) \propto \begin{cases} \text{constant if } 2 \leq k \leq N \\ \frac{1}{t^{2(k-N)}} \quad \text{if } N < k \leq 2N. \end{cases} \tag{5.23}$$

From this equation we see that $SALI \propto GALI_2 \propto$ constant, in accordance to (5.10).

5.3.2.1   Some Illustrative Paradigms

In what follows we illustrate the different behaviors of the $GALI_k$ by computing its evolution for some representative chaotic and regular orbits of various ND autonomous Hamiltonians and $2N$d symplectic maps. Before doing so let us note that for these systems the LEs comes in pairs of values having opposite signs

$$\lambda_i = -\lambda_{2N-i+1}, \quad i = 1, 2, \ldots, N, \tag{5.24}$$

while, moreover

$$\lambda_N = \lambda_{N+1} = 0 \tag{5.25}$$

for Hamiltonian systems [14, 41, 81].

Hamiltonian Systems

Initially, we consider the 2D Hamiltonian (5.4) which has a 4d phase space. For this system we can define the $GALI_k$ for $k = 2, 3$ and 4. Then, according to (5.24) and (5.25), the LEs satisfy the conditions $\lambda_1 = -\lambda_4$, $\lambda_2 = \lambda_3 = 0$. Thus, according to (5.22) the evolution of the GALIs for a chaotic orbit is given by

$$GALI_2(t) \propto e^{-\lambda_1 t}, \quad GALI_3(t) \propto e^{-2\lambda_1 t}, \quad GALI_4(t) \propto e^{-4\lambda_1 t}. \tag{5.26}$$

On the other hand, for a regular orbit (5.23) indicates that

$$GALI_2(t) \propto \text{constant}, \quad GALI_3(t) \propto \frac{1}{t^2}, \quad GALI_4(t) \propto \frac{1}{t^4}. \tag{5.27}$$

From the results of Fig. 5.5, where the time evolution of the $GALI_2$, the $GALI_3$ and the $GALI_4$ for a chaotic orbit (actually the one considered in Figs. 5.3a and 5.4) and a regular orbit are plotted, we see that the laws (5.26) and (5.27) describe quite accurately the obtained numerical data.

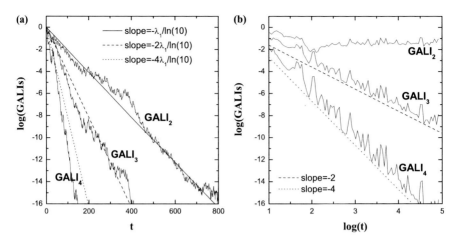

Fig. 5.5 The time evolution of the GALI$_2$, the GALI$_3$ and the GALI$_4$ for (**a**) a chaotic and (**b**) a regular orbit of the 2D Hamiltonian (5.4) for $H_2 = 0.125$. The chaotic orbit is the one considered in Fig. 5.3a, while the initial conditions of the regular orbit are $q_1 = 0$, $q_2 = 0$, $p_1 = 0.5$, $p_2 = 0$. The *straight lines* correspond in (**a**) to functions proportional to $\exp(-\lambda_1 t)$, $\exp(-2\lambda_1 t)$ and $\exp(-4\lambda_1 t)$, for $\lambda_1 = 0.047$ and in (**b**) to functions proportional to t^{-2} and t^{-4}. The slope of each line is mentioned in the legend. Note that the horizontal, time axis in (**a**) is linear, while in (**b**) is logarithmic (after [84])

For a 3D Hamiltonian like (5.5) the theoretical prediction (5.22) gives

$$\text{GALI}_2(t) \propto e^{-(\lambda_1 - \lambda_2)t}, \quad \text{GALI}_3(t) \propto e^{-(2\lambda_1 - \lambda_2)t}, \quad \text{GALI}_4(t) \propto e^{-(3\lambda_1 - \lambda_2)t},$$
$$\text{GALI}_5(t) \propto e^{-4\lambda_1 t}, \quad \text{GALI}_6(t) \propto e^{-6\lambda_1 t},$$
$$(5.28)$$

for a chaotic orbit, because, according to (5.24) and (5.25), $\lambda_1 = -\lambda_6$, $\lambda_2 = -\lambda_5$ and $\lambda_3 = \lambda_4 = 0$. On the other hand, a regular orbit lies on a 3d torus and according to (5.23) the GALIs should behave as

$$\text{GALI}_2(t) \propto \text{constant}, \quad \text{GALI}_3(t) \propto \text{constant}, \quad \text{GALI}_4(t) \propto \frac{1}{t^2},$$
$$\text{GALI}_5(t) \propto \frac{1}{t^4}, \quad \text{GALI}_6(t) \propto \frac{1}{t^6}.$$
$$(5.29)$$

In Fig. 5.6 we plot the time evolution of the various GALIs for a chaotic (Fig. 5.6a) and a regular (Fig. 5.6b) orbit of the 3D Hamiltonian (5.5). From the plotted results we see that the behaviors of the GALIs are very well approximated by (5.28) and (5.29). We note here that the constant values that the GALI$_2$ and the GALI$_3$ eventually attain in Fig. 5.6b are not the same. Actually, the limiting value of GALI$_3$ is smaller than the one of GALI$_2$.

As an example of evaluating the GALIs for multidimensional Hamiltonians we consider model (5.6) for $N = 8$ particles. This corresponds to an 8D Hamiltonian system H_8, having a 16d phase space, which allows the definition of several GALIs:

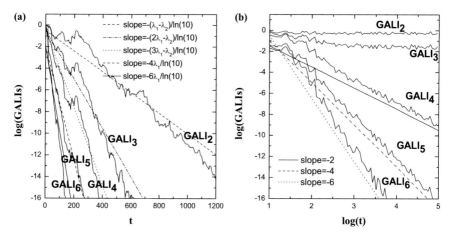

Fig. 5.6 The time evolution of the $GALI_k$, $k = 2, 3, \ldots, 6$ for (**a**) a chaotic and (**b**) a regular orbit of the 3D Hamiltonian (5.5) with $H_3 = 0.09$, $\omega_1 = 1$, $\omega_2 = \sqrt{2}$ and $\omega_3 = \sqrt{3}$. The initial conditions of the orbits are: (**a**) $q_1 = 0$, $q_2 = 0$, $q_3 = 0$, $E_1 = 0.03$, $E_2 = 0.03$, $E_3 = 0.03$, and (**b**) $q_1 = 0$, $q_2 = 0$, $q_3 = 0$, $E_1 = 0.005$, $E_2 = 0.085$, $E_3 = 0$, where the quantities E_1, E_2, E_3 (usually referred as the 'harmonic energies') are related to the momenta p_1, p_2, p_3 through $p_i = \sqrt{2E_i/\omega_i}$, $i = 1, 2, 3$. The *straight lines* in (**a**) correspond to functions proportional to $\exp[-(\lambda_1 - \lambda_2)t]$, $\exp[-(2\lambda_1 - \lambda_2)t]$, $\exp[-(3\lambda_1 - \lambda_2)t]$, $\exp(-4\lambda_1 t)$ and $\exp(-6\lambda_1 t)$ for $\lambda_1 = 0.03$, $\lambda_2 = 0.008$, which are accurate numerical estimations of the orbit's two largest LEs (see [84] for more details). The *straight lines* in (**b**) correspond to functions proportional to t^{-2}, t^{-4} and t^{-6}. The slope of each line is mentioned in the legend. The horizontal, time axis is linear in (**a**) and logarithmic in (**b**) (after [84])

starting from $GALI_2$ up to $GALI_{16}$. In Fig. 5.7 the time evolution of several of these indices are shown for a chaotic (Fig. 5.7a, b) and a regular (Fig. 5.7c, d) orbit. From these results we again conclude that the laws (5.22) and (5.23) are quite accurate in describing the time evolution of the GALIs.

The first seven indices, $GALI_2$ up to $GALI_8$, exhibit completely different behaviors for chaotic and regular motion: they tend exponentially fast to zero for a chaotic orbit (Fig. 5.7a, b), while they attain constant, positive values for a regular one (Fig. 5.7c). This characteristic makes them ideal numerical tools for discriminating between the two cases, as we see in Sect. 5.4.1.1 where some specific numerical examples are discussed in detail.

Although the constancy of the $GALI_k$, $k = 1, \ldots, 8$ for regular orbits is predicted from (5.23), nothing is yet said about the actual values of these constants. It is evident from Fig. 5.7c that these values decrease as the order k of the $GALI_k$ increases, something which was also observed in Fig. 5.6b for the 3D Hamiltonian (5.5). For the regular orbit of Fig. 5.7c we see that $GALI_8 \approx 10^{-7}$. One might argue that this very small value could be considered to be practically zero and that the orbit might be (wrongly) classified as chaotic. The flaw in this argumentation is that the possible smallness of $GALI_8 \approx 10^{-7}$ is of relative nature as this value should be compared with the values that the index reaches for actual chaotic

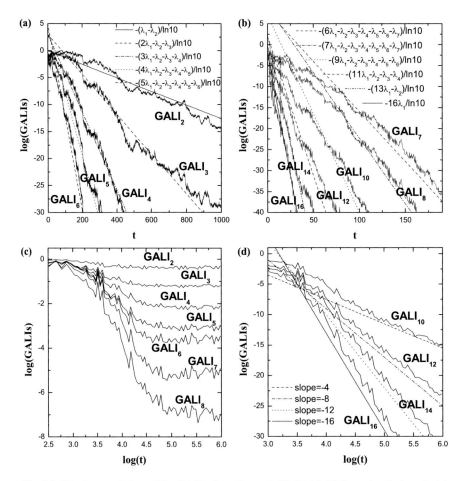

Fig. 5.7 The time evolution of the $GALI_k$, $k = 2, \ldots, 8, 10, 12, 14, 16$ for a chaotic (panels (**a**) and (**b**)) and a regular orbit (panels (**c**) and (**d**)) of the ND Hamiltonian (5.6) with $N = 8$ and $\beta = 1.5$. The initial conditions of the chaotic orbit are $Q_1 = Q_4 = 2$, $Q_2 = Q_5 = 1$, $Q_3 = Q_6 = 0.5$,

$$Q_7 = Q_8 = 0.1, P_i = 0 \text{ where } Q_i = \frac{2}{3} \sum_{j=1}^{8} q_j \sin\left(\frac{ij\pi}{9}\right), P_i = \frac{2}{3} \sum_{j=1}^{8} p_j \sin\left(\frac{ij\pi}{9}\right), i = 1, \ldots, 8$$

(see [85] for more details). The initial conditions of the regular orbit are $q_1 = q_2 = q_3 = q_8 = 0.05$, $q_4 = q_5 = q_6 = q_7 = 0.1$, $p_i = 0$, $i = 1, \ldots, 8$. The *straight lines* in (**a**) and (**b**) correspond to exponential functions of the form (5.22) for $\lambda_1 = 0.170$, $\lambda_2 = 0.141$, $\lambda_3 = 0.114$, $\lambda_4 = 0.089$, $\lambda_5 = 0.064$, $\lambda_6 = 0.042$, $\lambda_7 = 0.020$, which are estimations (obtained in [85]) of the orbit's seven largest LEs. The *straight lines* in (**d**) correspond to functions proportional to t^{-4}, t^{-8}, t^{-12} and t^{-16}. The slope of each line is mentioned in the legend. Note the huge range differences in the horizontal, time axes between panels (**a**) and (**b**), where the axes are linear, and panels (**c**) and (**d**) where the axes are logarithmic (after [85])

orbits. For instance, the chaotic orbit of Fig. 5.7b has $GALI_8 \approx 10^{-40}$, after only $t \approx 160$ time units! At the same time we get $GALI_8 \approx 10^{-1}$ for the regular orbit (Fig. 5.7c). In addition, extrapolating the results of $GALI_8$ for the chaotic orbit in

Fig. 5.7b to e.g. $t \approx 10^5$ we would obtain values extremely smaller than the value GALI$_8 \approx 10^{-7}$ archived for the regular orbit in Fig. 5.7c.

The necessity to determine an appropriate threshold value for the GALI$_k$, $2 \leq k \leq N$, below which orbits will be securely classified as chaotic, becomes evident from the above analysis. Since a theoretical, or even an empirical (numerical) relation between the order k of the GALI$_k$ and the constant value it reaches for regular orbits is still lacking, one efficient way to determine this threshold value is by computing the GALI$_k$ for some representative chaotic and regular orbits of each studied system. Then, a safe policy is to define this threshold to be a few orders of magnitude smaller than the minimum value obtained by the GALI$_k$ for the tested regular orbits. For example, based on the results of Fig. 5.6 for the 3D Hamiltonian (5.5) this threshold value for the GALI$_3$ could be set to be $\leq 10^{-8}$, while for the system of Fig. 5.7 a reliable threshold value for the GALI$_8$ could be $\leq 10^{-16}$.

The results of Fig. 5.7 verify the predictions of (5.22) and (5.23) that the GALIs of order $8 < k \leq 16$ tend to zero both for chaotic and regular orbits. Nevertheless, the completely different way they do so, i.e. they decay exponentially fast for chaotic orbits, while they follow a power law decay for regular ones, allows us again to develop a well-tailored strategy to discriminate between the two cases. The different decay laws result in enormous differences in the time the indices need to reach any predefined low value. Thus, the measurement of this time can be used to characterize the nature of the orbits, as we see in Sect. 5.4.1.2. For example, for the chaotic orbit of Fig. 5.7b GALI$_{16} \approx 10^{-30}$ after about $t \approx 25$ time units, while it reaches the same small value after about $t \approx 10^5$ time units for the regular orbit of Fig. 5.7d; a time interval which is larger by a factor ≈ 4000 with respect to the chaotic orbit!

Symplectic Maps

Although up to now our discussion concerned the implementation of the GALIs to Hamiltonian systems, the indices follow laws (5.22) and (5.23) also for symplectic maps (with the obvious substitution of the continuous time t by a discrete one which counts the map's iterations n) as the representative results of Figs. 5.8 and 5.9 clearly verify. In particular, in Fig. 5.8 we see the behavior of the GALIs for a chaotic (Fig. 5.8a) and a regular (Fig. 5.8b) orbit of the 4d map

$$
\begin{aligned}
x_1' &= x_1 + y_1' \\
y_1' &= y_1 + \frac{K}{2\pi} \sin(2\pi x_1) - \frac{\gamma}{2\pi} \sin[2\pi(x_2 - x_1)] \\
x_2' &= x_2 + y_2' \\
y_2' &= y_2 + \frac{K}{2\pi} \sin(2\pi x_2) - \frac{\gamma}{2\pi} \sin[2\pi(x_1 - x_2)],
\end{aligned}
\tag{5.30}
$$

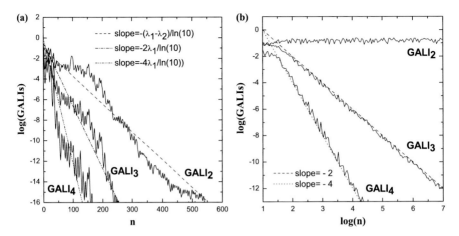

Fig. 5.8 The evolution of the $GALI_2$, the $GALI_3$ and the $GALI_4$ with respect to the number of iterations n for (**a**) a chaotic and (**b**) a regular orbit of the 4d map (5.30) with $K = 0.5$ and $\gamma = 0.05$. The initial conditions of the orbits are: (**a**) $x_1 = 0.55$, $y_1 = 0.1$, $x_2 = 0.005$, $y_2 = 0.01$, and (**b**) $x_1 = 0.55$, $y_1 = 0.1$, $x_2 = 0.54$, $y_2 = 0.01$. The *straight lines* in (**a**) correspond to functions proportional to $\exp[-(\lambda_1 - \lambda_2)n]$, $\exp(-2\lambda_1 n)$ and $\exp(-4\lambda_1 n)$ for $\lambda_1 = 0.07$, $\lambda_2 = 0.008$, which are the orbit's LEs obtained in [66]. The *straight lines* in (**b**) represent functions proportional to n^{-2} and n^{-4}. The slope of each line is mentioned in the legend. Note that the horizontal axis is linear in (**a**) and logarithmic in (**b**) (after [66])

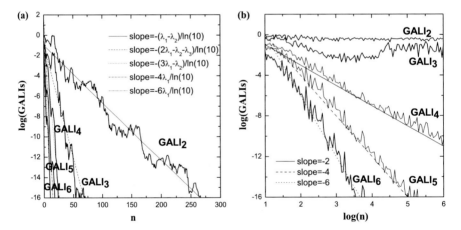

Fig. 5.9 The evolution of the $GALI_k$, $k = 2, 3, \ldots, 6$ with respect to the number of iterations n for (**a**) a chaotic (after [64]) and (**b**) a regular orbit of the 6d map (5.9) with $K = 3$ and $\gamma = 0.1$. The initial conditions of the orbits are: (**a**) $x_1 = x_2 = x_3 = 0.8$, $y_1 = 0.05$, $y_2 = 0.21$, $y_3 = 0.01$, and (**b**) $x_1 = x_2 = x_3 = 0.55$, $y_1 = 0.05$, $y_2 = 0.21$, $y_3 = 0$. The *straight lines* in (**a**) correspond to functions proportional to $\exp[-(\lambda_1 - \lambda_2)n]$, $\exp[-(2\lambda_1 - \lambda_2 - \lambda_3)n]$, $\exp[-(3\lambda_1 - \lambda_2)n]$, $\exp(-4\lambda_1 n)$ and $\exp(-6\lambda_1 n)$ for $\lambda_1 = 0.70$, $\lambda_2 = 0.57$, $\lambda_3 = 0.32$, which are the orbit's LEs obtained in [64]. The *straight lines* in (**b**) represent functions proportional to n^{-2}, n^{-4} and n^{-6}. The slope of each line is mentioned in the legend. Note that the horizontal axis is linear in (**a**) and logarithmic in (**b**)

obtained from (5.7) for $M = 2$ and $K_1 = K_2 = K$, while in Fig. 5.9 a chaotic
(Fig. 5.9a) and a regular (Fig. 5.9b) orbit of the 6d map (5.9) are considered.

These results illustrate the fact that the $GALI_k$ has the same behavior for
Hamiltonian flows and symplectic maps. For instance, even by simple inspection
we conclude that the GALIs behave similarly in Figs. 5.5 and 5.8, which refer to a
2D Hamiltonian and a 4d map respectively, as well as in Figs. 5.6 and 5.9, which
refer to a 3D Hamiltonian and a 4d map respectively.

5.3.2.2 The Case of 2d Maps

Equations (5.22) and (5.23) describe the behavior of the GALIs for ND Hamiltonian
systems and $2N$d symplectic maps with $N \geq 2$. What happens if $N = 1$? The case of
an 1D, time independent Hamiltonian is not very interesting because such systems
are integrable and chaos does not appear. But, this is not the case for 2d maps, which
can exhibit chaotic behavior.

In 2d maps only the $GALI_2$ (which, according to (5.17) is equivalent to the SALI)
is defined. For chaotic orbits the $GALI_2$ decreases exponentially to zero according
to (5.22), which becomes

$$GALI_2(n) \propto SALI(n) \propto \exp(-2\lambda_1 n), \qquad (5.31)$$

in this particular case, since, according to (5.24) $\lambda_1 = -\lambda_2 > 0$. Note that in
(5.31) we have substituted the continuous time t of (5.22) by the number n of map's
iterations. The agreement between the prediction (5.31) and actual, numerical data
can be seen for example in Fig. 5.10a where the evolution of the SALI ($\propto GALI_2$)
is plotted for a chaotic orbit of the 2d standard map

$$
\begin{aligned}
x_1' &= x_1 + y_1' \\
y_1' &= y_1 + \frac{K}{2\pi} \sin(2\pi x_1),
\end{aligned}
\qquad (5.32)
$$

obtained from (5.7) for $M = 1$. Thus, we conclude that (5.22) is also valid for 2d
maps.

But what happens in the case of regular orbits? Is (5.23) still valid for $k = 2$
and $N = 1$? First of all let us note that for these particular values of k and N only
the second branch of (5.23) is meaningful, and it provides the prediction that the
$GALI_2$ tends to zero as n^{-2}. This result is interesting, as this is the first case of
regular motion for which no GALI remains constant. But actually the vanishing
of the $GALI_2$ in this case is not surprising. Regular motion in 2d maps occurs on
1d invariant curves. So, any deviation vector from a regular orbit eventually falls
on the tangent space of this curve, which of course has dimension 1. Thus, the
two deviation vectors needed for the computation of the $GALI_2$ eventually becomes

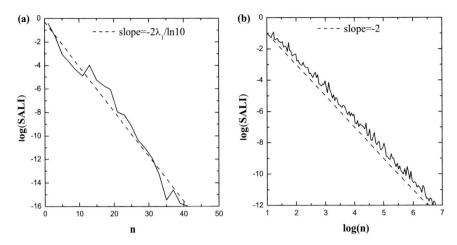

Fig. 5.10 The evolution of the SALI (which in practice is the $GALI_2$) with respect to the number of iterations n for (**a**) a chaotic and (**b**) a regular orbit of the 2d map (5.32) with $K = 2$. The initial conditions of the orbits are: (**a**) $x_1 = y_1 = 0.2$, and (**b**) $x_1 = 0.4$, $y_1 = 0.8$. The *straight line* in (**a**) corresponds to a function proportional to $\exp(-2\lambda_1 n)$ for $\lambda_1 = 0.438$, which is the orbit's mLE obtained in [65], while the line in (**b**) represents a function proportional to n^{-2}. The slope of each line is mentioned in the legend. Note that the horizontal axis is linear in (**a**) and logarithmic in (**b**) (after [65])

collinear and consequently $GALI_2 \to 0$. Actually the prediction obtained by (5.23), that for regular orbits of 2d maps

$$GALI_2(n) \propto SALI(n) \propto \frac{1}{n^2}, \qquad (5.33)$$

is correct, as for example the results of Fig. 5.10b show.

In conclusion we note that the behavior of the SALI/$GALI_2$ for chaotic and regular orbits in 2d maps is respectively given by (5.31) and (5.33), which are obtained from (5.22) and (5.23) for $k = 2$ and $N = 1$. The different behaviors of the index for chaotic (exponential decay) and regular motion (power law decay) were initially observed in [79], although the exact functional laws (5.31) and (5.33) were derived later [83, 84]. As was pointed out even from the first paper on the SALI [79], these differences allow us to use the SALI/$GALI_2$ to distinguish between chaotic and regular motion also in 2d maps (see for instance [65, 79]).

5.3.3 Regular Motion on Low Dimensional Tori

An important feature of the GALIs is their ability to identify regular motion on low dimensional tori. In order to explain this capability let us assume that a regular orbit lies on an sd torus, $2 \le s \le N$, in the $2Nd$ phase space on an ND Hamiltonian system or a $2N$d map with $N \geq 2$. Then, following similar arguments to the ones made in Sect. 5.3.2 for regular motion on an Nd torus, we conclude that the $GALI_k$ eventually remains constant for $2 \leq k \leq s$, because in this case the k deviation vectors will remain linearly independent when they eventually fall on the sd tangent space of the torus. On the other hand, any $s < k \leq 2N$ deviation vectors eventually become linearly dependent as there will be more vectors on the torus' tangent space than the space's dimension, and consequently the $GALI_k$ will vanish. In this case, the way the $GALI_k$ tends to zero depends not only on k and N, as in (5.23), but also on the dimension s of the torus. Actually, it was shown analytically in [27, 85] that for regular orbits on an sd torus the $GALI_k$ behaves as

$$
GALI_k(t) \propto \begin{cases} \text{constant if } 2 \leq k \leq s \\ \frac{1}{t^{k-s}} \quad \text{if } s < k \leq 2N - s \\ \frac{1}{t^{2(k-N)}} \quad \text{if } 2N - s < k \leq 2N. \end{cases} \tag{5.34}
$$

It is worth noting that for $s = N$ we retrieve (5.23) as the second branch of (5.34) becomes meaningless, while by setting $k = 2$, $s = 1$ and $N = 1$ we get (5.33).

The validity of (5.34) is supported by the results of Fig. 5.11 where two representative regular orbits of the H_8 Hamiltonian, obtained by setting $N = 8$ in (5.6), are considered (we note that Fig. 5.7 refers to the same model). The first orbit (Fig. 5.11a, b) lies on a 2d torus as the constancy of only $GALI_2$ indicates. The decay of the remaining GALIs is well reproduced by the power laws (5.34) for $N = 8$ and $s = 2$. The second orbit (Fig. 5.11c, d) lies on a 4d torus and consequently the $GALI_2$, the $GALI_3$ and the $GALI_4$ remain constant, while all other indices follow power law decays according to (5.34) for $N = 8$ and $s = 4$.

In Fig. 5.12 we see the evolution of some GALIs for regular motion on low dimensional tori of the 40d map obtained by (5.7) for $M = 20$. The results of Fig. 5.12a denote that the orbit lies on a 3d torus in the 40d phase space of the map, while in the case of Fig. 5.12b the motion takes place on a 6d torus. The plotted straight lines help us verify that for both orbits the behaviors of the decaying GALIs are accurately reproduced by (5.34) for $N = 20$, $s = 3$ (Fig. 5.12a) and $N = 20$, $s = 6$ (Fig. 5.12b).

5.3.3.1 Searching for Regular Motion on Low Dimensional Tori

Equation (5.34), as well as the results of Figs. 5.11 and 5.12 imply that the GALIs can be also used for identifying regular motion on low dimensional tori. From (5.34) we deduce that the dimension of the torus on which the regular motion occurs coincides with the largest order k of the GALIs for which the $GALI_k$ remains constant. Based on this remark we can develop a strategy for locating low dimensional tori in the phase space of a dynamical system. The $GALI_k$ of initial conditions resulting in motion on an sd torus eventually will remain constant for $2 \leq k \leq s$, while it will decay to zero following the power law (5.34) for $k > s$. So,

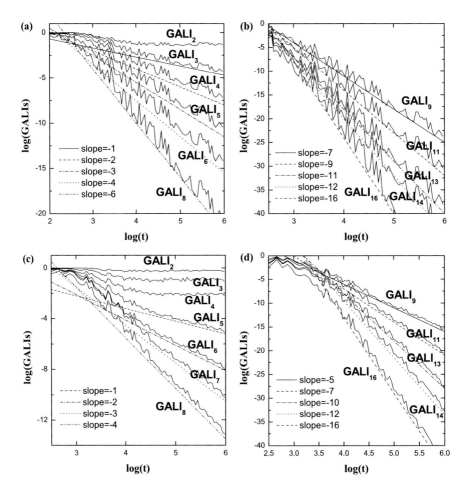

Fig. 5.11 The time evolution of the $GALI_k$, $k = 2, \ldots, 9, 11, 13, 14, 16$ for a regular orbit lying on a 2d torus (panels (**a**) and (**b**)) and for another one lying on a 4d torus (panels (**c**) and (**d**)) of the 8D Hamiltonian H_8 considered in Fig. 5.7. The initial conditions of the first orbit are $Q_1 = 2$, $P_1 = 0$, $Q_i = P_i = 0$, $i = 2, \ldots, 8$ (the definition of these variables is given in the caption of Fig. 5.7). The initial conditions of the second orbit are $q_i = 0.1$, $p_i = 0$, $i = 1, \ldots, 8$. The plotted *straight lines* correspond to the power law predictions (5.34) for $N = 8$, $s = 2$ (panels (**a**) and (**b**)) and for $N = 8$, $s = 4$ (panels (**c**) and (**d**)). The slope of each line is mentioned in the legend (after [85])

after some relatively long time interval, all the GALIs of order $k > s$ will have much smaller values than the ones of order $k \leq s$. Thus, in order to identify the location of sd tori, $2 \leq s \leq N$, in the $2Nd$ phase space of a dynamical system we evaluate at first various GALIs for several initial conditions and then find the initial conditions which result in large $GALI_k$ values for $k \leq s$ and small values for $k > s$.

As was mentioned in Sect. 5.3.2.1, the constant, final values of the GALIs for regular motion decrease with the order of the GALI (see Figs. 5.6b, 5.7c, 5.9b, 5.11c

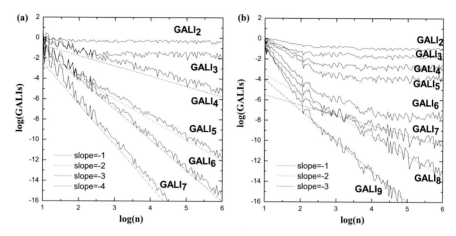

Fig. 5.12 The evolution of several GALIs for a regular orbit lying (**a**) on a 3d torus and (**b**) on a 6d torus of the 40d map obtained by setting $M = 20$ in (5.7). In (**a**) the initial conditions of the orbit are $x_{11} = 0.65$, $x_{12} = 0.55$, $x_i = 0.5$ $\forall i \neq 11, 12$, and $y_i = 0$, $i = 1, \ldots, 20$, while the parameters of the map are set to $\gamma = 0.001$ and $K_i = K = 2$, $i = 1, \ldots, 20$. In (**b**) $\gamma = 0.00001$ and K_i are set in triplets of $-1.35, -1.45, -1.55$ (i.e. $K_1 = -1.35$, $K_2 = -1.45$, $K_3 = -1.55$, $K_4 = -1.35, \ldots, K_{20} = -1.45$), while the orbit's exact initial conditions can be found in [21]. The plotted *straight lines* correspond to the power law predictions (5.34) for (**a**) $N = 20$, $s = 3$ and (**b**) $N = 20$, $s = 6$. The slope of each line is mentioned in the legend (after [21])

and 5.12). Since this decrease has not been quantified yet, a good computational approach in the quest for low dimensional tori is to 'normalize' the values of the GALIs for each individual orbit by dividing them by the largest GALI_k value, $\max(\text{GALI}_k)$, obtained by all orbits in the studied ensemble at the end time $t = t_e$ of the integration. In this way we define the 'normalized GALI_k'

$$g_k(t) = \frac{\text{GALI}_k(t)}{\max[\text{GALI}_k(t_e)]}. \tag{5.35}$$

Then, by coloring each initial condition according to its $g_k(t_e)$ value we can construct phase space charts where the position of low dimensional tori is easily located.

To illustrate this method we present (following [38]) the search for low dimensional tori in a subspace of the 8d phase space of the 4D Hamiltonian system H_4 obtained by setting $N = 4$ and $\beta = 1.5$ in (5.6). In order to facilitate the visualization of the whole procedure we restrict our search in the subspace (q_3, q_4) by setting the other initial conditions of the studied orbits to $q_1 = q_2 = 0.1$, $p_1 = p_2 = p_3 = 0$, while $p_4 > 0$ is evaluated so that $H_4 = 0.010075$. In Fig. 5.13 we color each permitted initial condition in the (q_3, q_4) plane according to its g_2, g_3 and g_4 value at $t = t_e = 10^6$ time units (panels (a), (b) and (c) respectively).

For this particular Hamiltonian we can have regular motion on 2d, 3d and 4d tori. Let us see now how we can exploit the results of Fig. 5.13 to locate such tori.

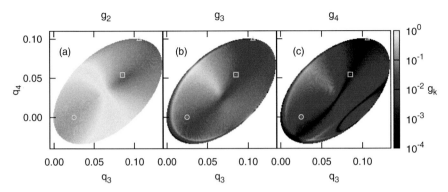

Fig. 5.13 Regions of different g_k (5.35) values for (**a**) $k = 2$, (**b**) $k = 3$, (**c**) $k = 4$, in the subspace (q_3, q_4) of the 4D Hamiltonian H_4 obtained from (5.6) for $N = 4$ and $\beta = 1.5$. The remaining coordinates of the considered initial conditions are set to $q_1 = q_2 = 0.1$, $p_1 = p_2 = p_3 = 0$, while $p_4 > 0$ is evaluated so that $H_4 = 0.010075$. *White regions* correspond to forbidden initial conditions. The color scales shown at the *right* of the panels are used to color each point according to the orbit's g_k value at $t = 10^6$. The points with coordinates $q_3 = 0.106$, $q_4 = 0.0996$ (marked by a *triangle*), $q_3 = 0.085109$, $q_4 = 0.054$ (marked by a *square*) and $q_3 = 0.025$, $q_4 = 0$ (marked by a *circle*) correspond to regular orbits on a 2d, a 3d and a 4d torus respectively (after [38])

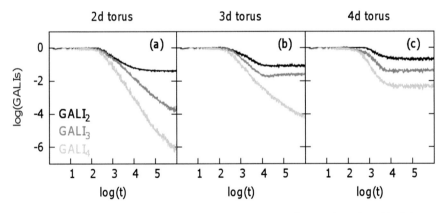

Fig. 5.14 The time evolution of the $GALI_2$, the $GALI_3$ and the $GALI_4$ of regular orbits lying on a (**a**) 2d, (**b**) 3d, (**c**) 4d torus of the 4D Hamiltonian considered in Fig. 5.13. The initial conditions of these orbits are respectively marked by a triangle, a square and a circle in Fig. 5.13 (after [38])

Motion on 2d tori results in large final g_2 values and to small g_3 and g_4. So, such tori should be located in regions colored in yellow or light red in Fig. 5.13a and in black in Fig. 5.13b, c. A region which satisfies these requirements is located at the upper border of the colored areas in Fig. 5.13. The evolution of the GALIs of an orbit with initial conditions in that region (denoted by a triangle in Fig. 5.13) is shown in Fig. 5.14a and it verifies that the motion takes place on a 2d torus, as only the $GALI_2$ remains constant.

Extending the same argumentation to higher dimensions we see that motion on a 3d torus can occur in regions colored in yellow or light red in both Fig. 5.13a, b and in black in Fig. 5.13c. The initial condition of an orbit of this kind is marked by a small square in Fig. 5.13. The evolution of this orbit's GALIs (Fig. 5.14b) verifies that the orbit lies on a 3d torus, because only the GALI$_2$ and the GALI$_3$ remain constant. Orbits on 4d tori is the most common situation of regular motion for this 4D Hamiltonian system. This is evident from the results of Fig. 5.13 because most of the permitted area of initial conditions correspond to high g_2, g_3 and g_4 values. A randomly chosen initial condition in this region (marked by a circle in Fig. 5.13) results indeed to regular motion on a 4d torus as the constancy of its GALI$_k$, $k = 2, 3, 4$ in Fig. 5.14c clearly indicates.

We note that initial conditions leading to chaotic motion in this system would correspond to very small g_2, g_3 and g_4 values (due to the exponential decay of the associated GALIs) and consequently would be colored in black in *all* panels of Fig. 5.13. The lack of such regions in Fig. 5.13 signifies that all considered initial conditions lead to regular motion. This happens because regions of chaotic motion occupy a tiny fraction of the system's phase space, because its nonlinearity strength is very small. Therefore, chaotic motion is not captured by the grid of initial conditions of Fig. 5.13.

5.3.4 Behavior of the GALI for Periodic Orbits

Let us now discuss the behavior of the GALIs for periodic orbits of period T; i.e. orbits satisfying the condition $\mathbf{x}(t + T) = \mathbf{x}(t)$, with $\mathbf{x}(t)$ being the coordinate vector in the system's phase space. In the presentation of this topic we mainly follow the analysis performed in [67]. The linear stability of periodic orbits is defined by the eigenvalues of the so-called monodromy matrix, which is obtained by the solution of the variational equations (for Hamiltonian systems) or by the evolution of the tangent map (for symplectic maps) for one period T (see for example [22, 80] and Sect. 3.3 of [53]). When all eigenvalues lie on the unit circle in the complex plane the orbit is characterized as elliptic, while otherwise it is called hyperbolic (unstable). For a detailed presentation of the various stability types of periodic orbits the reader is referred for example to [22, 40, 44, 45, 80].

The presence of periodic orbits influence significantly the dynamics. In most systems we observe that the majority of non-periodic orbits in the vicinity of an elliptic one are regular. So, although initial conditions near an elliptic orbit can lead to chaos, regular orbits exhibiting a time evolution similar to the elliptic orbit itself prevail. If one assumes that the elliptic orbit is integrable and in its vicinity the Kolmogorov–Arnold–Moser (KAM) theorem (see for example Sect. 3.2 of [53] and references therein) can be applied (for which one needs to check a non-degeneracy condition which is typically satisfied), then there is large measure of orbits on KAM tori nearby. In Hamiltonian systems of dimension larger than 2 the phenomenon of Arnold diffusion (see for example Chap. 6 of [53] and references therein) typically would lead to an escape of orbits from the neighborhood of the elliptic orbit. However, it is generally believed that Arnold diffusion occurs on a slow time scale, and we do not expect interference with the GALI method. Of course, regular behavior on nearby KAM tori does not imply that the elliptic orbit itself is stable (e.g. Appendix of [34]). On the other hand, in chaotic Hamiltonian systems and symplectic maps orbits in the vicinity of an unstable periodic orbit typically behave chaotically and diverge from the periodic one exponentially fast. This divergence is characterized by LEs (with at least one of them being positive) which are determined by the eigenvalues of the monodromy matrix (e.g. [13, 84] and Sect. 5.2b of [53]). Thus, following arguments similar to the ones developed in Sect. 5.3.2 for chaotic orbits, we easily see that the GALI$_k$ of unstable periodic orbits decreases to zero following the exponential law (5.22), i. e.

$$\text{GALI}_k(t) \propto \exp\left\{-\left[(\lambda_1 - \lambda_2) + (\lambda_1 - \lambda_3) + \cdots + (\lambda_1 - \lambda_k)\right]t\right\}, \qquad (5.36)$$

where λ_i, $i = 1, \ldots, k$ are the periodic orbit's k largest LEs.

In Fig. 5.15a we see that the evolution of the GALIs for an unstable periodic orbit of the 2D Hamiltonian (5.4) is well approximated by (5.36) for $\lambda_1 = 0.084$. This value is the orbit's mLE determined by the eigenvalues of the corresponding monodromy matrix (see [67] for more details). We also note that according to (5.24) and (5.25) we set $\lambda_1 = -\lambda_4$, and $\lambda_2 = \lambda_3 = 0$ in (5.36). The agreement between the numerical data and the theoretical prediction (5.36) is lost after about $t \approx 350$ time units. This happens because the numerically computed orbit eventually deviates

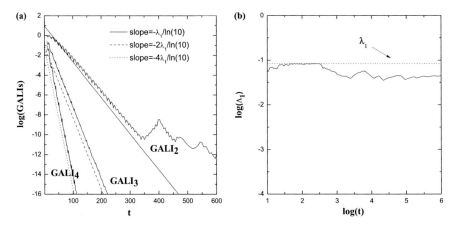

Fig. 5.15 The time evolution of (**a**) the GALI$_2$, the GALI$_3$, the GALI$_4$ and (**b**) the finite time mLE Λ_1 of an unstable periodic orbit of the 2D Hamiltonian (5.4) for $H_2 = 0.125$. The initial conditions of the orbit are $q_1 = 0, q_2 = 0.2083772012, p_1 = 0.4453146996, p_2 = 0.1196065752$. The *straight lines* in (**a**) correspond to functions proportional to $\exp(-\lambda_1 t)$, $\exp(-2\lambda_1 t)$ and $\exp(-4\lambda_1 t)$, for $\lambda_1 = 0.084$, which is the mLE of the periodic orbit. The slope of each line is mentioned in the legend. The *horizontal dotted line* in (**b**) indicates the value $\lambda_1 = 0.084$ (after [67])

from the unstable periodic one due to unavoidable computational inaccuracies and enters the chaotic region around the periodic orbit. In general, this region is characterized by different LEs with respect to the ones of the periodic orbit. The effect of this behavior on the orbit's finite time mLE Λ_1 (5.1) is seen in Fig. 5.15b. The computed Λ_1 deviates from the value $\lambda_1 = 0.084$ (marked by a horizontal dotted line) at about the same time the GALI$_2$ changes its decreasing rate in Fig. 5.15a. Eventually, Λ_1 stabilizes at another positive value, which characterizes the chaoticity of the region around the periodic orbit.

On the other hand, the case of stable periodic orbits is a bit more complicated, because the GALIs behave differently for Hamiltonian flows and symplectic maps. In [67] it was shown analytically that for stable periodic orbits of ND Hamiltonian systems, with $N \geq 2$, the GALIs decay to zero following the following power laws

$$\text{GALI}_k(t) \propto \begin{cases} \frac{1}{t^{k-1}} & \text{if } 2 \leq k \leq 2N - 1 \\ \frac{1}{t^{2N}} & \text{if } k = 2N. \end{cases} \tag{5.37}$$

We observe that this equation can be derived from (5.34), which describes the behavior of the GALIs for motion on an sd tori, by setting $s = 1$. We note that the first branch of (5.34) is meaningless for $s = 1$, while the other two branches take the forms appearing in (5.37). The connection between (5.34) and (5.37) is not surprising if we notice that a periodic orbit is nothing more than an 1d closed curve in the system's phase space, having the some dimension with an 1d torus.

Small, random perturbations from the stable periodic orbit generally results in regular motion on an Nd torus. So, the GALIs of the perturbed orbit will follow (5.23). Thus, in general, the GALIs of regular orbits in the vicinity of a stable periodic orbit behave differently with respect to the indices of the periodic orbit itself (except from the GALI$_{2N}$ and the GALI$_{2N-1}$, which respectively follow the laws $\propto t^{-2N}$ and $\propto t^{-(2N-2)}$ in both cases). The most profound change happens for the GALIs of order $2 \leq k \leq N$ because, according to (5.23), they remain constant in the neighborhood of the periodic orbit, while they decay to zero following the power law (5.37) for the periodic orbit.

The correctness of (5.37) becomes evident from the results of Fig. 5.16a, where the time evolution of the GALIs of a stable periodic orbit of the 2D Hamiltonian (5.4) is shown. In particular, we see that the indices decay to zero following the power laws GALI$_2 \propto t^{-1}$, GALI$_3 \propto t^{-2}$, GALI$_4 \propto t^{-4}$ predicted from (5.37). According to (5.23) the GALIs of regular orbits in the neighborhood of the stable periodic orbit should behave as GALI$_2 \propto$ constant, GALI$_3 \propto t^{-2}$ and GALI$_4 \propto t^{-4}$. Thus, only the GALI$_2$ is expected to behave differently for regular orbits in the vicinity of the periodic orbit of Fig. 5.16a. The results of Fig. 5.16b show that this is actually true. The GALI$_2$ of the neighboring regular orbits initially follows the same power law decay of the periodic orbit (GALI$_2 \propto t^{-1}$), but later on it stabilizes to a constant positive value. We see that the further the orbit is located from the periodic one the sooner the GALI$_2$ deviates from the power law decay.

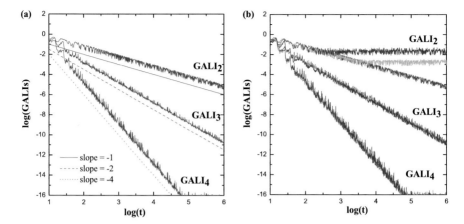

Fig. 5.16 (a) The time evolution of the $GALI_2$, the $GALI_3$ and the $GALI_4$ for a stable periodic orbit of the 2D Hamiltonian (5.4) for $H_2 = 0.125$. The orbit's initial conditions are $q_1 = 0 = q_{10}$, $q_2 = 0.35207 = q_{20}$, $p_1 = 0.36427 = p_{10}$, $p_2 = 0.14979 = p_{20}$. The *straight lines* correspond to functions proportional to t^{-1}, t^{-2} and t^{-4}. The slope of each line is mentioned in the legend. (b) The same plot as in (a) where apart from the GALIs of the stable periodic orbit (*red curves*) the indices of two neighboring, regular orbits are also plotted. Their initial conditions are $q_1 = q_{10}$, $p_2 = p_{20}$ for both of them, while $q_2 = q_{20} + 0.00793$ (*green curves*), and $q_2 = q_{20} + 0.02793$ (*blue curves*). In both cases the $p_1 > 0$ initial condition is set so that $H_2 = 0.125$. Note that the curves of the $GALI_3$ and the $GALI_4$ for all three orbits overlap each other (after [67])

These differences of the $GALI_2$ values can be used to identify the location of stable periodic orbits in the system's phase space, although the index was not developed for this particular purpose.[9] This becomes evident from the result of Fig. 5.17 where the values of the $GALI_2$ at $t = 10^5$ for several orbits of the Hénon-Heiles system (5.4) are plotted as a function of the q_2 coordinate of the orbits' initial conditions. The remaining coordinates are $q_1 = p_2 = 0$, while $p_1 > 0$ is set so that $H_2 = 0.125$. Actually these initial conditions lie on the symmetry line of the subspace defined by $q_1 = 0$, $p_1 > 0$, i.e. the horizontal line $p_2 = 0$ in Figs. 5.19 and 5.20 below. This line passes through the initial condition of some periodic orbits of the system. For the construction of Fig. 5.17 we considered an ensemble of 7000 orbits whose q_2 coordinates are equally distributed in the interval $-0.1 \le q_2 \le 0.6$. The data points are line connected, so that the changes of the $GALI_2$ values become easily visible.

In Fig. 5.17 regions of relatively large $GALI_2$ values ($\gtrsim 10^{-4}$) correspond to regular (periodic or quasiperiodic) motion. Chaotic orbits and unstable periodic orbits have very small $GALI_2$ values ($\lesssim 10^{-12}$), while domains with intermediate values ($10^{-12} \lesssim GALI_2 \lesssim 10^{-4}$) correspond to sticky chaotic orbits. An interesting

---

[9]It is worth mentioning here that other chaos indicators, like the Orthogonal Fast Lyapunov Indicator (OFLI) and its variations [7, 8], are quite successful in performing this task as they were actually designed for this purpose.

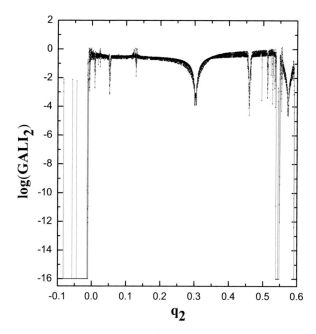

Fig. 5.17 The values of the GALI$_2$ at $t = 10^5$ for several orbits of the 2D Hamiltonian (5.4) as a function of the q_2 coordinate of the orbits' initial conditions. The remaining coordinates are $q_1 = p_2 = 0$, while $p_1 > 0$ is set so that $H_2 = 0.125$. Actually these initial conditions lie on the $p_2 = 0$ line of Figs. 5.19 and 5.20. The numerical data (*black points*) are line connected (*grey line*) in order to facilitate the visualization of the value changes (after [67])

feature of Fig. 5.17 is the appearance of some relatively narrow regions where the GALI$_2$ decreases abruptly obtaining values $10^{-4} \lesssim$ GALI$_2 \lesssim 10^{-1}$; the most profound one being in the vicinity of $q_2 \approx 0.3$. These regions correspond to the immediate neighborhoods of stable periodic orbits, with the periodic orbit itself been located at the point with the smallest GALI$_2$ value.

The creation of these characteristic 'pointy' shapes is due to the behavior depicted in Fig. 5.16b: the GALI$_2$ has relatively small values on the stable periodic orbit, for which it decreases as $\propto t^{-1}$, while it attains constant, positive values for regular orbits in the vicinity of the periodic orbit. These constant values increase as the orbit's initial conditions depart further away from the periodic orbit. So, more generally, the appearance of such 'pointy' formations in GALI$_k$ plots ($2 \leq k \leq N$) provide good indications for the location of stable periodic orbits.

Let us now turn our attention to maps. In 2*N*d symplectic maps stable periodic orbits of period l correspond to l distinct points (the so-called stable fixed points of order l). Any deviation vector from the periodic orbit rotates around each fixed

point. This behavior can be easily seen in the case of 2d maps where the tori around
a stable fixed point correspond to closed invariant curves which can be represented,
through linearization, by ellipses (see for example Sect. 3.3b of [53]). Thus, any
k initially distinct deviation vectors needed for the computation of the $GALI_k$ will
rotate around the fixed point keeping on average the angles between them constant.
Consequently the volume of the parallelepiped they define, i.e. the value of the
$GALI_k$, will remain practically constant. Thus, in the case of stable periodic orbits
of 2Nd maps, with $N \geq 1$ we have

$$GALI_k(t) \propto \text{constant}, \quad \text{for } 2 \leq k \leq 2N. \quad (5.38)$$

This behavior is clearly seen in Fig. 5.18a where the evolution of the $GALI_2$, the
$GALI_3$ and the $GALI_4$ for a stable periodic orbit of period 7 of the 4d map (5.30) is
plotted.

Again small perturbations of the periodic orbit's initial conditions generally
result in motion on an Nd tori. Then, the evolution of the corresponding GALIs
is provided by (5.23) for $N \geq 2$, while the $GALI_2$ will decrease to zero according
to (5.33) for 2d maps. So, the most striking difference between the behavior of the
$GALI_k$ of a stable periodic orbit and of a neighboring, regular orbit appears for
$k > N$, because in this case the $GALI_k$ remains constant for the periodic orbit, while
it decays to zero for the neighboring one. Differences of this kind can be observed
in Fig. 5.18b.

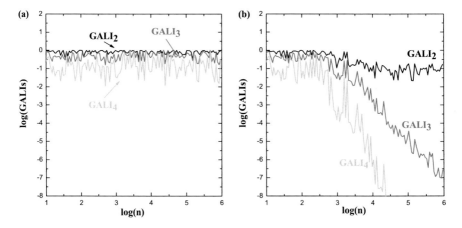

Fig. 5.18 The evolution of the $GALI_2$, the $GALI_3$ and the $GALI_4$ with respect to the number of
iterations n for (**a**) a stable periodic orbit and (**b**) a nearby regular orbit, of the 4d map (5.30) with
$K = 0.9$ and $\gamma = 0.05$. The initial conditions of the orbits are: (**a**) $x_1 = 0.23666$, $y_1 = 0.0$,
$x_2 = 0.23666$, $y_2 = 0.0$, and (**b**) $x_1 = 0.23$, $y_1 = 0.0$, $x_2 = 0.236$, $y_2 = 0.0$

5.4   Applications

The ability of the SALI and the GALI methods to efficiently discriminate between chaotic and regular motion was described in detail in the previous sections, where some exemplary Hamiltonian systems and symplectic maps were considered. In what follows we present applications of this ability to various dynamical systems originating from different research fields.

5.4.1   Global Dynamics

In Sect. 5.3.2 we discussed how one can use the various GALIs to reveal the chaotic or regular nature of individual orbits in the $2Nd$ phase space of a dynamical system. Additionally, in Sect. 5.3.4 we saw how the measurement of the $GALI_2$ values for an ensemble of orbits can facilitate the uncovering of some dynamical properties of the studied system, in particular the pinpointing of stable periodic orbits (Fig. 5.17), while in Sect. 5.3.3.1 we described how a more general search can help us locate motion on low dimensional tori.

Now we see how one can use the GALIs in order to study the global dynamics of a system. For simplicity we use in our analysis the 2D Hamiltonian system (5.4), but the methods presented below can be (and actually have already been) implemented to higher-dimensional systems.

5.4.1.1   Investigating Global Dynamics by the $GALI_k$ with $2 \leq k \leq N$

According to (5.22) and (5.23) the $GALI_k$, with $2 \leq k \leq N$, behaves in a completely different way for chaotic (exponential decay) and regular (remains practically constant) orbits. Thus, by coloring each initial condition of an ensemble of orbits according to its $GALI_k$ value at the end of a fixed integration time we can produce color plots where regions of chaotic and regular motion are easily seen. In addition, by choosing an appropriate threshold value for the $GALI_k$, below which the orbit is characterized as chaotic (see Sect. 5.3.2.1 on how to set up this threshold), we can efficiently determine the 'strength' of chaos by calculating the percentage of chaotic orbits in the studied ensemble. Then, by performing the same analysis for different parameter values of the system we can determine its physical mechanisms that increase or suppress chaotic behavior.

A practical question arises though: which index should one use for this kind of analysis? The obvious advantage of the $GALI_2$/SALI is its easy computation according to (5.8), which requires the evolution of only two deviation vectors. On the other hand, evaluating the GALIs of order up to $k = N$ is more CPU-time consuming as the computation of the index from (5.20) requires the evolution of more deviation vectors, as well as the implementation of the SVD algorithm. An advantage of these higher order indices is that they tend to zero faster than the GALI$_2$/SALI for chaotic orbits. So, reaching their threshold value which characterizes an orbit as chaotic, requires in general, less computational effort. This feature is particularly useful when we want to estimate the percentage of chaotic orbits, as there is no need to continue integrating orbits which have been characterized as chaotic (see Sect. 5.2 of [84] for an example of this kind). Thus, we conclude that the reasonable choices for such global studies are the GALI$_2$/SALI and the GALI$_N$.

In order to illustrate this process, let us consider the 2D Hénon-Heiles system (5.4), for which GALI$_N$ $\equiv$ GALI$_2$, since $N = 2$. In Fig. 5.19 we see color plots of its Poincaré surface of section defined by $q_1 = 0$ (a concise description of the construction of a surface of section can be found for instance in Sect. 1.2b of [53]). The remaining initial conditions of each orbit are its coordinates on the (q_2, p_2) plane of Fig. 5.19, while $p_1 > 0$ is set so that $H_2 = 0.125$. For each panel of Fig. 5.19 a 2d grid of approximately 350,000 equally distributed initial conditions is considered. Each point on the (q_2, p_2) plane is colored according to its log(GALI$_2$) value at $t = 2000$, while white regions denote not permitted initial conditions. Regions colored in yellow or light red correspond to regular orbits, while dark blue and black domains contain chaotic ones. Intermediate colors at the borders between these two regions indicate sticky chaotic orbits.

This kind of color plots can reveal fine details of the underlying dynamics, like for example the small yellow 'islands' of regular motion inside the large, black chaotic 'sea', as well as allow the accurate estimation of the percentage of chaotic or regular orbits in the studied ensemble. Naturally the denser the used grid is, the finer the uncovered details become, but unfortunately the higher the needed computational effort gets. In an attempt to speed up the whole process the following procedure was followed in [3] where the dynamics of the Hénon-Heiles system (5.4) was studied. The final GALI$_2$/SALI value and the corresponding color was assigned not only to the initial condition of the studied orbit, but also to all intersection points of the orbit with the surface of section. This assignment can be extended even further by additionally taking into account the symmetry of Hamiltonian (5.4) with respect to the q_2 variable, which results in structures symmetric with respect to the $p_2 = 0$ axis in Fig. 5.19. Consequently, points symmetric to this axis should have the same GALI$_2$/SALI value. So, orbits with initial conditions on grid points to which a color has already been assigned, as they were intersection points with the surface of section of previously computed orbits, are not computed again and so the construction of color plots like the ones of Fig. 5.19 is speeded up significantly. In [3] it was shown that this approach achieves very accurate estimations of the percentages of chaotic orbits with respect to the ones obtaining by coloring each and every initial condition according to the index's value at the end of the integration time (this is actually how Fig. 5.19 was produced).

Let us now discuss the differences between panels (a) and (b) of Fig. 5.19. In both figures the chaotic regions are practically the same. Nevertheless, in the yellow and light red colored domains, where regular motion occurs, some 'spurious' structures appear in Fig. 5.19a, which are not present in Fig. 5.19b. For example, inside the large stability island with $0 \lesssim q_2 \lesssim 0.5$ at the right side of Fig. 5.19a we observe an almost horizontal formation colored in light red, while similar colored 'arcs' appear

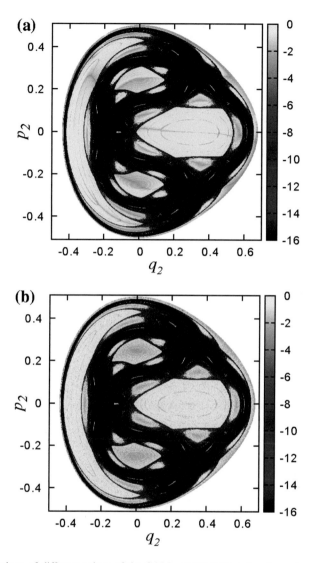

Fig. 5.19 Regions of different values of the GALI$_2$ on the Poincaré surface of section defined by $q_1 = 0$ of the 2D Hamiltonian (5.4) for H_2=0.125. A set of approximately 350,000 equally spaced initial conditions on the grid $(q_2, p_2) \in [-0.5, 0.7] \times [-0.5, 0.5]$ is used. *White regions* correspond to forbidden initial conditions. The color scales shown at the *right* of the panels are used to color each point according to the orbit's log(GALI$_2$) value at $t = 2000$. In (**a**) the same set of initial orthonormal deviation vectors was used for the computation of the GALI$_2$ of each initial condition, while in (**b**) a different, randomly produced set of vectors was used for each orbit

inside many other islands of regular motion. These artificial features emerge when one uses exactly the same set of orthonormal, initial deviation vectors for every studied orbit, as we did in Fig. 5.19a. The appearance of such features in color plots of other chaos detection methods has already been reported in the literature [9]. A simple way to avoid them is to use a different, random set of initial, orthonormal vectors for the computation of the $GALI_2$, as we did in Fig. 5.19b. By doing so, these spurious features disappear and only structures related to the actual dynamics of the system remain, like for instance the cyclical 'chain' of the light red colored, elongated regions inside the big stability island at the right side of Fig. 5.19b. This structure indicates the existence of some higher order stability islands, which are surrounded by an extremely thin chaotic layer. This layer is not visible for the resolution used in Fig. 5.19b. A magnification, and a much finer grid would reveal this tiny chaotic region.

5.4.1.2   Investigating Global Dynamics by the $GALI_k$ with $N < k \leq 2N$

As was clearly explained in Sect. 5.3.2.1 the GALIs of order $N < k \leq 2N$ tend to zero both for chaotic and regular orbits, but with very different time rates as (5.22) and (5.23) state. This deference can be also used to investigate global dynamics, but following an alternative approach to the one developed in Sect. 5.4.1.1. Since these GALIs decay to zero exponentially fast for chaotic orbits, but follow a much slower power law decay for regular ones, the time t_{th} they need to reach an appropriately chosen, small threshold value will be significantly different for the two kinds of orbits. We note that both the exponential and the power law decays become faster with increasing order k of $GALI_k$. Consequently, the creation of huge differences in the $GALI_k$ values, which allow the discrimination between chaotic and regular motion, will appear earlier for larger k values. So, in general, the overall required computational time decreases significantly by using a higher order $GALI_k$, despite the integration of more deviation vectors, since this integration will be terminated earlier. Thus, the best choice in investigations of this kind is to use the $GALI_{2N}$.

Let us illustrate this approach by computing the $GALI_4$ for the 2D Hénon-Heiles system (5.4), at a grid in its $q_1 = 0$ surface of section. The outcome of this procedure is seen in Fig. 5.20, where each initial condition is colored according to the time t_{th} needed for its $GALI_4$ to become $\leq 10^{-12}$. Each orbit is integrated up to $t = 500$ time units and if its $GALI_4$ value at the end of the integration is larger than the threshold value 10^{-12} the corresponding t_{th} value is set to $t_{th} = 500$ and the initial condition is colored in blue according to the color scales seen below the panel of Fig. 5.20. Regions of regular motion correspond to large t_{th} values and are colored in blue, while all the remaining colored domains contain chaotic orbits. Again, white regions correspond to forbidden initial conditions. This approach yields a very detailed chart of the dynamics, analogous to the one seen in Fig. 5.19.

An advantage of the current approach is its ability to clearly reveal various 'degrees' of chaotic behavior in regions not colored in blue. Strongly chaotic orbits are colored in red and yellow as their $GALI_4$ becomes $\leq 10^{-12}$ quite fast. Orbits with larger t_{th} values correspond to chaotic orbits which need more time in order to show their chaotic nature, while the 'sticky' chaotic regions are characterized by even higher t_{th} values and are colored in light blue. We note that for every initial condition

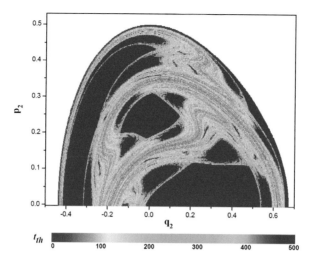

Fig. 5.20 Regions of different values of the time t_{th} needed for the GALI$_4$ to become less than 10^{-12} on the $q_1 = 0$ surface of section of the 2D Hénon-Heiles system (5.4). Each orbit is integrated up to $t = 500$ time units. *White regions* correspond to forbidden initial conditions. The color scales shown below the panel are used to color each point according to the orbit's t_{th} value (after [84])

we used a different, random set of orthonormal deviation vectors in order to avoid the appearance of possible 'spurious' structures, like the ones seen in Fig. 5.19a.

5.4.2   Studies of Various Dynamical Systems

The SALI and the GALI methods have been used broadly for the study of the phase space dynamics of several models originating from different scientific fields. These studies include the characterization of individual orbits as chaotic or regular, as well as the consideration of large ensembles of initial conditions along the lines presented in Sect. 5.4.1, whenever a more global understanding of the underlying dynamics was needed.

In this section we present a brief, qualitative overview of such investigations. For this purpose we focus mainly on the outcomes of these studies avoiding a detailed presentation of mathematical formulas and equations for each studied model.

5.4.2.1   An Accelerator Map Model

Initially, let us discuss two representative applications of the SALI. The first one concerns the study of a 4d symplectic map which describes the evolution of a charged particle in an accelerator ring having a localized thin sextupole magnet.

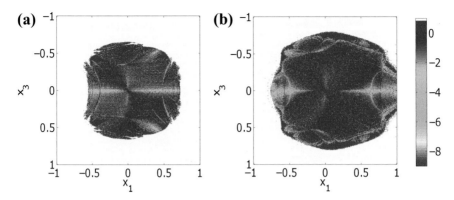

Fig. 5.21 Regions of different SALI values of (**a**) the 4d uncontrolled accelerator map studied in [19] and (**b**) the controlled map constructed in [16]. The coordinates x_1, x_3 respectively describe horizontal and vertical deflections of a charged particle from the ideal circular orbit passing from $x_1 = x_3 = 0$ in some appropriate units (see [19] for more details). 16,000 uniformly distributed initial conditions on the grid $(x_1, x_3) \in [-1, 1] \times [-1, 1]$ were evolved for 10^5 iterations of each map and colored according to the orbit's log(SALI) value, using the color scales shown at the *right* of the panels. The *white colored regions* correspond to orbits that escape in less than 10^5 iterations. *Red points* denote chaotic orbits, while regular ones are colored in *blue*. The increase of the stability region around the point $x_1 = x_3 = 0$ is evident (after [17])

The specific form of this map can be found in [19] where the SALI method was used for the construction of phase space color charts where regions of chaotic and regular motion were clearly identified, as well as for evaluating the percentage of chaotic orbits.

Later on, in [16, 17] this map was used to test the efficiency of chaos control techniques for increasing the stability domain (the so-called 'dynamic aperture') around the ideal circular orbit of this simplified accelerator model. These techniques turned out to be quite successful, as the addition of a rather simple control term, which potentially could be approximated by real multipole magnets, increased the stability region of the map as can be seen in Fig. 5.21.

5.4.2.2 A Hamiltonian Model of a Bose-Einstein Condensate

Let us now turn our attention to a 2D Hamiltonian system describing the interaction of three vortices in an atomic Bose-Einstein condensate, which was studied in [52]. By means of SALI color plots the extent of chaos in this model was accurately measured and its dependence on physically important parameters, like the energy and the angular momentum of the vortices, were determined.

In real experiments, from which the study of this model was motivated, the life time of Bose-Einstein condensates is limited. For this reason the time in which the chaotic nature of orbits is uncovered played a significant role in the analysis presented in [52]. Actually, different 'degrees of chaoticity' are revealed by

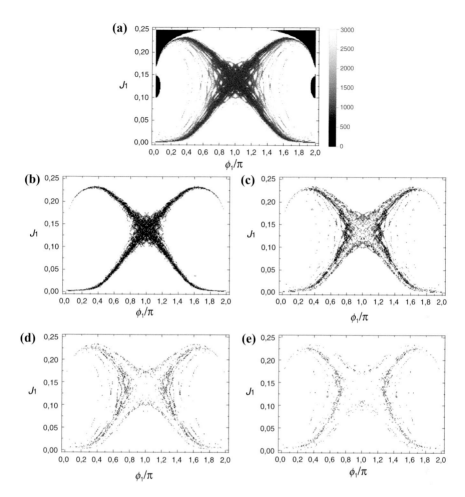

Fig. 5.22 (**a**) Regions of different values of the time t_{th} needed for the SALI to become less than 10^{-12} for a 2D Hamiltonian describing the interaction of three vortices in an atomic Bose-Einstein condensate. The explicit definition of the coordinates J_1 and ϕ_1/π can be found in [52] where this model was studied in detail. Each orbit is integrated up to $t = 3000$ time units. *White regions* correspond to regular orbits, while *black areas* at the upper two corners, as well as in the middle of the vertical axes at both sides of the plot, denote not permitted initial conditions. The color scales shown at the *right* of the panel are used to color each point according to the orbit's t_{th} value. The initial conditions of (**a**) are decomposed in four different sets according to their t_{th} value: (**b**) $140 \leq t_{th} \leq 500$, (**c**) $500 < t_{th} \leq 1000$, (**d**) $1000 < t_{th} \leq 1500$ and (**e**) $1500 < t_{th} \leq 2000$ (after [52])

registering the time t_{th} that the SALI of a chaotic orbit requires in order to become $\leq 10^{-12}$ (Fig. 5.22). This approach is similar to the one presented in Sect. 5.4.1.2, and allows the identification of regions with different strengths of chaos.

The chaotic orbits of Fig. 5.22a are decomposed in Figs. 5.22b–e in four different sets according to their t_{th} value: $t_{th} \in [140, 500]$ (Fig. 5.22b), $t_{th} \in (500, 1000]$ (Fig. 5.22c), $t_{th} \in (1000, 1500]$ (Fig. 5.22d) and $t_{th} \in (1500, 2000]$ (Fig. 5.22e), where time is measured in some appropriate units (see [52] for more details). From these results we see that, as the initial conditions move further away from the center of the x-shaped region of Fig. 5.22a the orbits need more time to show their chaotic nature and consequently, some of them can be considered as regular from a practical (experimental) point of view. For instance, in real experiments one would expect to detect chaotic motion in regions shown in Fig. 5.22b where orbits have relatively small t_{th} values. Thus, an analysis of this kind can provide practical information about where one should look for chaotic behavior in actual experimental set ups.

5.4.2.3 Further Applications of the SALI and the GALI Methods

The SALI and the GALI methods have been successfully employed in studies of various physical problems and mathematical toy models, as well as for the investigation of fundamental aspects of nonlinear dynamics (e.g. see [30]). In what follows we briefly present some of these studies

In [65] the SALI/GALI$_2$ method was used for the global study of the standard map (5.32). By considering large ensembles of initial conditions the percentage of chaotic motion was accurately computed as a function of the map's parameter K. This work revealed the periodic re-appearance of small (even tiny) islands of stability in the system's phase space for increasing values of K. Subsequent investigations of the regular motion of the standard map in [62] led to the clear distinction between typical islands of stability and the so-called accelerator modes, i.e. motion resulting in an anomalous enhancement of the linear in time orbits' diffusion. Typically, this motion is highly superdiffusive and is characterized by a diffusion exponent ≈ 2.

In [21] the GALI was used for the detection of chaotic orbits in many dimensions, the prediction of slow diffusion, as well as the determination of quasiperiodic motion on low dimensional tori in the system (5.7) of many coupled standard maps. Additional applications of the SALI in studying maps can be found in [73], where the index was used for shedding some light in the properties of accelerator models, while in [76] a coupled logistic type predator-prey model describing population growths in biological systems was considered. Further studies of 2d and 4d maps based on the SALI method were performed in [35].

Models of dynamical astronomy and galactic dynamics are considered to be the spearhead of the chaos detection methods [31]. Actually, many of these methods have been used, or often even constructed, to investigate the properties of such systems. Several applications of the SALI to systems of this kind can be found in the literature. In [18, 88, 89] the stability properties of orbits in a particular few-body problem, the so-called the Sitnikov problem, were studied, while in [94] the long term stability of two-planet extrasolar systems initially trapped in the 3:1 mean motion resonance was investigated. The SALI was also used to study the dynamics of the Caledonian symmetric four-body problem [90], as well as the circular restricted three-body problem [75].

In systems modeling the dynamics of galaxies special care should be taken with respect to the determination of the star motion's nature, because this has to be done as fast as possible and in physically relevant time intervals (e.g. smaller than the age of the universe). Hence, in order to check the adequacy of a proposed galactic model, in terms of being able to sustain structures resembling the ones seen in observations of real galaxies, the detection of chaotic and regular motion for rather small integration times is imperative. The SALI and the GALI methods have proved to be quite efficient tools for such studies, as they allow the fast characterization of orbits. This ability reduces significantly the required computational burden, as in many cases the determination of the orbits' nature is achieved before the predefined, final integration time.

In particular, the SALI method has been used successfully in studying the chaotic motion and spiral structure in self-consistent models of rotating galaxies [93], the dynamics of self-consistent models of cuspy triaxial galaxies with dark matter haloes [23], the orbital structure in N body models of barred-spiral galaxies [42], the secular evolution of elliptical galaxies with central masses [50], the chaotic component of cuspy triaxial stellar systems [25], as well as the chaoticity of non-axially symmetric galactic models [97] and of models with different types of dark matter halo components [96].

The SALI was used in [65] for investigating the dynamics of 2D and 3D Hamiltonian models of rotating bared galaxies. This work was extended in [60] by using the GALI for studying the global dynamics of different galactic models of this type. In particular, the effects of several parameters related to the shape and the mass of the disk, the bulge and the bar components of the models, as well as the rotation speed of the bar, on the amount of chaos appearing in the system were determined. Moreover, the implementation of the $GALI_3$ in the 3D Hamiltonians allowed the detection of regular motion on low (2d) dimensional tori, although these systems support, in general, 3d orbits. The astronomical significance of these orbits was discussed in detail in [60].

Implementations of the SALI to nuclear physics systems can be found in [56–58, 86, 87] where the chaotic behavior of boson models is investigated, as well as in [5] where the dynamics of a Hamiltonian model describing a confined microplasma was studied. Recently the SALI and the GALI methods, together with other chaos indicators, were reformulated in the framework of general relativity, in order to become invariant under coordinate transformation [54].

The SALI and the GALI have been also used to study the dynamics of nonlinear lattice models. Applications of these indices to the Fermi–Pasta–Ulam model can be found in [1, 2, 4, 27–29, 71, 85] where the properties of regular motion on low dimensional tori, the long term stability of orbits, as well as the interpretation of Fermi–Pasta–Ulam recurrences were studied. In [63] the GALI method managed to capture the appearance of a second order phase transition that the Hamiltonian Mean Field model exhibits at a certain energy density. The index successfully verified also other characteristics of the system, like the sharp transition from weak to strong chaos. Further applications of the SALI method to other models of nonlinear lattices can be found in [4, 72].

In addition, the SALI was further used in studying the chaotic and regular nature of orbits in non-Hamiltonian dynamical systems [6, 47], some of which model chaotic electronic circuits [46, 48, 49].

5.4.3 Time Dependent Hamiltonians

The applications presented so far concerned autonomous dynamical systems. However, there are several phenomena in nature whose modeling requires the invocation of parameters that vary in time. Whenever these phenomena are described according to the Hamiltonian formalism, the corresponding Hamiltonian function is not an integral of motion as its value does not remain constant as time evolves.

The SALI and the GALI methods can be also used to determine the chaotic or regular nature of orbits in time dependent systems as long as, their phase space does not shrink ceaseless or expand unlimited, with respect to its initial volume, during the considered times. This property allows us to utilize the time evolution of the volume defined by the deviation vectors, as in the case of the time independent models, and estimate accurately its possible decay for time intervals where the total phase space volume has not changed significantly.

In conservative time independent Hamiltonians orbits can be periodic (stable or unstable), regular (quasiperiodic) or chaotic and their nature does not change in time. Sticky chaotic orbits may exhibit a change in their orbital morphologies from almost quasiperiodic to completely chaotic behaviors, but in reality their nature does not change as they are weakly chaotic orbits. On the other hand, in time dependent models, individual orbits can display abrupt transitions from regular to chaotic behavior, and vice versa, during their time evolution. This is an intriguing characteristic of these systems which should be captured by the used chaos indicator. Such transitions between chaotic and regular behaviors can be seen for example in N body simulations of galactic models. For this reason, time dependent analytic potentials trying to mimic the evolution of N body galactic systems, are expected to exhibit similar transitions.

An analytic time dependent bared galaxy model consisting of a bar, a disk and a bulge component, whose masses vary linearly in time was studied in [68]. The time dependent nature of the model influences drastically the location and the size of stability islands in the system's phase space, leading to a continuous interplay between chaotic and regular behaviors. The GALI was able to capture subtle changes in the nature of individual orbits (or ensemble of orbits) even for relatively small time intervals, verifying that it is an ideal diagnostic tool for detecting dynamical transitions in time dependent systems.

Although both 2D and 3D time dependent Hamiltonian models were studied in [68], we further discuss here only the 3D model in order to illustrate the procedure followed for detecting the various dynamical epochs in the evolution of an orbit. The main idea for doing that is the re-initialization of the computation of the $GALI_k$, with $2 \leq k \leq N$, whenever the index reaches a predefined low value (which signifies chaotic behavior) by considering k new, orthonormal deviation vectors resetting $GALI_k = 1$.

Let us see this procedure in more detail. In [68] the evolution of the $GALI_3$ was followed for each studied orbit. The three randomly chosen, initial orthonormal deviation vectors set $GALI_3 = 1$ in the beginning of the numerical simulation $(t = 0)$. These vectors were evolved according to the dynamics induced by the 3D, time dependent Hamiltonian up to the time $t = t_d$ that the $GALI_3$ became smaller than 10^{-8} for the fist time. At that point the time $t = t_d$ was registered and three new, random, orthonormal vectors were considered resetting $GALI_3 = 1$. Afterwards, the evolution of these vectors was followed until the next, possible occurrence of $GALI_3 < 10^{-8}$. Then the same process was repeated.

Why was this procedure implemented? What is the reason behind this strategy? In order to reveal this reason let us assume that an orbit initially behaves in a chaotic way and later on it drifts to a regular behavior. The volume formed by the deviation vectors will shrink exponentially fast, becoming very small during the initial chaotic epoch and will remain small throughout the whole evolution in the regular epoch, unless one re-initializes the deviation vectors and the volume they define. In this way the deviation vectors will be able to 'feel' the new, current dynamics.

An example case of this kind is shown in Fig. 5.23. In particular, in Fig. 5.23a we see that the evolution of the finite time mLE Λ_1 is not able to provide valid information about the different dynamical epochs that the studied orbit experiences. This is due to the index's averaging nature which takes into account the whole history of the evolution. On the other hand, the re-initialized $GALI_3$ (whose time evolution is shown in Fig. 5.23b) clearly succeeds in depicting the transitions between regular epochs, where it oscillates around positive values (such time intervals are denoted by I and III in Fig. 5.23a, b), and chaotic ones, where it exhibits repeated exponential decays to very small values (epoch II). From the results of Fig. 5.23a it becomes evident that the computation of the mLE cannot be used as a reliable criterion for determining the chaotic or regular nature of the orbit in these three time intervals.

Another way to visualize the results of Fig. 5.23b is through the measurement of the time t_d needed for the repeated re-initializations of the $GALI_3$, or in other words, of the time needed for the $GALI_3$ to decrease from $GALI_3 = 1$ to $GALI_3 \leq 10^{-8}$. In Fig. 5.23c we present t_d as a function of the evolution time of the orbit. From the results of this figure we see that during the time interval $7500 \lesssim t \lesssim 14{,}000$ the value of t_d is rather small, indicating strong chaotic motion. For smaller times, $t \lesssim 7500$, the $GALI_3$ takes a long time to become small, suggesting the presence of regular motion or of (relatively) weaker chaotic motion. The upwardly pointing arrow, after $t \gtrsim 15{,}000$, shows that the $GALI_3$ no longer falls to zero, which again indicates the appearance of a regular epoch.

After the first, successful application of the GALIs to time dependent Hamiltonians in [68], the same approach was followed for the study of a more sophisticated time dependent galactic model in [61]. This analytic Hamiltonian model succeeded

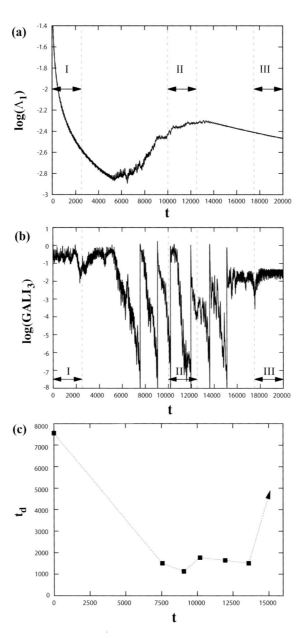

Fig. 5.23 Time evolution of (**a**) the finite time mLE Λ_1, (**b**) the re-initialized $GALI_3$, and (**c**) the time t_d needed for the re-initialized $GALI_3$ to decrease from $GALI_3 = 1$ to $GALI_3 \leq 10^{-8}$ for a particular orbit of the 3D time dependent galactic model studied in [68]. The orbit changes its dynamical nature from regular to chaotic and again to regular. Three characteristic epochs are located between the *vertical dashed gray lines* in (**a**) and (**b**) and are denoted by I (regular), II (chaotic) and III (regular). The *arrow* at the right end of (**c**) indicates that after $t \gtrsim 15,000$ the $GALI_3$ in (**b**) does not fall back to zero (until of course, the final integration time $t = 20,000$), which is a clear indication that in this time interval the orbit is regular (after [68])

to incorporate the evolution of the basic morphological features of an actual N body simulation, by allowing all the relevant parameters of its dynamical components to vary in time.

5.5   Summary

In this chapter we presented how the SALI and the various GALIs can be used to study the chaotic behavior of dynamical systems.

Following the history of the evolution of these indices, we initially presented in Sect. 5.2 the underlying idea behind the introduction of the SALI: the index actually quantifies the possible alignment of two initially distinct deviation vectors. The natural generalization of this idea, by considering more than two deviation vectors and checking if they become linearly dependent, led later on, to the introduction of the GALI, as we explained in Sect. 5.3. The close relation between the two indices was also pointed out, as according to (5.17) the GALI$_2$ and the SALI practically coincide

$$\text{GALI}_2 \propto \text{SALI}.$$

Avoiding the presentation of mathematical proofs (which the interested reader can find in the related references), we formulated in Sect. 5.3 the laws that the indices follow for chaotic and regular orbits, providing also several numerical results which demonstrate their validity.

In particular, for ND Hamiltonian systems ($N \geq 2$) and $2N$d symplectic maps ($N \geq 1$) the GALI$_k$ tends exponentially to zero for chaotic orbits and unstable periodic orbits following (5.22)

$$\text{GALI}_k(t) \propto \exp\left\{-[(\lambda_1 - \lambda_2) + (\lambda_1 - \lambda_3) + \cdots + (\lambda_1 - \lambda_k)]\,t\right\},$$

while for regular motion on as sd torus, with $2 \leq s \leq N$, the evolution of the GALI$_k$ is given by (5.34)

$$\text{GALI}_k(t) \propto \begin{cases} \text{constant} & \text{if } 2 \leq k \leq s \\ \frac{1}{t^{k-s}} & \text{if } s < k \leq 2N - s \\ \frac{1}{t^{2(k-N)}} & \text{if } 2N - s < k \leq 2N. \end{cases}$$

The latter formula is quite general as (a) for $s = N$ it provides (5.23), which describes the behavior of the GALI$_k$ for motion on an Nd torus, i.e. the most common situation of regular motion in the $2N$d phase space of the system, (b) for $k = 2$, $s = 1$ and $N = 1$ it gives (5.33), which describes the power law decay of the GALI$_2$ in the case of a 2d map (the GALI$_2$ is the only possible GALI in this case), and (c) for $s = 1$ it becomes (5.37), which provides the power law decay of the GALI_k for stable periodic orbits of Hamiltonian systems (we remind that in the case of stable periodic in maps all the GALIs remain constant (5.38)).

In our presentation, we paid much attention to issues concerning the actual computation of the indices. In Sect. 5.3.1 we explained in detail an efficient way to evaluate the GALI_k, which is based on the SVD procedure (5.20), while in the Appendix we provide pseudo-codes for the computation of the SALI and the GALI. In Sect. 5.3.3.1 we discussed a numerical strategy for the detection of regular motion on low dimensional tori (see Figs. 5.13 and 5.14), while in Sect. 5.3.4 we showed how the evaluation of the GALI for an ensemble of orbits can lead to the location of stable periodic orbits (see Figs. 5.17 and 5.18). In addition, the effect of the choice of the initial deviation vectors on the color plots depicting the global dynamics of a system, was discussed in Sect. 5.4.1.1, where specific strategies to avoid the appearance of spurious structures in these plots were presented (see Fig. 5.19).

One of the main advantages of the SALI and the GALI methods is their ability to discriminate between chaotic and regular motion very efficiently. The GALI_k with $2 \leq k \leq N$ tends exponentially fast to zero for chaotic orbits, while it attains positive values for regular ones. Due to these different behaviors these indices, and in particular the $\text{GALI}_2/\text{SALI}$ and the GALI_N, can reveal even tiny details of the underlying dynamics, if one follows the procedure presented in Sect. 5.4.1.1. Implementing the numerical strategies developed in Sect. 5.4.1.2 we can also use the completely different time rates with which the GALI_k with $N < k \leq 2N$, tends to zero (exponentially fast for chaotic orbits and power law decay for regular ones) in order to study the dynamics globally. Finally, in Sect. 5.4.3 a particular numerical method, the re-initialization of the GALI_k, proved to be the suitable approach to reveal even brief changes in the dynamical nature of orbits in time dependent Hamiltonians.

The SALI and the GALI have already proven their usefulness in chaos studies as their many applications to a variety of dynamical systems show (see Sect. 5.4.2). Nevertheless, several other chaos indicators have been developed over the years. A few, sporadic comparisons between some of these methods have been performed in studies of particular dynamical systems (e.g. [9, 75, 79, 83]). Recently, detailed and systematic comparisons between many chaos indicators based on the evolution of deviation vectors were conducted [33, 59], and the SALI method was added in the software package LP-VIcode [24], which includes several of these indicators. The main outcome of these comparative studies was that the use of more than one chaos indicators is useful, if not imperative, for revealing the dynamics of a system.

Acknowledgements Many of the results described in this chapter were obtained in close collaboration with Prof. T. Bountis, Dr. Ch. Antonopoulos and Dr. E. Gerlach. This work was partially supported by the European Union (European Social Fund - ESF) and Greek national funds through the Operational Program "Education and Lifelong Learning" of the National Strategic Reference Framework (NSRF)—Research Funding Program: 'THALES'. Ch. S. would like to thank the Research Office of the University of Cape Town for the Research Development Grant which funded part of this study, as well as the Max Planck Institute for the Physics of Complex Systems in Dresden for its hospitality during his visit in December 2014–January 2015, when part of this work was carried out. In addition, Ch. S. thanks T. van Heerden for the careful reading of the manuscript and for his valuable comments. We are also grateful to the three anonymous referees whose constructive remarks helped us improve the content and the clarity of the chapter.

Appendix: Pseudo-Codes for the Computation of the SALI and the GALI$_k$

We present here pseudo–codes for the numerical computation of the SALI (Table 5.1) and the GALI$_k$ (Table 5.2) methods, according to the algorithms presented in Sects. 5.2 and 5.3.1 respectively.

Table 5.1 Numerical computation of the SALI

Input:	1. Hamilton equations of motion and variational equations, or equations of the map and of the tangent map.
	2. Initial condition for the orbit $\mathbf{x}(0)$.
	3. Initial *orthonormal* deviation vectors $\mathbf{w}_1(0)$, $\mathbf{w}_2(0)$.
	4. Renormalization time τ.
	5. Maximum time: T_M and small threshold value of the SALI: S_m.
Step 1	**Set** the stopping flag, $SF \leftarrow 0$, the counter, $i \leftarrow 1$, and the orbit characterization variable, $OC \leftarrow$ 'regular'.
Step 2	**While** ($SF = 0$) **Do**
	Evolve the orbit and the deviation vectors from time $t = (i-1)\tau$ to $t = i\tau$, i.e. **Compute** $\mathbf{x}(i\tau)$ and $\mathbf{w}_1(i\tau)$, $\mathbf{w}_2(i\tau)$.
Step 3	**Normalize** the two vectors, i.e. **Set** $\mathbf{w}_1(i\tau) \leftarrow \mathbf{w}_1(i\tau)/\|\mathbf{w}_1(i\tau)\|$ and $\mathbf{w}_2(i\tau) \leftarrow \mathbf{w}_2(i\tau)/\|\mathbf{w}_2(i\tau)\|$.
Step 4	**Compute** and **Store** the current value of the SALI: $\mathrm{SALI}(i\tau) = \min\{\|\mathbf{w}_1(i\tau) + \mathbf{w}_2(i\tau)\|, \|\mathbf{w}_1(i\tau) - \mathbf{w}_2(i\tau)\|\}$.
Step 5	**Set** the counter $i \leftarrow i + 1$.
Step 6	**If** $[\mathrm{SALI}((i-1)\tau) < S_m]$ **Then** **Set** $SF \leftarrow 1$ and $OC \leftarrow$ 'chaotic'. **End If**
Step 7	**If** $[(i\tau > T_M)]$ **Then** **Set** $SF \leftarrow 1$. **End If** **End While**
Step 8	**Report** the time evolution of the SALI and the nature of the orbit.

The algorithm for the computation of the SALI according to Eq. (5.8). The program computes the evolution of the SALI with respect to time t up to a given upper value of time $t = T_M$ or until the index becomes smaller than a low threshold value S_m. In the latter case the studied orbit is considered to be chaotic

Table 5.2 Numerical computation of the GALI$_k$

Input:	1. Hamilton equations of motion and variational equations, or equations of the map and of the tangent map.
	2. Order k of the desired GALI.
	3. Initial condition for the orbit $\mathbf{x}(0)$.
	4. Initial *orthonormal* deviation vectors $\mathbf{w}_1(0)$, $\mathbf{w}_2(0)$, ..., $\mathbf{w}_k(0)$.
	5. Renormalization time τ.
	6. Maximum time: T_M and small threshold value of the GALI: G_m.
Step 1	**Set** the stopping flag, $SF \leftarrow 0$, the counter, $i \leftarrow 1$, and the orbit characterization variable, $OC \leftarrow$ 'regular'.
Step 2	**While** $(SF = 0)$ **Do**
	Evolve the orbit and the deviation vectors from time $t = (i-1)\tau$ to $t = i\tau$, i.e. **Compute** $\mathbf{x}(i\tau)$ and $\mathbf{w}_1(i\tau)$, $\mathbf{w}_2(i\tau)$, ..., $\mathbf{w}_k(i\tau)$.
Step 3	**Normalize** the vectors:
	Do for $j = 1$ to k
	Set $\mathbf{w}_j(i\tau) \leftarrow \mathbf{w}_j(i\tau)/\|\mathbf{w}_j(i\tau)\|$.
	End Do
Step 4	**Compute** and **Store** the current value of the GALI$_k$:
	Create matrix $\mathbf{A}(i\tau)$ having as rows the deviation vectors $\mathbf{w}_1(i\tau)$, $\mathbf{w}_2(i\tau)$, ..., $\mathbf{w}_k(i\tau)$.
	Compute the singular values $z_1(i\tau)$, $z_2(i\tau)$, ..., $z_k(i\tau)$ of matrix $\mathbf{A}^T(i\tau)$ by applying the SVD algorithm.
	GALI$_k(i\tau) = \prod_{j=1}^{k} z_j(i\tau)$.
Step 5	**Set** the counter $i \leftarrow i + 1$.
Step 6	**If** $[\text{GALI}_k((i-1)\tau) < G_m]$ **Then**
	Set $SF \leftarrow 1$ and $OC \leftarrow$ 'chaotic'.
	End If
Step 7	**If** $[(i\tau > T_M)]$ **Then**
	Set $SF \leftarrow 1$.
	End If
	End While
Step 8	**Report** the time evolution of the GALI$_k$ and the nature of the orbit.

The algorithm for the computation of the GALI$_k$ according to Eq. (5.20). The program computes the evolution of the GALI$_k$ with respect to time t up to a given upper value of time $t = T_M$ or until the index becomes smaller than a low threshold value G_m. In the latter case the studied orbit is considered to be chaotic

References

1. Antonopoulos, Ch., Bountis, T.: Detecting order and chaos by the Linear Dependence Index (LDI) method. ROMAI J. **2**(2), 1–13 (2006)
2. Antonopoulos, Ch., Christodoulidi, H.: Weak chaos detection in the Fermi-Pasta-Ulam-α system using q-Gaussian statistics. Int. J. Bifurcat. Chaos **21**, 2285 (2011)
3. Antonopoulos, Ch., Manos, A., Skokos, Ch.: SALI: an efficient indicator of chaos with application to 2 and 3 degrees of freedom Hamiltonian systems. In: Tsahalis, D.T. (ed.) From Scientific Computing to Computational Engineering. Proceedings of the 1st International Conference, vol. III, pp. 1082–1088. Patras University Press, Patras (2005)
4. Antonopoulos, Ch., Bountis, T., Skokos, Ch.: Chaotic dynamics of N-degree of freedom Hamiltonian systems. Int. J. Bifurcat. Chaos **16**, 1777–1793 (2006)
5. Antonopoulos, Ch., Basios, V., Bountis, T.: Weak chaos and the 'Melting Transition' in a confined microplasma system. Phys. Rev. E **81**, 016211 (2010)
6. Antonopoulos, C., Basios, V., Demongeot, J, Nardone, P., Thomas, R.: Linear and nonlinear arabesques: a study of closed chains of negative 2-element circuits. Int. J. Bifurcat. Chaos **23**, 1330033 (2013)
7. Bario, R.: Sensitivity tools vs. Poincaré sections. Chaos Solitons Fractals **25**, 711–726 (2005)
8. Bario, R.: Painting chaos: a gallery of sensitivity plots of classical problems. Int. J. Bifurcat. Chaos **16**, 2777–2798 (2006)
9. Barrio, R., Borczyk, W., Breiter, S.: Spurious structures in chaos indicators maps. Chaos Solitons Fractals **40**, 1697–1714 (2009)
10. Benettin, G., Galgani, L.: Lyapunov characteristic exponents and stochasticity. In: Laval, G., Grésillon, D. (eds.) Intrinsic Stochasticity in Plasmas, pp. 93–114. Edit. Phys., Orsay (1979)
11. Benettin, G., Galgani, L., Strelcyn, J.M.: Kolmogorov entropy and numerical experiments. Phys. Rev. A **14**, 2338–2344 (1976)
12. Benettin, G., Galgani, L., Giorgilli, A., Strelcyn, J.M.: Tous les nombres caractéristiques sont effectivement calculables. C. R. Acad. Sc. Paris Sér. A **286**, 431–433 (1978)
13. Benettin, G., Froeschlé, C., Scheidecker, J.P.: Kolmogorov entropy of a dynamical system with an increasing number of degrees of freedom. Phys. Rev. A **19**, 2454–2460 (1979)
14. Benettin, G., Galgani, L., Giorgilli, A., Strelcyn, J.M.: Lyapunov characteristic exponents for smooth dynamical systems and for Hamiltonian systems; A method for computing all of them. Part 1: theory. Meccanica (March) 9–20 (1980)
15. Benettin, G., Galgani, L., Giorgilli, A., Strelcyn, J.M.: Lyapunov characteristic exponents for smooth dynamical systems and for Hamiltonian systems; A method for computing all of them. Part 2: Numerical application. Meccanica (March) 21–30 (1980)
16. Boreux, J., Carletti, T., Skokos, Ch., Vittot, M.: Hamiltonian control used to improve the beam stability in particle accelerator models. Commun. Nonlinear Sci. Numer. Simul. **17**, 1725–1738 (2012)
17. Boreux, J., Carletti, T., Skokos, Ch., Papaphilippou, Y., Vittot, M.: Efficient control of accelerator maps. Int. J. Bifurcat. Chaos **22**(9), 1250219 (2012)
18. Bountis, T., Papadakis, K.E.: The stability of vertical motion in the N-body circular Sitnikov problem. Celest. Mech. Dyn. Astron. **104**, 205–225 (2009)
19. Bountis, T., Skokos, Ch.: Application of the SALI chaos detection method to accelerator mappings. Nucl. Instrum. Methods Phys. Res. A **561**, 173–179 (2006)
20. Bountis, T.C., Skokos, Ch.: Complex Hamiltonian Dynamics. Springer, Berlin (2012)
21. Bountis, T., Manos, T., Christodoulidi, H.: Application of the GALI Method to localization dynamics in nonlinear systems. J. Comp. Appl. Math. **227**, 17–26 (2009)
22. Broucke, R.A.: Periodic orbits in the elliptic restricted three–body problem. NASA, Jet Propulsion Laboratory, Tech. Rep. 32-1360 (1969)
23. Capuzzo-Dolcetta, R., Leccese, L., Merritt, D., Vicari, A.: Self-consistent models of cuspy triaxial galaxies with dark matter haloes. Astrophys. J. **666**, 165–180 (2007)

24. Carpintero, D.D., Maffione, N., Darriba, L.: LP–VIcode: a program to compute a suite of variational chaos indicators. Astron. Comput. **5**, 19–27 (2014)
25. Carpintero, D.D., Muzzio, J.C., Navone, H.D.: Models of cuspy triaxial stellar systems –III. The effect of velocity anisotropy on chaoticity. Mon. Not. R. Astron. Soc. **438**, 2871–2881 (2014)
26. Casati, G., Chirikov, B.V., Ford, J.: Marginal local instability of quasi-periodic motion. Phys. Lett. A **77**, 91–94 (1980)
27. Christodoulidi, H., Bountis, T.: Low-dimensional quasiperiodic motion in Hamiltonian systems. ROMAI J. **2**(2), 37–44 (2006)
28. Christodoulidi, H., Efthymiopoulos, Ch.: Low-dimensional q-tori in FPU lattices: dynamics and localization properties. Physica D **261**, 92 (2013)
29. Christodoulidi, H., Efthymiopoulos, Ch., Bountis, T.: Energy localization on q-tori, long-term stability, and the interpretation of Fermi-Pasta-Ulam recurrences. Phys. Rev. E **81**, 016210 (2010)
30. Cincotta, P.M., Efthymiopoulos, C., Giordano, C.M., Mestre, M.F.: Chirikov and Nekhoroshev diffusion estimates: bridging the two sides of the river. Physica D **266**, 49–64 (2014)
31. Contopoulos, G.: Order and Chaos in Dynamical Astronomy. Springer, Berlin, Heidelberg (2002)
32. Contopoulos, G., Galgani, L., Giorgilli, A.: On the number of isolating integrals in Hamiltonian systems. Phys. Rev. A **18**, 1183–1189 (1978)
33. Darriba, L.A., Maffione, N.P., Cincotta, P.M., Giordano, C.M.: Comparative study of variational chaos indicators and ODEs' numerical integrators. Int. J. Bifurcat. Chaos **22**, 1230033 (2012)
34. Dullin, H.R., Meiss, J.D., Sterling, D.: Generic twistless bifurcations. Nonlinearity **13**, 203–224 (2000)
35. Faranda, D., Mestre, M.F., Turchetti, G.: Analysis of round off errors with reversibility test as a dynamical indicator. Int. J. Bifurcat. Chaos **22**, 1250215 (2012)
36. Fermi, E., Pasta, J., Ulam, S.: Studies of nonlinear problems. I. Los Alamos Rep LA-1940 (1955)
37. Froeschlé, C., Gonczi, R., Lega, E.: The fast Lyapunov indicator: a simple tool to detect weak chaos. Application to the structure of the main asteroidal belt. Planet. Space Sci. **45**, 881–886 (1997)
38. Gerlach, E., Eggl, S., Skokos, Ch.: Efficient integration of the variational equations of multi–dimensional Hamiltonian systems: application to the Fermi–Pasta–Ulam lattice. Int. J. Bifurcat. Chaos **22**, 1250216 (2012)
39. Gottwald, G.A., Melbourne, I.: A new test for chaos in deterministic systems. Proc. R. Soc. Lond. A **460**, 603–611 (2004)
40. Hadjidemetriou, J.: The stability of periodic orbits in the three-body problem. Celest. Mech. **12**, 255–276 (1975)
41. Haken, H.: At least one Lyapunov exponent vanishes if the trajectory of an attractor does not contain a fixed point. Phys. Lett. A **94**, 71–72 (1983)
42. Harsoula, M., Kalapotharakos, C.: Orbital structure in N-body models of barred-spiral galaxies. Mon. Not. R. Astron. Soc. **394**, 1605–1619 (2009)
43. Hénon, M., Heiles, C.: The applicability of the third integral of motion: Some numerical experiments. Astron. J. **69**, 73–79 (1964)
44. Howard, J.E., Dullin, H.R.: Linear stability of natural symplectic maps. Phys. Lett. A **246**, 273–283 (1998)
45. Howard, J.E., MacKay, R.S.: Linear stability of symplectic maps. J. Math. Phys. **28**, 1036–1051 (1987)
46. Huang, G., Cao, Z.: Numerical analysis and circuit realization of the modified LÜ chaotic system. Syst. Sci. Control Eng. **2**, 74–79 (2014)
47. Huang, G-Q, Wu, X.: Analysis of Permanent-Magnet Synchronous Motor Chaos System. Lecture Notes in Computer Science, vol. 7002, pp. 257–263. Springer, Berlin, Heidelberg (2011)

48. Huang, G.Q., Wu, X.: Analysis of new four-dimensional chaotic circuits with experimental and numerical methods. Int. J. Bifurcat. Chaos **22**, 1250042 (2012)
49. Huang, G., Zhou, Y.: Circuit simulation of the modified Lorenz system. J. Inf. Comput. Sci. **10**, 4763–4772 (2013)
50. Kalapotharakos, C.: The rate of secular evolution in elliptical galaxies with central masses. Mon. Not. R. Astron. Soc. **389**, 1709–1721 (2008)
51. Kantz, H., Grassberger, P.: Internal Arnold diffusion and chaos thresholds in coupled symplectic maps. J. Phys. A **21**, L127–133 (1988)
52. Kyriakopoulos, N., Koukouloyannis, V., Skokos, Ch., Kevrekidis, P.: Chaotic behavior of three interacting vortices in a confined Bose-Einstein condensate. Chaos **24**, 024410 (2014)
53. Lichtenberg, A.J., Lieberman, M.A.: Regular and Chaotic Dynamics, 2nd edn. Springer, Berlin (1992)
54. Lukes-Gerakopoulos, G.: Adjusting chaotic indicators to curved spacetimes. Phys. Rev. D **89**, 043002 (2014)
55. Lyapunov, A.M.: The general problem of the stability of motion. Taylor and Francis, London (1992) (English translation from the French: Liapounoff, A.: Problème général de la stabilité du mouvement. Annal. Fac. Sci. Toulouse **9**, 203–474 (1907). The French text was reprinted in Annals Math. Studies vol. 17. Princeton University Press (1947). The original was published in Russian by the Mathematical Society of Kharkov in 1892)
56. Macek, M., Stránský, P., Cejnar, P., Heinze, S., Jolie, J., Dobeš, J.: Classical and quantum properties of the semiregular arc inside the Casten triangle. Phys. Rev. C **75**, 064318 (2007)
57. Macek, M, Dobeš, J., Cejnar, P.: Occurrence of high-lying rotational bands in the interacting boson model. Phys. Rev. C **82**, 014308 (2010)
58. Macek, M, Dobeš, J., Stránský, P., Cejnar, P.: Regularity-induced separation of intrinsic and collective dynamics. Phys. Rev. Lett. **105**, 072503 (2010)
59. Maffione, N.P., Darriba, L.A., Cincotta, P.M., Giordano, C.M.: A comparison of different indicators of chaos based on the deviation vectors: Application to symplectic mappings. Celest. Mech. Dyn. Astron. **111**, 285–307 (2011)
60. Manos, T., Athanassoula, E.: Regular and chaotic orbits in barred galaxies - I. Applying the SALI/GALI method to explore their distribution in several models. Mon. Not. R. Astron. Soc. **415**, 629–642 (2011)
61. Manos, T., Machado, R.E.G.: Chaos and dynamical trends in barred galaxies: bridging the gap between N-body simulations and time-dependent analytical models. Mon. Not. R. Astron. Soc. **438**, 2201–2217 (2014)
62. Manos, T., Robnik, M.: Survey on the role of accelerator modes for the anomalous diffusion: The case of the standard map. Phys. Rev. E **89**, 022905 (2014)
63. Manos, T., Ruffo, S.: Scaling with system size of the Lyapunov exponents for the Hamiltonian mean field model. Transp. Theory Stat. Phys. **40**, 360–381 (2011)
64. Manos, T., Skokos, Ch., Bountis, T.: Application of the Generalized Alignment Index (GALI) method to the dynamics of multi-dimensional symplectic maps. In: Chandre C., Leoncini, X., Zaslavsky, G. (eds.) Chaos, Complexity and Transport: Theory and Applications. Proceedings of the CCT 07, pp. 356–364. World Scientific, Singapore (2008)
65. Manos, T., Skokos, Ch., Athanassoula, E., Bountis, T.: Studying the global dynamics of conservative dynamical systems using the SALI chaos detection method. Nonlinear Phenomen. Complex Syst. **11**(2), 171–176 (2008)
66. Manos, T., Skokos, Ch., Bountis, T.: Global dynamics of coupled standard maps. In: Contopoulos, G., Patsis, P.A. (eds.) Chaos in Astronomy, Astrophysics and Space Science Proceedings, pp. 367–371. Springer, Berlin, Heidelberg (2009)
67. Manos, T., Skokos, Ch., Antonopoulos, Ch.: Probing the local dynamics of periodic orbits by the generalized alignment index (GALI) method. Int. J. Bifurcat. Chaos **22**, 1250218 (2012)
68. Manos, T., Bountis, T., Skokos, Ch.: Interplay between chaotic and regular motion in a time-dependent barred galaxy model. J. Phys. A Math. Theor. **46**, 254017 (2013)
69. Nagashima, T., Shimada, I.: On the C-system-like property of the Lorenz system. Prog. Theor. Phys. **58**, 1318–1320 (1977)

70. Oseledec, V.I.: A multiplicative ergodic theorem. Ljapunov characteristic numbers for dynamical systems. Trans. Moscow Math. Soc. **19**, 197–231 (1968)
71. Paleari, S., Penati, T.: Numerical Methods and Results in the FPU Problem. Lecture Notes in Physics, vol. 728, pp. 239–282. Springer, Berlin (2008)
72. Panagopoulos, P., Bountis, T.C., Skokos, Ch.: Existence and stability of localized oscillations in 1-dimensional lattices with soft spring and hard spring potentials. J. Vib. Acoust. **126**, 520–527 (2004)
73. Petalas,, Y.G., Antonopoulos, C.G., Bountis, T.C., Vrahatis, M.N.: Evolutionary methods for the approximation of the stability domain and frequency optimization of conservative maps. Int. J. Bifurcat. Chaos **18**, 2249–2264 (2008)
74. Press, W.H., Teukolsky, S.A., Vetterling, W.T., Flannery, B.P.: Numerical Recipes in Fortran 77, 2nd edn. The Art of Scientific Computing. Cambridge University Press, Cambridge (1992)
75. Racoveanu, O.: Comparison of chaos detection methods in the circular restricted three-body problem. Astron. Nachr. **335**, 877–885 (2014)
76. Saha, L.M., Sahni, N.: Chaotic evaluations in a modified coupled logistic type predator-prey model. Appl. Math. Sci. **6**(139), 6927–6942 (2012)
77. Sándor, Zs., Érdi, B., Széll, A., Funk, B.: The relative Lyapunov indicator: an efficient method of chaos detection. Celest. Mech. Dyn. Astron. **90**, 127–138 (2004)
78. Shimada, I., Nagashima, T.: A numerical approach to ergodic problem of dissipative dynamical systems. Prog. Theor. Phys. **61**, 1605–1615 (1979)
79. Skokos, Ch.: Alignment indices: a new, simple method for determining the ordered or chaotic nature of orbits. J. Phys. A **34**, 10029–10043 (2001)
80. Skokos, Ch.: On the stability of periodic orbits of high dimensional autonomous Hamiltonian systems. Physica D **159**, 155–179 (2001)
81. Skokos, Ch.: The Lyapunov Characteristic Exponents and their Computation. Lecture Notes in Physics, vol. 790, pp. 63–135. Springer, Berlin, Heidelberg (2010)
82. Skokos, Ch., Antonopoulos, Ch., Bountis, T.C., Vrahatis, M.N.: How does the smaller alignment index (SALI) distinguish order from chaos? Prog. Theor. Phys. Suppl. **150**, 439–443 (2003)
83. Skokos, Ch., Antonopoulos, Ch., Bountis, T.C., Vrahatis, M.N.: Detecting order and chaos in Hamiltonian systems by the SALI method. J. Phys. A **37**, 6269–6284 (2004)
84. Skokos, Ch., Bountis, T.C., Antonopoulos, Ch.: Geometrical properties of local dynamics in Hamiltonian systems: the generalized alignment index (GALI) method. Physica D **231**, 30–54 (2007)
85. Skokos, Ch., Bountis, T.C., Antonopoulos, Ch.: Detecting chaos, determining the dimensions of tori and predicting slow diffusion in Fermi-Pasta-Ulam lattices by the generalized alignment index method. Eur. Phys. J. Spec. Top. **165**, 5–14 (2008)
86. Stránský, P., Cejnar, P., Macek, M.: Order and chaos in the Geometric Collective Model. Phys. At. Nucl. **70**(9), 1572–1576 (2007)
87. Stránský, P., Hruška, P., Cejnar, P.: Quantum chaos in the nuclear collective model: classical-quantum correspondence. Phys. Rev. E **79**, 046202 (2009)
88. Soulis, P., Bountis, T., Dvorak, R.: Stability of motion in the Sitnikov 3-body problem. Celest. Mech. Dyn. Astron. **99**, 129–148 (2007)
89. Soulis, P.S., Papadakis, K.E., Bountis, T.: Periodic orbits and bifurcations in the Sitnikov four-body problem. Celest. Mech. Dyn. Astron. **100**, 251–266 (2008)
90. Széll, A., Érdi, B., Sándor, Z., Steves, B.: Chaotic and stable behavior in the Caledonian Symmetric Four-Body problem. Mon. Not. R. Astron. Soc. **347**, 380–388 (2004)
91. Voglis, N., Contopoulos, G., Efthymiopoulos, C.: Method for distinguishing between ordered and chaotic orbits in four-dimensional maps. Phys. Rev. E **57**, 372–377 (1998)
92. Voglis, N., Contopoulos, G., Efthymiopoulos, C.: Detection of ordered and chaotic motion using the dynamical spectra. Celest. Mech. Dyn. Astron. **73**, 211–220 (1999)
93. Voglis, N., Harsoula, M., Contopoulos, G.: Orbital structure in barred galaxies. Mon. Not. R. Astron. Soc. **381**, 757–770 (2007)

94. Voyatzis, G.: Chaos, order, and periodic orbits in 3:1 resonant planetary dynamics. Astrophys. J. **675**, 802–816 (2008)
95. Wolf, A., Swift, J.B., Swinney, H.L., Vastano, J.A.: Determining Lyapunov exponents from a time series. Physica D **16**, 285–317 (1985)
96. Zotos, E.E.: Classifying orbits in galaxy models with a prolate or an oblate dark matter halo component. Astron. Astrophys. **563**, A19 (2014)
97. Zotos, E.E., Caranicolas, N.D.: Order and chaos in a new 3D dynamical model describing motion in non-axially symmetric galaxies. Nonlinear Dyn. **74**, 1203–1221 (2013)

Chapter 6
The Relative Lyapunov Indicators: Theory and Application to Dynamical Astronomy

Zsolt Sándor and Nicolás Maffione

Abstract A recently introduced chaos detection method, the Relative Lyapunov Indicator (RLI) is investigated in the cases of symplectic mappings and continuous Hamiltonian systems. It is shown that the RLI is an efficient numerical tool in determining the true nature of individual orbits, and in separating ordered and chaotic regions of the phase space of dynamical systems. A comparison between the RLI and some other variational indicators are presented, as well as the recent applications of the RLI to various problems of dynamical astronomy.

6.1 Introduction

One of the most important questions of investigating a dynamical system with $i > 1$ degrees of freedom is to identify the ordered or chaotic behaviour of its orbits. If the dynamical system is governed by ordered orbits its time evolution is predictable. On the contrary, if the dynamical system evolves through chaotic orbits, its long-term behaviour cannot be predicted. In this paper we are considering a special class of dynamical systems called Hamiltonian systems. The phase space of a Hamiltonian-system usually contains both regions for ordered (predictable) and chaotic (unpredictable) motion, therefore the informations about the locations and extent of these regions are of high interest in investigating the evolution of such systems. A typical class of Hamiltonian systems are the planetary systems such as the Solar System, or extrasolar planetary systems.

Z. Sándor (✉)
Computational Astrophysics Group, Konkoly Observatory of the Hungarian Academy of Sciences, Budapest, Hungary
e-mail: zssandor@konkoly.hu

N. Maffione
Grupo de Caos en Sistemas Hamiltonianos, Facultad de Ciencias Astronómicas y Geofísicas, Universidad Nacional de la Plata, Buenos Aires, Argentina
e-mail: nmaffione@fcaglp.unlp.edu.ar

© Springer-Verlag Berlin Heidelberg 2016
Ch. Skokos et al. (eds.), *Chaos Detection and Predictability*, Lecture Notes in Physics 915, DOI 10.1007/978-3-662-48410-4_6

The ordered behaviour of an orbit or trajectory[1] is strongly related to its stability. By the term *stability* we mean that the trajectories are located in a bounded region of the phase space. If the region of chaotic motion is not bounded in the phase space, the trajectories could leave that domain through chaotic diffusion. In this case the chaotic trajectories become *unstable*. Thus one way to perform stability investigations of dynamical systems is the detection of ordered or chaotic behaviour of their orbits. The stability of the planets or asteroids in the Solar System is an outstanding question of dynamical astronomy. The ongoing discovery of exoplanetary systems made the stability investigations of planetary systems even more important.

In recent years there has been a growing interest in development and application of different chaos detection methods. Beside the "traditional" tools such as the largest Lyapunov Characteristic Exponent (LCE) or Lyapunov Characteristic Number (LCN; [5]) and the frequency analysis [26], several new methods have been developed, which can be used to detect the ordered and chaotic nature of individual orbits, or to separate regions of ordered and chaotic motions in the phase space of a dynamical system. These methods are the Fast Lyapunov Indicator (FLI; [17], [20]), the Orthogonal Fast Lyapunov Indicator (OFLI; [16]), the Mean Exponential Growth factor of Nearby Orbits (MEGNO; [9, 10]), the Spectral Distance (SD; [49]), the Smaller ALignment Index (SALI; [40, 42]), the Generalized ALignment Index (GALI; [43, 44]), and finally, the Relative Lyapunov Indicator (RLI; [36, 37]), whose analysis is the main scope of this paper. We note that all of these methods are based on the time evolution of the infinitesimally small tangent vector to the orbit, which is provided by the variational equations. Thus these chaos detection methods can be classified as *variational* indicators.

In what follows, after recalling the definition of the RLI, we present its behaviour in symplectic mappings and continuous Hamiltonian systems. The efficiency of the RLI is presented by a comparative study with the already mentioned variational indicators. This paper closes with a chapter presenting the recent applications of the RLI.

6.1.1 *Definition of the RLI*

The ordered or the chaotic nature of a trajectory can be characterized most precisely by the calculation of the LCE:

$$L^1(\mathbf{x}^*) = \lim_{t \to \infty} \frac{1}{t} \log \frac{||\boldsymbol{\xi}(t)||}{||\boldsymbol{\xi}_0||} \, ,$$

---

[1]In the case of continuous dynamical systems the *trajectory* is a continuous curve in the phase space given by the points representing the time evolution of an initial state. In discrete dynamical systems the set of the discrete points representing the time evolution of the system is called as *orbit*.

where $\mathbf{x}^* \in \mathbb{R}^n$ is the initial state of the system, or in other words, the starting point of the trajectory, and $\boldsymbol{\xi}(t)$ is the image of an initial infinitesimally small deviation vector $\boldsymbol{\xi}_0$ between two nearby trajectories after time t. The time-evolution of $\boldsymbol{\xi}$ is given by the equations of motions and their linearized equations:

$$\frac{d\mathbf{x}(t)}{dt} = f[\mathbf{x}(t)] , \qquad \frac{d\boldsymbol{\xi}(t)}{dt} = Df[\mathbf{x}(t)]\boldsymbol{\xi} ,$$

where $Df[\mathbf{x}(t)]$ is the Jacobian matrix of the function $\mathbf{f} : \mathbb{R}^n \rightarrow \mathbb{R}^n$ evaluated at $\mathbf{x}(t) \in \mathbb{R}^n$, and $\mathbf{x} : \mathbb{R} \rightarrow \mathbb{R}^n$, $\boldsymbol{\xi} : \mathbb{R} \rightarrow \mathbb{R}^n$ are vector-valued functions, too. In Hamiltonian systems if $L^1(\mathbf{x}^*) = 0$, the orbit emanating from the initial state $\mathbf{x}^*$ is ordered, if $L^1(\mathbf{x}^*) > 0$ it is chaotic. A serious disadvantage of the calculation of the LCE is that it can be obtained as a limit, thus its value can only be extrapolated, which makes the identification of weakly chaotic orbits unreliable.

In practice, one calculates only the finite-time approximation of the LCE, called the finite-time Lyapunov Indicator (LI):

$$L(\mathbf{x}, t) = \frac{1}{t} \log \frac{||\boldsymbol{\xi}(t)||}{||\boldsymbol{\xi}_0||} .$$

It is obvious that by calculating the LI for short time, the true nature of individual orbits cannot be identified. However, the basic features of the phase space of a system (the existence and approximate position of regular regions and extended chaotic domains) can be discovered very quickly by calculating a large number of LIs for short time. Let $\mathbf{x}$ be a vector variable taken along a line, which is going through both regular and chaotic regions of the phase space of a dynamical system. Then by fixing the integration time t_{int}, one can calculate the curve $L(\mathbf{x}, t_{\text{int}})$. In the case of regular regions (KAM tori, islands of stability) the curve $L(\mathbf{x}, t_{\text{int}})$ varies smoothly, while in the case of an extended chaotic region it shows large fluctuations [11]. However, in the case of weak chaos the fluctuations of the curve $L(\mathbf{x}, t_{\text{int}})$ are not large enough to decide the true nature of the investigated region. In order to measure the fluctuations of the curve of the finite-time LI at $\mathbf{x}^*$, we introduce the quantity:

$$\Delta L(\mathbf{x}^*, t) = |L(\mathbf{x}^* + \Delta\mathbf{x}^*, t) - L(\mathbf{x}^*, t)| , \qquad (6.1)$$

which is the difference between the finite-time LI of two neighbouring orbits, and $\Delta\mathbf{x}^*$ is the distance between the two initial condition vectors. This quantity has been introduced and called RLI in [36, 37]. Definition (6.1) contains $\Delta\mathbf{x}^*$ as a free parameter, which should be chosen small enough to reflect the local properties of the phase space. In our numerical investigations we have experienced that the arbitrary choice of $\Delta\mathbf{x}^*$ in a quite large interval $||\Delta\mathbf{x}^*|| \in [10^{-14}, 10^{-7}]$ does not modify essentially the behaviour of the RLI as a function of the time. For ordered orbits the RLI shows linear dependence on $||\Delta\mathbf{x}^*||$, while for chaotic orbits the RLI practically is invariant with respect to the choice of $||\Delta\mathbf{x}^*||$.

Although there is not developed a strict mathematical theory describing the time behaviour of the RLI so far, the results of numerical simulations clearly show its power in separating ordered and chaotic orbits. An intuitive explanation could be that in the case of ordered orbits the time evolution of the two LI curves ($L(\mathbf{x}^*, t)$ and $L(\mathbf{x}^* + \Delta\mathbf{x}^*, t)$) practically cannot be distinguished meaning that they converge with the same (or very similar) rate to the LCE $= 0$ limit. On the other hand, the convergence rate of the LI of two close chaotic orbits (separated in the phase space by the vector $\Delta\mathbf{x}^*$) could be very different, which is reflected in the time evolution of the RLI. In the next sections of the paper we shall investigate through extensive numerical experiments the completely different behaviour of the RLI as a function of time in the cases of ordered and chaotic orbits, which makes it a suitable tool of chaos detection.

6.1.2 Properties of the RLI in Chaos Detection

In order to eliminate the high frequency fluctuations of the curve $\Delta L(\mathbf{x}, t)$ for a fixed $\mathbf{x} \in \mathbb{R}^n$, we suggested the following smoothing

$$\langle \Delta L(\mathbf{x}) \rangle (t) = \frac{1}{t} \sum_{i=1}^{[t/\Delta t]} \Delta L(\mathbf{x}, i \cdot \Delta t) \,,$$

where Δt is the stepsize. In numerical experiments we always use the above smoothed value of the RLI.

The different behaviour of the RLI for ordered and chaotic orbits are first presented for discrete Hamiltonian systems, such as the following 2D:

$$\begin{cases} x_1' = x_1 + x_2 \\ x_2' = x_2 - v \cdot \sin(x_1 + x_2) \quad \mathrm{mod}\ (2\pi) \,, \end{cases} \tag{6.2}$$

and 4D symplectic mapping:

$$\begin{cases} x_1' = x_1 + x_2 \\ x_2' = x_2 - v \cdot \sin(x_1 + x_2) - \mu \cdot [1 - \cos(x_1 + x_2 + x_3 + x_4)] \\ x_3' = x_3 + x_4 \\ x_4' = x_4 - \kappa \cdot \sin(x_3 + x_4) - \mu \cdot [1 - \cos(x_1 + x_2 + x_3 + x_4)] \quad \mathrm{mod}\ (2\pi) \,, \end{cases} \tag{6.3}$$

where v and κ are the non-linearity parameters, and μ is the coupling parameter of the 4D mapping.

In the case of the 2D symplectic mapping (6.2) the initial conditions of the ordered orbit are $x_1 = 2$, $x_2 = 0$, while the initial conditions of the chaotic orbit are $x_1 = 3$, $x_2 = 0$. In both cases $v = 0.5$. The phase plots of these orbits are

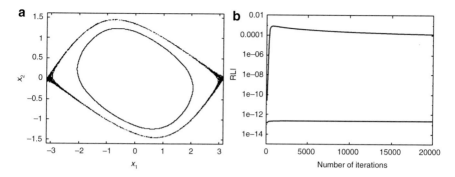

Fig. 6.1 (**a**) (*left*) The phase plot of an ordered and a chaotic orbit in the mapping (6.2); (**b**) (*right*) the behaviour of the RLI as the function of time for a chaotic orbit (*upper curve*) and for an ordered orbit (*lower curve*)

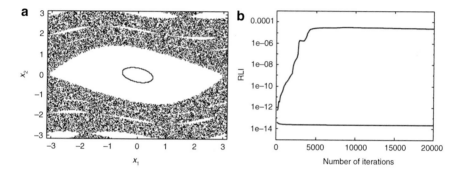

Fig. 6.2 (**a**) (*left*) The phase plot of an ordered and a chaotic orbit in the mapping (6.3); (**b**) (*right*) the behaviour of the RLI as the function of time for a chaotic orbit (*upper curve*) and for an ordered orbit (*lower curve*)

shown in Fig. 6.1a and the corresponding time behaviour of the RLI is displayed in Fig. 6.1b. In the case of the 4D mapping (6.3) the following initial conditions are used: $x_1 = 0.5$, $x_2 = 0$, $x_3 = 0.5$, $x_4 = 0$ for the ordered, and $x_1 = 3$, $x_2 = 0$, $x_3 = 0.5$, $x_4 = 0$ for the chaotic orbit. In both cases the parameters are $\nu = 0.5$, $\kappa = 0.1$ and $\mu = 0.001$. The projections of these orbits onto the $x_1 - x_2$ plane are shown in Fig. 6.2a. The behaviour of the RLI as the functions of time of the ordered and chaotic orbits are plotted in Fig. 6.2b. Studying Figs. 6.1b and 6.2b one can see that the RLI for an ordered orbit is almost constant. The RLI of a chaotic orbit grows very rapidly, and after reaching a maximum value it decreases very slowly.

The maximum value of the RLI of a chaotic orbit is much higher (in the examples shown by 9–10 orders of magnitude) than the almost constant value of the RLI of an ordered orbit. It can be seen that by using the RLI, the ordered or chaotic nature of orbits can be identified after a few hundred iterations of the investigated mapping.

A crucial test for a chaos detection method is whether it separates the weakly chaotic orbits (also called "sticky" orbits) from the ordered orbits. In order to demonstrate this property of the RLI we used the 4D symplectic mapping and initial conditions for an ordered and a weakly chaotic orbit as [40, 49]:

$$\begin{cases} x_1' = x_1 + x_2' \\ x_2' = x_2 + (K/2\pi) \sin(2\pi x_1) - (\beta/\pi) \sin[2\pi (x_3 - x_1)] \\ x_3' = x_3 + x_4' \\ x_4' = x_4 + (K/2\pi) \sin(2\pi x_3) - (\beta/\pi) \sin[2\pi (x_1 - x_3)] \mod (1), \end{cases} \tag{6.4}$$

where K is the non-linearity and β is the coupling parameter. According to [49] the orbit with the parameters $K = 3$, $\beta = 0.1$ and with initial conditions $x_1 = 0.55$, $x_2 = 0.1$, $x_3 = 0.62$, $x_4 = 0.2$ is ordered, while the orbit with the same initial conditions and parameter K, but with $\beta = 0.3051$ is slightly chaotic tending to a very small LCE. The projection of the weakly chaotic orbit on the $x_1 - x_2$ plane is shown in Fig. 6.3a, and it seems to be an ordered orbit on a torus. Using the RLI (Fig. 6.3b) one can see that the chaotic nature of this orbit can be detected after about $N \sim 5 \times 10^6$ iterations.

Finally, we should discuss the role of the initial separation between the two orbits, which is one of the free parameters of the RLI (the other one is the length of the time needed to calculate the RLI, as will be mentioned later on). In what follows, we give an evidence that $||\Delta \mathbf{x}^*||$ can be chosen arbitrarily from a quite large interval $[10^{-14}, 10^{-7}]$. The smallest value of this interval is due to the finite representation of numbers by computers. On the other hand, the largest value of the above interval should also be small enough in order to the RLI reflect the local property of the phase space around the orbit under study.

Figure 6.4 shows the dependence of the RLI (obtained after a fixed number of iterations, which in this particular case was 2×10^4), on $||\Delta \mathbf{x}^*||$ for the 4D ordered orbit of mapping (6.3), shown in Fig. 6.2a. In Fig. 6.4 both the horizontal and vertical axes are scaled logarithmically. Studying Fig. 6.4a one can see that for the ordered orbit the RLI changes linearly with respect to $||\Delta \mathbf{x}^*||$. In Fig. 6.4b we display the

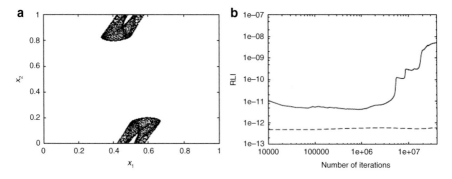

Fig. 6.3 (**a**) (*left*) The phase plot of a weakly chaotic orbit in the mapping (6.4); (**b**) (*right*) the behaviour of the RLI as the function of time for the weakly chaotic orbit seen in the *left panel* (*upper curve*) and for an ordered orbit (*lower curve*)

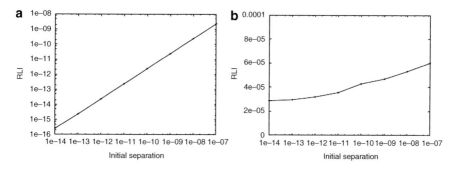

Fig. 6.4 (**a**) (*left*) The linear dependence of the final value of the RLI on the initial separation for an ordered orbit; (**b**) (*right*) the final value of the RLI versus the initial separation for a chaotic orbit

dependence of the RLI (after 2×10^4 iterations) on $||\Delta \mathbf{x}^*||$ for the chaotic 4D orbit shown in Fig. 6.2a. One can see that in the chaotic case the final value of the RLI practically does not depend on the choice of $||\Delta \mathbf{x}^*||$. Thus in the above sense the RLI is invariant with respect to the choice of the initial separation. This invariance property can be explained by the fact that the RLI of a chaotic orbit practically measures the average width of the oscillation of the LI curve (as a function of time), which does not depend heavily on the choice of the initial separation.

6.2 A Short Comparison of the RLI to Other Methods of Chaos Detection

In this section we compare the performances of the RLI with the variational indicators mentioned in Sect. 6.1: the LI, the FLI and the OFLI, the MEGNO, the SD, the SALI and the GALI.

In Sect. 6.2.1 we analyze the dependence of the RLI on its free parameters and in the following one we compare the typical behaviours of the RLI with other techniques in the well-known Hénon–Heiles model [21] (hereinafter HH model). In Sect. 6.2.3 we apply the RLI and other indicators to study dynamical systems of different complexity: the 4D symplectic mapping (6.3) presented in Sect. 6.1.2 and a rather complex 3D potential resembling a Navarro, Frenk and White triaxial halo (hereinafter NFW model; [48]). We compare the phase space portraits given by the RLI and the other methods to decide whether the results are comparable. In Sect. 6.2.4 we briefly discuss the dependence of the RLI on the computing times.

The Bulirsch–Stoer integrator is used throughout this section.

6.2.1 The Dependence of the RLI on the Free Parameters

The RLI has two free parameters: (a) the initial separation (Δx^*) between the basis orbit and its "shadow" (Sect. 6.1.1) and (b) its threshold (the threshold is a value that separates chaotic from regular motion and it is related with the length of the time needed to calculate the RLI). These free parameters are "user-choice" quantities. Thus, it is of interest to study the dependence of the RLI on both of them.

The following experiments are undertaken on the HH model:

$$\mathcal{H} = \frac{1}{2} \left(p_x^2 + p_y^2 \right) + \frac{1}{2} \left(x^2 + y^2 \right) + x^2 \cdot y - \frac{1}{3} y^3 ,$$

where $\mathcal{H}$ is the Hamiltonian and x, y, p_x, p_y are the usual phase space variables.

6.2.1.1 The Initial Separation Parameter

The dependence of the RLI on the initial separation parameter is strongly related to the type of orbit under study (see Sect. 6.1.1). Therefore, we take four different types of orbits with initial conditions located on the line defined by $x = p_y = 0$ and $y \in [-0.1 : 0.1]$ and the energy surface $E = 0.118$, namely, a regular orbit close to a stable periodic orbit (r-sp); a quasiperiodic orbit (r-qp); a regular orbit close to an unstable periodic orbit (r-up); and a chaotic orbit inside a stochastic layer (c-sl). The initial conditions are taken from [10]. The integration time is 10^4 units of time (hereinafter u.t.), which is enough time to provide a reliable characterization of the orbits and the stepsize of the numerical integration is 0.01. We note that these values have been used in the following numerical experiments, too.

In Fig. 6.5a, we present the final values (i.e. the values of the indicator at the end of the integration time) of the RLI as a function of the initial separation parameter[2] for the orbits introduced earlier. We show that the initial separation parameter does not significantly affect the RLI when we apply the indicator to the chaotic orbit "c-sl", but it does when we apply it to the regular orbits "r-sp", "r-qp" and "r-up" (and confirming the results shown in Sect. 6.1.2). This dependence of the RLI on its free parameter has severe implications in the selection of the threshold. For instance, if we start the computation with an initial separation of 10^{-14}, the relation shown in Fig. 6.5a will indicate that a good candidate for the threshold to distinguish between the chaotic orbit "c-sl" (RLI~ 0.1) and the regular orbit "r-sp" (RLI$\sim 10^{-13.5}$) can be 10^{-12}. Then, the orbits with values of the RLI higher than 10^{-12} will be classified as chaotic orbits. However, this choice of the threshold leads to a misclassification of the regular orbits "r-up" (RLI$\sim 10^{-10}$) and "r-qp" (RLI$\sim 10^{-12}$). Furthermore, since the correspondence between the RLI and the initial separation parameter for

---

[2]The values for the parameter have been taken from the interval suggested in Sect. 6.1.2.

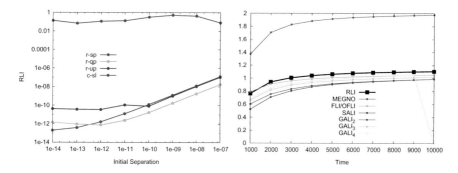

Fig. 6.5 (**a**) (*left*) The RLI values as a function of the initial separation parameter for the orbits "r-sp", "r-qp", "r-up" and "c-sl"; (**b**) (*right*) the normalized approximation rates for several chaos indicators, including the RLI (see text for further details)

the regular orbits of the sample tends to be linear (see Sect. 6.1.2), the threshold 10^{-12} does not work at all for an initial separation greater than 10^{-11}.

In order to determine a reliable threshold for the RLI, the relationship with the initial separation parameter should be done by computing the indicator for a group of orbits known to be regular but with some level of instability (e.g. regular orbits close to a hyperbolic object such as an unstable periodic orbit, see [27]).

6.2.1.2 The Threshold

In this section we investigate how the RLI and other indicators (listed above) depend on their thresholds. For the following experiment in the HH model we have adopted a sample of 125751 initial conditions in the region defined by $x = 0$, $y \in [-0.1 : 0.1]$, $p_y \in [-0.05 : 0.05]$ and on the energy surface defined by $E = 0.118$. The thresholds of the LI, the RLI, the MEGNO, the SALI, the FLI/OFLI and the GALIs are shown in Table 6.1, where t is the time (see [13]). From here, the initial separation parameter will be 10^{-12}, unless stated otherwise. The threshold used for the RLI has been computed following the remarks discussed in the previous section.

To proceed with the experiment we define the approximation rate as the rate of convergence with a final percentage of chaotic orbits. This rate will show a combination of the reliability of the indicator and the accuracy of the selected threshold if the final percentage of chaotic orbits approaches the "true percentage" of chaotic orbits in the system. Therefore, as we require a reliable final percentage of chaotic orbits, we consider the percentage of chaotic orbits given by the LI by 10^4 u.t.: ∼39.92 %, the "true percentage" of chaotic orbits in the system. Both the overwhelming number of papers claiming the reliability of the LI as a chaos indicator and the experimental evidence showing that 10^4 u.t. seems to be a reliable convergent time for all the indicators in the experiment (see, for instance, Sect. 6.2.2,

Table 6.1 Thresholds for several indicators, including the RLI

Indicator	Threshold
LI	$ln(t)/t$
RLI	10^{-10}
MEGNO	2
SALI	10^{-4}
FLI/OFLI	t
GALI$_2$	t^{-1}
GALI$_3$	t^{-2}
GALI$_4$	t^{-4}

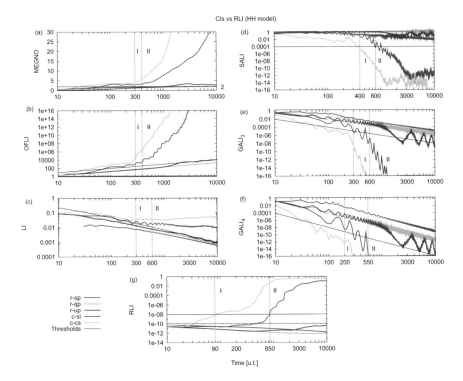

Fig. 6.6 Behaviours of (**a**) the MEGNO, (**b**) the OFLI, (**c**) the LI, (**d**) the SALI, (**e**) the GALI$_3$, (**f**) the GALI$_4$ and (**g**) the RLI for orbits "r-sp", "r-qp", "r-up", "c-sl" and "c-cs". The thresholds as well as the lines "I" and "II" are included (see text for further details)

Fig. 6.6) supporting this statement. Thus, we normalize the time evolution of the percentage of chaotic orbits given by the methods with this "true percentage". Hence, the values of the normalized approximation rates higher than 1 show percentages of chaotic orbits higher than the "true percentage".

We test the reliability of the thresholds given in Table 6.1 according to the above mentioned rates. The results are indicated in Fig. 6.5b. The convergence towards a constant rate of the RLI and the FLI/OFLI is faster than that of the other indicators of the sample. Despite this rapid tendency towards a constant value for both indicators, the final percentages of chaotic orbits given by the RLI and the FLI/OFLI are higher than that of the LI, which means that the values of the rates are above 1. Nevertheless, this slight difference in the final percentages of chaotic orbits can always be fixed with a small adjustment of the corresponding thresholds. Since the results for the MEGNO show substantial disagreement between the percentages of the chaotic orbits given by the indicator and the LI, a significant empirical adjustment of the MEGNO's threshold should be made to avoid an overestimation in the number of chaotic orbits. The final percentages given by the SALI and the GALIs are in perfect agreement with the "true percentage". However, their tendency towards a stable percentage of chaotic orbits is slower than the one showed by the RLI or the FLI/OFLI.

Thus, among the above mentioned CIs, the RLI and the FLI/OFLI show the best approximation rates, i.e. the best combination of the reliability of the indicators and the accuracy of their thresholds. For further details on the experiment, refer to [13].

By 10^4 u.t. the threshold taken for the $GALI_4$ reaches the computer's precision (10^{-16}) and thus, every chaotic orbit lies beyond such precision. Therefore, the last point for $GALI_4$ in Fig. 6.5b falls apart from the tendency.

Now that we have finished studying the importance of a wise selection of the free parameters of the RLI, we calibrated the indicator following the suggestions mentioned here and continue comparing its performance with other indicators.

6.2.2 Expected Behaviour of the Indicators in the HH Model

In this experiment, done in the HH model, our goal is to compare the typical behaviours of the RLI to the other techniques and show its advantages and disadvantages.

We take the orbits of Sect. 6.2.1.1 and a chaotic orbit inside the chaotic sea (c-cs) with initial conditions on the line defined by $x = p_y = 0$ and $y \in [-0.1 : 0.1]$ and the energy surface $E = 0.118$ (the initial conditions are taken from [10]). The final integration time is 10^4 u.t. and the stepsize is 0.01.

Figure 6.6 shows the time evolution curves of several indicators for the five different types of orbits introduced at the beginning of the section. Some of the main features of a chaos indicator are the speed of convergence and the resolving power. The former is the time it takes to distinguish between chaotic and regular motion. In order to visualize this quantity, in Fig. 6.6 we introduce the vertical lines "I" and "II". The first one shows the time after which the orbit "c-cs" is clearly identified as a chaotic orbit with the indicator and the second one plays the same role as "I" for the orbit "c-sl".

On panel (a) we present the typical behaviours of the MEGNO (see [9]). The values of the indicator for the three regular orbits tend towards the theoretical asymptotic threshold, 2, in different ways (see the right bottom of panel a). The values for the chaotic orbits increase linearly with time. At ~300 u.t. (see line "I"), the separation between orbit "c-cs" and the threshold line is already significant. Hence, orbit "c-cs" is clearly identified as a chaotic orbit then. Only 100 u.t. later (line "II") the same happens with orbit "c-sl".

On panel (b) we show the time evolution curves of the OFLI [16]. The values of the indicator for two of the regular orbits increase linearly with time (see the threshold and its expression in Table 6.1) while the values for the chaotic orbits increase exponentially fast with time. This distinction between both tendencies can be made at the same times that have been shown by the MEGNO. Besides, orbit "r-sp" has an almost constant value because it is very close to a stable periodic orbit.

On panel (c) we present the LI (see e.g. [41]). The values of the indicator for the three regular orbits decrease with time (see the threshold and its expression in Table 6.1) while the values for the chaotic orbits tend towards a constant value which depends on the chaoticity of the orbit. The distinction between the regular orbits and the chaotic orbit "c-sl" is made by the LI later than by the previous indicators: the orbit leaves the linear tendency of the threshold around $\sim$600 u.t. (line "II").

On panel (d) we present the time evolution curves of the SALI [40]. The values of the indicator for the regular orbits "r-qp" and "r-up" oscillate within the interval (0,2), while the orbit "r-sp" tends towards 0 following a power law behaviour. The chaotic orbits decrease exponentially fast with time. The time needed for the SALI to clearly identify the chaotic orbits is the time used by the chaotic orbits to reach the threshold (see its value in Table 6.1 and lines "I" and "II" in the figure to locate the times). The indicator delays making this distinction (in fact, it does so later than the LI) because for smaller values of the integration time, the chaotic orbits decrease with a power law as the regular orbit "r-sp" does.

On panels (e) and (f) we show the time evolution curves of the GALI$_3$ and the GALI$_4$, respectively (the GALI$_2$ and the SALI have almost identical behaviours and the former is not included). Their theoretical thresholds (Table 6.1; see [8, 29, 43, 44] for further details) yield good estimations of the time needed for the indicators to distinguish the chaotic orbits from the regular orbits. The GALI$_3$ makes this distinction in the same time as the LI did (see lines "I" and "II"). Once again, the reason for this delay is that the GALI$_3$ decreases with a power law for regular orbits as well as for chaotic orbits at the beginning of the integration interval. Nevertheless, the higher the order of the GALI, the faster its tendency towards 0 for the chaotic orbits. Then, the GALI$_4$ has registered the best time so far to distinguish the chaotic orbit "c-cs": $\sim$200 u.t.

On panel (g) we present the time evolution curves of the RLI (Sect. 6.1.2). The values of the indicator for the three regular orbits are in the interval $(10^{-12}, 10^{-10})$ according to the initial separation of 10^{-12} (see Fig. 6.5a, in Sect. 6.2.1.1). Thus, as the value 10^{-10} (depicted with a dotted line in the figure) that have been selected in Sect. 6.2.1.2 is in the limit of the interval, it is not reliable as a threshold any more. Therefore, we selected the value 10^{-8} for the threshold. The characterization of the regular orbits does not clearly differentiate among them as the MEGNO, the OFLI or the SALI. The values for the chaotic orbits increase with time until they reach a constant value. On the one hand, orbit "c-cs" is clearly identified as a chaotic orbit by $\sim$80 u.t. (line "I") when the orbit reaches the threshold. This is the fastest characterization of the chaotic orbit "c-cs". On the other hand, the RLI identifies orbit "c-sl" as a chaotic orbit around $\sim$850 u.t. (line "II"), which is the slowest characterization.

All the indicators delay in making a reliable characterization of the chaotic orbit "c-sl", which shows that the chaotic orbit "c-cs" has a larger LI than orbit "c-sl".

Finally, the characterization of the five representative orbits made by the RLI as well as its speed of convergence is similar to the other techniques. Thus, the RLI is most welcome to the group of fast variational indicators.

6.2.3 Performances of the Indicators Under Different Scenarios

We have seen in the previous section how similar are the performances of the RLI and the other fast indicators in the rather simple HH model. Here we will focus on experiments in scenarios that are different from the HH model to determine whether the RLI is a reliable technique for studying different or more complex systems than the HH model.

6.2.3.1 The 4D Symplectic Mapping

The time evolution curves of the indicators (used in Sect. 6.2.2) are not efficient to analyze a large number of orbits. The appropriate way to gather information in these cases is in terms of the final values of the methods. Thus, let us now turn our attention to the study of the resolving power of the techniques using their final values.

The following study will be conducted in the 4D mapping (6.3) presented in Sect. 6.1.2 by adopting different samples of initial conditions and 10^5 iterations. The version of the MEGNO considered here is the MEGNO(2,0), whose threshold value is 0.5 (see [10]).

The large number of iterations used in the experiments deserves a further explanation. In Fig. 6.7, we present the RLI mappings for 10^3 (left panel), 5×10^3 (middle panel) and 10^4 (right panel) iterations. The RLI mapping corresponding to 10^3 iterations presents a very noisy phase space portrait probably due to a combination of a poor election of the initial deviation vectors (see for instance [1, 2]) and the short number of iterations. It is also clear from the figure that the phase space portrait presents a stable picture after 5×10^3 iterations. Furthermore, increasing the number of iterations helps to resolve very sticky orbits but no further advantage is observed. Thus, we iterate the map 10^5 times in order to distinguish the most sticky regions.

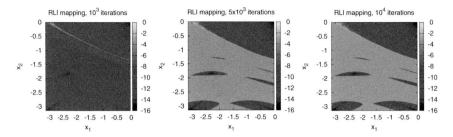

Fig. 6.7 The RLI mapping on gray-scale plots composed of 10^6 initial conditions, for 10^3 (*left*), 5×10^3 (*middle*) and 10^4 (*right*) iterations

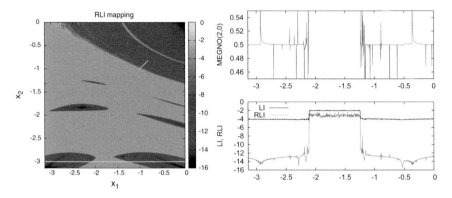

Fig. 6.8 (**a**) (*left*) The RLI mapping on gray-scale plot composed of 10^6 initial conditions, for 10^5 iterations (in logarithmic scale); (**b**) (*right*) the MEGNO(2,0), the LI and the RLI final values for 10^3 initial conditions along the line $x_2 = -3$, for 10^5 iterations (the LI and the RLI final values are in logarithmic scale)

Initial Conditions Inside High-Order Resonances

In Fig. 6.8a we present the RLI mapping for a region of the phase space corresponding to the 4D mapping and is composed of 10^6 initial conditions. The main resonance as well as the high-order resonances are clearly depicted in dark gray (i.e. small values of the RLI) while the stochastic layers inside the main stability island and the chaotic sea are depicted in light gray (large values of the RLI). We show an horizontal line of initial conditions ($x_1 \in [-\pi, 0]$, $x_2 = -3$, $x_3 = 0.5$ and $x_4 = 0$) used in the following experiment to compare the performances of the RLI with the mostly used variational indicator, the LI, and with the MEGNO(2,0), which is faster than the LI and which is also a reliable indicator. With a diagonal segment we depict the initial conditions ($x_1 = x_2 \in [-1.03, -0.8]$, $x_3 = 0.5$ and $x_4 = 0$) used in the experiment of next section.

In Fig. 6.8b, we compare the performances of the RLI, the LI and the MEGNO(2,0) on 10^3 equidistant initial conditions lying on the horizontal line that crosses the high-order resonances in Fig. 6.8a. This figure clearly shows that the RLI unzips the hidden structure inside the high-order resonances better than the LI. Furthermore, the RLI and the MEGNO(2,0) reveal similar structures (the Y-range of the MEGNO(2,0) has been centered on the threshold and shortened to amplify the details of the revealed structure). For further discussions on the experiment, refer to [27].

On behalf of the previous experiments the RLI is not only more reliable than the LI to reveal small scale structures, but also an accurate indicator to describe a large array of initial conditions.

Sticky Orbits

Sticky orbits are the most difficult type of orbit to characterize by a variational indicator. Thus, we further analyze the identification of this type of orbits (Sect. 6.1.2) to study the performance of the RLI.

In Fig. 6.9, we show the sticky region enclosed in the interval $(-1.03, -0.8)$ (and depicted earlier in Fig. 6.8a with a diagonal segment) in terms of the final values of the same indicators previously used. We also point out the final values of three representative orbits, two chaotic orbits (one of them which is sticky chaotic) and a regular orbit. The thresholds are also included.

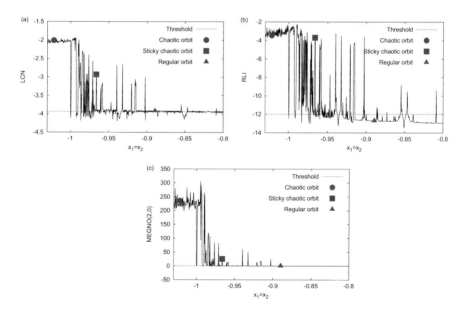

Fig. 6.9 Sticky region inside the interval $(-1.03, -0.8)$ for 10^5 iterations and for three indicators: (**a**) the LI, (**b**) the RLI and (**c**) the MEGNO(2,0). The representative orbits are depicted with points of different types. The thresholds are depicted with a *dashed line* (the LI and the RLI final values are in logarithmic scale). See text for details. The figures were taken from [27] and slightly modified

In Fig. 6.9a, the LI perfectly identifies the three representative orbits and their domains. In Fig. 6.9b, some sticky and chaotic orbits share the same RLI final values ($\sim 10^{-3.5}$) which hide the different levels of chaoticity (such is the case of the representative chaotic and sticky chaotic orbits, both of them have similar RLI values). This is not the case for the other indicator shown in the figure: the MEGNO(2,0) on panel (c). The MEGNO(2,0) has completely different final values for the sticky and the chaotic orbits.

The RLI shows a reliable performance revealing the global characteristics of the region, such as the regular domain (where we find the representative regular orbit) and some small high-order resonances (e.g. $x_1 = x_2 \sim -0.85$). Nevertheless, it does not distinguish the sticky from the chaotic orbit in the experiment (see [27] for further details).

In the following section, we follow with the experiments in a scenario of completely different nature and much more complex than the HH model or the 4D mapping.

6.2.3.2 The 3D NFW Model

According to the current paradigm of a hierarchically clustering universe, large galaxies formed through the accretion and mergers of smaller objects. The imprints of such events should be well preserved in the outer stellar haloes where dynamical mixing processes are not significant in relatively short times (for instance, many stellar streams have been identified in the outer regions of the Milky Way [4, 22, 28]). Furthermore, this galaxy formation paradigm predicts that the centres of the accreted component of stellar haloes should contain the oldest products of accretion events [12] in the formation history of the galaxy (such as the Milky Way), and therefore, this substructure might be hidden in its inner regions (e.g. close to the Solar neighbourhood). However, in order to study the phase space portraits of stars in small volumes in the inner regions of the stellar haloes to quantify and classify the substructure, we need a model of the Dark Matter (DM) halo that hosts the galaxy.

In [33, 34] the authors introduced a universal density profile for DM haloes (i.e. haloes with masses ranging from dwarf galaxy haloes to those of major clusters): the NFW profile. However, in Cold Dark Matter (CDM) cosmologies DM haloes are not spherical. Furthermore, numerical simulations suggest that their shape vary with radius. In [48] the authors have built a triaxial extension of the NFW profile, resembling a triaxial DM halo; the corresponding potential is the so-called NFW model. Therefore, the NFW model is a triaxial potential used in galactic dynamics associated with equilibrium density profiles of DM haloes in CDM cosmologies.

In a forthcoming paper we study the phase space portraits of stellar particles inside solar neighbourhood-like volumes to gain insights about the formation history of Milky Way-like galaxies. Hereinafter, we use some results from that investigation to demonstrate the reliability of the RLI on a rather complex model.

The solar neighbourhood volume is a sphere of 2.5 Kpc of radius located at 8 Kpc from the center of the NFW model (denoted by Φ_N in the following equation):

$$\Phi_N = -\frac{A}{r_p} \ln\left(1 + \frac{r_p}{r_s}\right) ,$$

where A is the constant:

$$A = \frac{G M_{200}}{\ln\left(1 + C_{NFW}\right) - C_{NFW}/\left(1 + C_{NFW}\right)} ,$$

with G, the gravitational constant, M_{200}, the virial mass of the DM halo and C_{NFW}, the concentration parameter used to describe the shape of the density profile. Besides, r_p follows the relation:

$$r_p = \frac{(r_s + r)r_e}{r_s + r_e} ,$$

where r_s is the scale radius defined by dividing the virial radius of the DM halo by C_{NFW}. The scale radius represents a transition scale between an ellipsoidal and a near spherical shape of the Φ_N. The r_e is an ellipsoidal radius:

$$r_e = \sqrt{\left(\frac{x}{a}\right)^2 + \left(\frac{y}{b}\right)^2 + \left(\frac{z}{c}\right)^2} ,$$

where b/a and c/a are the principal axial ratios with a the major axis and where we require $a^2 + b^2 + c^2 = 3$ (see [48]). The values of the constants used in the following experiment are taken at redshift $z = 0$ and listed in Table 6.2 (see [19] and references therein for further details on the model).

In order to begin the study of the (6 dimensional) regions of interest in the phase space of the NFW model, we needed to restrain some of the variables that defined the original sample of 22,500 initial conditions. We fixed the positions of the particles to the centre of the solar neighbourhood. Then, the stellar particles had the following positions at the beginning of the simulation: $x = 8$, $y = 0$ and $z = 0$ (in [Kpc]). The initial velocity in the polar axis (v_z) was restrained to the value -250 in [km s^{-1}]. The energy (E) was restrained to the interval (E_{mb}, E_{lb}) with $E_{mb} \sim -195{,}433$ the energy of the most bound particle, and $E_{lb} \sim -59{,}293$ (in [M$_\odot$ Kpc^2 Gyr^{-2}] with [M$_\odot$] the mass in solar mass units) the energy of the

Table 6.2 Constants used for the Φ_N potential

A	4158670.1856267899
r_s	19.044494521343964
a	1.3258820840000000
b	0.86264540200000000
c	0.70560584600000000

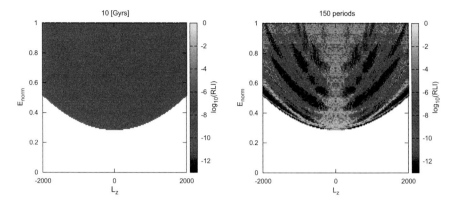

Fig. 6.10 Gray-scale plots of the RLI mapping for the velocity surface $v_z = -250$. (*left*) For a fixed integration time of 10 Gyrs. (*right*) For a fixed number (150) of radial periods. The values of the indicator are in logarithmic scale

least bound particle of the sample. The angular momentum (L_z) was restrained to the interval $(-2000, 2000)$ in [Kpc km s^{-1}]. We integrated the initial conditions for different integration times (1 u.t. corresponds to $\sim$1 Gyr).

In Fig. 6.10, we present the phase space portraits given by the RLI for two different choices of the integration times. On the left panel, we integrate the orbits for a fixed time interval of the order of the Hubble time, i.e. 10 Gyrs. On the right panel, we integrate the orbits for a fixed number of radial periods. The radial period of the stellar orbits in galactic potentials such as the NFW model scales as $\sim E^{-3/2}$. Then, we integrate the orbits for 150 radial periods in order to have a stable portrait of the phase space. It is evident that 10 Gyrs (left panel) is not enough to classify properly the orbits with the RLI (or any other indicator). Then, on the right panel the time interval used was $[\sim 57,750]$ Gyrs where 750 Gyrs is enough time to set reliable values of the RLI for the least bound particle of the sample. However, the most sticky regions are not clearly depicted yet. Therefore, in the following experiment we choose to scale the integration time linearly with the energy of the orbit. The linear relation between the computed integration times and the energy overestimates the former for the most bound particles. Indeed, the time interval used for the experiment was $[\sim 204,750]$ Gyrs where the integration times are clearly larger than those applied with the $\sim E^{-3/2}$ scale. The larger integration times given by the linear scale improve the identification of the most sticky regions which helps to evaluate the performance of the indicators.

On the left panels of Figs. 6.11 and 6.12 we present the gray-scale plots of the final values of four different chaos indicators in the (E_{norm}, L_z) plane[3]: the RLI and the MEGNO; panels (a) and (b1) of Fig. 6.11, respectively; the OFLI and the 1/GALI$_3$: panels (c1) and (d1) of Fig. 6.12, respectively. The GALI$_3$ is inverted in

---

[3]The E_{norm} is the normalized energy: $(E - E_{mb}) / (E_{lb} - E_{mb})$.

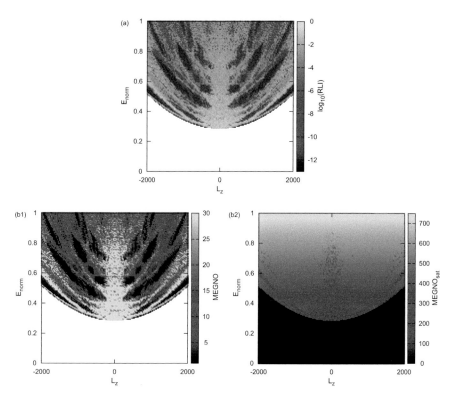

Fig. 6.11 Gray-scale plots for the velocity surface $v_z = -250$. (**a**) The RLI mapping (the values of the indicator are in logarithmic scale), (**b1**) the MEGNO mapping and (**b2**) the MEGNO$_{sat}$ mapping

the gray-scale plots to make the comparison of the portraits with the other indicators easier.

In many situations it is useful to define a saturation value by which the chaos indicator "saturates". For instance, the OFLI and the GALI$_3$ for chaotic motion increases or decreases exponentially fast, respectively. Then, if the chaotic nature of an orbit is well characterized by the OFLI when the indicator reaches 10^{16} or by the GALI$_3$ when it reaches the computer's precision (10^{-16}), the computation should be stopped. Hence, the values 10^{16} and 10^{-16} can be used as saturation values for the OFLI and the GALI$_3$, respectively. Another example is the MEGNO: the MEGNO has an asymptotic value for regular orbits, 2 (see Table 6.1), and increases linearly for chaotic orbits. Then, if the MEGNO reaches the value 30, the orbit is undoubtfully chaotic and it is worthless to continue the computation of the indicator. Then, the value 30 can be used as a saturation value for the MEGNO. The time of saturation, that is the time by which the indicator saturates, it is a quantity useful in recovering the chaoticity levels in the chaotic domain: the smaller the value of the time of saturation, the more chaotic the orbit (see [27, 43]). Finally, if the indicator

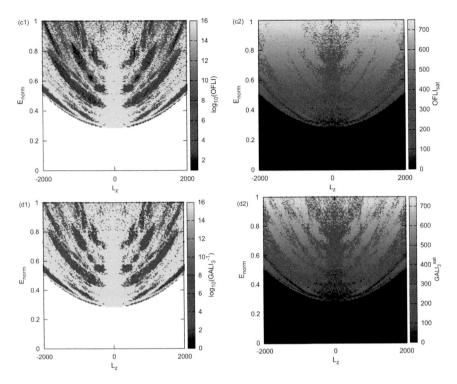

Fig. 6.12 Gray-scale plots for the velocity surface $v_z = -250$. (**c1**) The OFLI mapping and (**c2**) the OFLI$_{sat}$ mapping, (**d1**) the GALI$_3$ mapping and (**d2**) the GALI$_3^{sat}$ mapping (the final values of the indicators are in logarithmic scale)

saturates (i.e. the indicator reaches its saturation value), the integration times will be the times of saturation, but, if the indicator does not saturate, the integration times will be the final integration times.

On the right panels of Figs. 6.11 and 6.12 we present gray-scale plots of the integration times used for three of the four indicators above mentioned. On panel (b2) in Fig. 6.11 we present the integration times for the MEGNO or MEGNO$_{sat}$ and in Fig. 6.12: panels (c2) and (d2), the integration times for the OFLI and the GALI$_3$, or OFLI$_{sat}$ and GALI$_3^{sat}$, respectively.

The discussion below is not intended to analyze the dynamics of the system, which is the aim of a future work. Our goal here is to demonstrate that the performance of the RLI in this complex system is as good as the performances of the other three wide–spread indicators.

On the left panels of Figs. 6.11 and 6.12, we can clearly see the great level of coincidence among the phase space portraits of the four indicators. The regular component is composed of symmetrical structures around the L_z axis and the four indicators represent these structures with very similar shapes, sizes and shades of gray. This shows that the indicators do not only agree in the location, extension and shape of the domains of regular motion, but also in the description of these domains.

The chaotic domain is described equivalently by the four indicators and their corresponding times of saturation. For instance, the RLI (Fig. 6.11a) shows that the chaoticity (light gray) is inversely proportional to E_{norm} and does not depend on the L_z. That is, more bound orbits are more chaotic orbits. We arrive at the same (trivial) conclusion with the information given by the integration times of the other three indicators (Figs. 6.11b2 and 6.12c2, d2). The MEGNO (Fig. 6.11b1) shows an uniform and almost white color for the chaotic domain. This implies that the saturation value (30) has been reached by these chaotic orbits and thus, no further structure is revealed. However, the MEGNO$_{sat}$ (Fig. 6.11b2) shows such structure (surrounded by orbits that did not saturate).[4] This structure revealed by the MEGNO$_{sat}$ shows that the times of saturation are directly proportional to E_{norm} or, once again, that chaoticity is inversely proportional to E_{norm} and also does not depend on the L_z. Similar conclusions can be drawn from Fig. 6.12 for the OFLI (panel c1) and the OFLI$_{sat}$ (panel c2) and the GALI$_3$ (panel d1) and GALI$_3^{sat}$ (panel d2). However, the region composed of orbits that have reached the associated saturation values (10^{16} and 10^{-16} for the OFLI and the GALI$_3$, respectively) within the interval of integration is now much extended. On the one hand, this region in Fig. 6.12c1,d1 is depicted in an uniform and almost white color and thus, the structure cannot be revealed. On the other hand, the times of saturation in Fig. 6.12c2,d2 fulfill the missing information.

In the next experiment, we follow two orbits in the NFW model for a time–span of 1000 Gyrs (i.e. $\sim$77 Hubble times) in order to have convergent final values of all the indicators in the study. We compute the time evolution curves of several indicators, including the RLI, and present the results for a chaotic and a regular orbit ("cha" and "reg", respectively) in Fig. 6.13. In order to distinguish efficiently the chaotic orbit from the regular one, we proceed as in Sect. 6.2.2 and use the same thresholds used there for the MEGNO, the OFLI, the LI, the SALI and the RLI. The threshold for the GALI$_3$ in the NFW model will be the same constant used for the SALI (see [8]) because the model is a 3 degree of freedom (hereinafter d.o.f.) system. The threshold for the GALI$_5$ will be t^{-4}, with t the time (see [13] for further details). In Fig. 6.13, we mark with the vertical line "I" the time after which the orbit "cha" is clearly identified as a chaotic orbit.

Figure 6.13 shows that the accurate identification of the orbit "cha" as a chaotic orbit by the MEGNO (panel a), the OFLI (panel b), the GALI$_5$ (panel f) and the RLI (panel g) is made within a Hubble time ($\sim$13 Gyrs). The above mentioned indicators show that the orbit will behave as a chaotic orbit within a physical meaningful time–span (i.e. the age of the Universe) which is important to understand the dynamics of a real system like a galaxy.

---

[4]Remember that the integration time varies with the E_{norm}, which explains the transition from dark to light gray in the background of the plots on the left side of Figs. 6.11 and 6.12. Also the time of integration is fixed to 0 where there are not initial conditions.

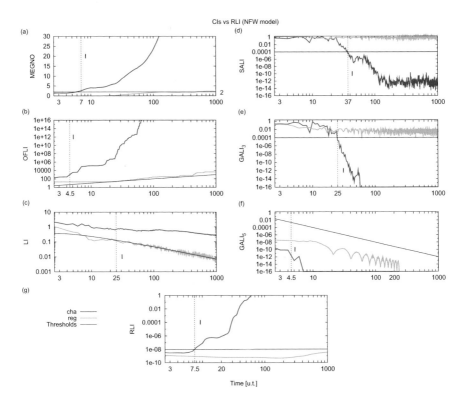

Fig. 6.13 Behaviours of (**a**) the MEGNO, (**b**) the OFLI, (**c**) the LI, (**d**) the SALI, (**e**) the GALI$_3$, (**f**) the GALI$_5$ and (**g**) the RLI for orbits "cha" and "reg". The thresholds as well as the line "I" are included (see text for further details)

In this section, we present results that support the RLI as a reliable indicator. This technique shows a phase space portrait very similar to those shown by the MEGNO, the OFLI or the GALI$_3$. Furthermore, the RLI identified the chaotic nature of the orbit "cha" very fast in the second experiment (the fastest being the OFLI and the GALI$_5$), within a Hubble time. These results put the RLI on an equal footing with the other fast variational indicators.

6.2.4 A Short Discussion on the Computing Times

The computing times of the indicators are specially crucial for time-consuming simulations and their estimation helps for an efficient usage of the computational resources. However, such estimation is not an easy task. The computing times depend on a wide variety of factors such as the complexity of the model and that of the indicator's algorithm, its numerical implementation, the hardware, etc.

Although making a detailed study of the computing times is not our concern (see [13] for further information on the subject), we would like to point out the fact that the easy algorithm of the RLI helps for a rather fast computation of the indicator. Furthermore, we are not dealing with the computing times themselves. We are going to register the ratios between the computing times of the different techniques and the computing time of the LIs[5] for orbits in two of the systems previously studied in the chapter, the HH and the NFW models. If the ratio is above 1, then the computing time of the indicator is larger than the computing time of the LIs.

In this experiment, we used the following hardware/software configuration: an Intel Core i5 with four cores, CPU at 2.67 GHz, 3 GB of RAM, an OS of 32 bits, and the gfortran compiler of gcc version 4.4.4, without any optimizations. The code to compute the indicators is the LP-VIcode, the acronym for "La Plata Variational Indicators code". The alpha version of the LP-VIcode was introduced in [14] and currently, it is a fully operational code that computes a suite of ten variational indicators (see [6]).[6] The initial setup of the LP-VIcode is the following: the integration time is 1000 u.t. (or 1000 Gyrs in the NFW model), the step of integration is 0.001 u.t. (or 1 Myr in the NFW model).

We registered the ratios for two pairs of orbits and for every indicator. Both pairs have one chaotic and one regular orbit. One of the pairs of orbits is located in the HH model and the other in the NFW model. The computing time of the LIs for the HH model (i.e. 4 LIs) is 0m15.204s., and for the NFW model (i.e. 6 LIs) is 3m27.703s. The energy conservation is $\Delta E \sim 10^{-13}$.

The results are shown in Table 6.3. The value "N" is the number of d.o.f. of the system. For instance, in the HH model (second column in Table 6.3), we compute 4 LIs and 3 GALIs: the $GALI_2$, the $GALI_3$ and the $GALI_4$.

The RLI is not the least consuming indicator so far, according to the information given in Table 6.3. The FLI/OFLI, the MEGNO, the SALI and the SD might be more desirable options (in that order). Nevertheless, the easy implementation of the

Table 6.3 Ratios of the computing times for several indicators, including the RLI, for the HH and the NFW models

Indicator(s)	Ratios (HH model)	Ratios (NFW model)
(2N-1) GALIs	∼2.026	∼1.024
(2N) LIs	1	1
RLI	∼0.892	∼0.443
SD	∼0.613	∼0.375
SALI	∼0.582	∼0.372
MEGNO	∼0.53	∼0.174
FLI/OFLI	∼0.432	∼0.151

---

[5]Here, the LIs are the numerical approximations of the spectra of Lyapunov Characteristic Exponents.

[6]Further information on the LP-VIcode can be found at the following url: http://www.fcaglp.unlp.edu.ar/LP-VIcode/.

RLI helps to reduce significantly the computing time of the LIs, and its ratio is not much larger than those of the SD or the SALI.

6.3 Application of the RLI to Planetary Systems

The ongoing discovery of exoplanetary systems is certainly the most rapidly growing field of astronomy. Up to now the number of known planetary systems is almost 200. On the other hand, the masses and orbital elements of the planets detected can be determined only with uncertainties. Therefore it is of high importance to provide stability estimates of these systems. Before the launch of the space missions devoted to detect terrestrial planets (CoRoT, Kepler), only the massive giant planets were discovered mainly by the radial velocity method. One of the major applications of the RLI was to study the stability of the still hypothetical terrestrial planets in planetary systems containing at least one giant planet [38]. Another possible application area of the RLI is to study the stability of different orbital solutions of resonant exoplanetary systems provided by radial velocity observations. In this review we shortly describe the stability studies done for the resonant planetary system HD 73526 [39]. Finally, we present the applicability of the RLI to map the high order resonances in the restricted three-body problem, which might have relevance when studying the behavior of Kuiper belt objects [15]. We note that close relatives to our investigations are the works of [50] computing the stability maps of the system 55 Cancri by using various indicators (LI, SALI and FLI), and of [7] studying the dynamical stability of the Kuiper-Belt using the LI indicator.

In all of the above problems the mean motion resonances (MMR) play an essential role, therefore we shortly summarize their properties. A MMR occurs between two bodies orbiting a more massive body if their orbital periods can approximately be expressed as a ratio of two positive integer numbers, $T_1/T_2 = (p + q)/p$, where T_1 and T_2 denotes the orbital periods of the two bodies, respectively. A MMR can be characterized by studying the behaviour of the resonant angle, which in the model of the restricted three-body problem for an inner MMR is

$$\theta = (p + q)\lambda' - p\lambda - q\varpi, \tag{6.5}$$

while for an outer MMR can be written as

$$\theta = (p + q)\lambda - p\lambda' - q\varpi, \tag{6.6}$$

where λ and ϖ are the mean orbital longitude, and longitude of the pericenter of the massless body, while λ' is the mean orbital longitude of the massive body. If θ librates with a certain amplitude, the two bodies are engulfed in the $(p + q) : p$ MMR. In this way almost the same orbital configuration of the bodies involved in the given MMR is repeated. Depending on the relative positions of the bodies, this configuration can be protective, or can result in unstable orbits, see for more details in [32].

6.3.1 Stability Catalogue of the Habitable Zones of Exoplanetary Systems

The main idea behind the stability catalogue was to map the regions of a planetary system that can host dynamically stable terrestrial planets [38]. The dynamical stability of a terrestrial planet is one of the strongest requirements for a habitable planetary climate. The most important requirement for the *habitability* of a planet is to contain water in liquid phase on its surface. A region around a star, in which an Earth-mass planet could be habitable in the above sense is called as the *habitable zone* (HZ), see [23] and [25] for more details.

6.3.1.1 Used Models and Initial Conditions

The stability of terrestrial planets can be studied by using different approaches: (i) by detecting the stable and unstable regions of the parameter space of each exoplanetary system separately or (ii) by using stability maps computed in advance for a large set of orbital parameters. In the stability catalogue we presented such stability maps also showing how to apply them to the exoplanetary systems under study. This second approach has the advantage that the stability properties of a terrestrial planet can be easily reconsidered when the orbital parameters of the giant planet of an exoplanetary system are modified. This is very often the case, since the orbital parameters of the giant planets are quite uncertain, and due to the accumulation and improvement of the observational data, they are subject to change quite frequently. Instead of the re-exploration of the phase space of each individual exoplanetary system after possible modification of the orbital parameters of the giant planet, the stability properties of the investigated planetary system can be easily re-established from the already existing stability maps. These stability maps, which form a stability catalogue, can also be used to study the stability properties of the habitable zones of known exoplanetary systems.

The majority of planetary systems which are detected so far consists of a star and a giant planet revolving in an eccentric orbit. Therefore, we used a simple dynamical model, the *elliptic restricted three-body problem* in which there are two massive bodies (the primaries) moving in elliptic orbits about their common center of mass, and a third body of negligible mass moving under their gravitational influence (for details see [45]). In our particular case the primaries are the star and the giant planet, and the third body is a small Earth-like planet, being regarded as massless. We note that among the extrasolar planetary systems there is a high rate of multiple planet systems. Thus, a more convenient model for the stability maps would be the restricted N-body problem (with N-2 giant planets, $N \geq 4$). The presence of additional giant planets certainly enhance the instabilities induced by just one massive planet, turning the HZ of the system more unstable. The main source of instabilities are the *mean motion resonances* (MMR) between the massive giant planet and the Earth-like planet. Thus, by mapping these resonances, we can find the possible regions of the instabilities in the HZs. On the other hand, the dynamical model with one giant planet also offers the most convenient way to display the most important MMRs as a function of the mass ratio of the star and the giant planet, and of the eccentricity of the giant planet.

In the catalogue of dynamical stability one important quantity is the mass parameter of the problem $\mu = m_1/(m_0 + m_1)$, where m_0 is the stellar, and m_1 is the planetary mass. The mass parameter has been changed between broad limits $(10^{-4} - 10^{-2})$ with various steps of $\Delta\mu$, in total the different stability maps have been calculated for 23 values of μ. The giant planet was placed around the star in an elliptic orbit, with semi-major axis a_1, eccentricity e_1, argument of periastron ω_1 and mean anomaly M_1. The semimajor axis a_1 was taken as unit distance $a_1 = 1$ during all simulations. The eccentricity e_1 was changed between 0.0 and 0.5 with a stepsize of 5×10^{-3}. The argument of periastron was fixed at $\omega_1 = 0°$, while the mean anomaly M_1 was changed between $0°$ and $360°$ with $\Delta M_1 = 45°$. The test planet was started in the orbital plane of the giant planet with an initial eccentricity $e = 0$, argument of periastron $\omega = 0°$, and mean anomaly $M = 0°$. The semi-major axis a of the test planet was changed in two different intervals: (i) for orbits of the test planet 'inside' the orbit of the giant planet between 0.1 and 0.9 with a stepsize of $\Delta a = 10^{-3}$ and (ii) for 'outside' orbits between 1.1 and 4.0 with a stepsize of $\Delta a = 3.625 \times 10^{-3}$. Further details and the complete catalogue can be found at http://astro.elte.hu/exocatalogue/index.html.

6.3.1.2 Stability Maps

Due to the very good visibility of the outer MMRs, we first display the case when the semi-major axis of the test planet is larger than the giant planet's semi-major axis $(a > 1)$. Additionally, in the stability map shown in Fig. 6.14 the values $\mu = 0.001$ and $M = 0°$ were kept fixed. For each value of the RLI a gray shade has been assigned. White regions correspond to small RLI values, thus they are very stable. The MMRs appear either as light strips in the dark, strongly chaotic regions or as the well-known "V"-shaped structures representing the separatrices between resonant and non-resonant motion. The inner regions of the resonances may be lighter than the lines of the bounding separatrices indicating regular motion in a protective resonance (e.g. the 2:5 MMR). Near the separatrices the motion is always chaotic, moreover at some MMRs even the inner part of the resonance is chaotic as indicated in the case of the 1:3 MMR, for instance. By increasing the giant planet's eccentricity e_1 many resonances overlap giving rise to strongly chaotic and thus very unstable behaviour. The reason of this phenomenon is that by increasing

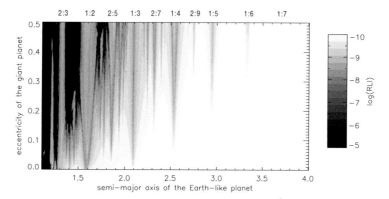

Fig. 6.14 Stability map of the outer MMRs for the Earth-like planets in the elliptic restricted three-body problem for $\mu = 0.001$ and $M = 0°$. *White colour* denotes ordered motion, *light grey strips* and "V" shapes the different resonances, while *black* the strongly chaotic regions

e_1 the apocenter distance of the giant planet also increases, therefore the giant planet perturbs more strongly the outer test planet.

In stability maps displayed in Fig. 6.15 the semi-major axis of the test planet is smaller than that of the giant planet ($a < 1$). The two panels for the mass parameter $\mu = 0.001$ show the stability maps for the two different starting positions of the giant planet, $M_1 = 0°$ (upper panel) and $M_1 = 180°$ (lower panel), respectively. Between the test planet and the giant planet, a large number of inner MMRs can be found, which dominate the stability maps. Inside the resonances the stable or chaotic behaviour of the test planet depends on the initial angular positions of the two planets. This is clearly visible by comparing the two panels. On the other hand the location of the MMRs is not altered, since this depends on the ratio of the semi-major axes of the two planets. In the lower panel of Fig. 6.15 several MMRs (5:2, 5:3, 3:2) are stable, which is not the case in the upper panel of Fig. 6.15. This is due to the fact that the relative initial positions determine the places of conjunctions of the two planets. If they meet regularly near the pericenter of the giant planet, the motion of the test planet becomes chaotic, while it can remain regular if the conjunctions take place near the apocenter of the giant planet. The effect of the initial phase difference between the planets is important, therefore a bunch of stability maps have been prepared for more initial values of the mean anomaly M_1 of the giant planet. These stability maps can be found in the online exocatalogue (http://astro.elte.hu/exocatalogue/index.html). By increasing the value of the mass parameter μ it becomes clearly visible that the larger mass of the giant planet results in stronger perturbations, and therefore more enhanced chaotic region.

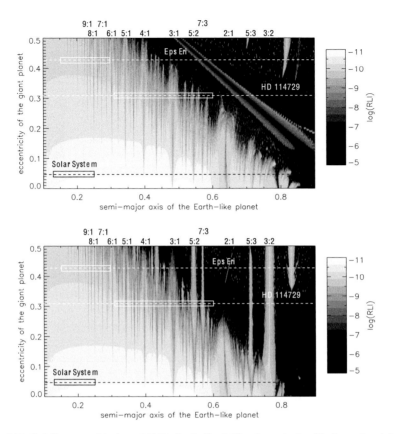

Fig. 6.15 Stability map of the inner MMRs for the Earth-like planets in the elliptic restricted three-body problem for $\mu = 0.001$, $M = 0°$ (*upper panel*), and $M = 180°$ (*lower panel*), respectively. We note the different character of the 5:2, 7:3, and 3:2 MMRs depending on the orbital positions between the test particle, and the perturbing body. *White colour* denotes ordered motion, *light grey strips* and "V" shapes the different resonances, while *black* the strongly chaotic regions. We displayed the properly scaled HZs of three exoplanetary systems in the stability maps. Studying the figures one can conclude that the HZ of the Solar system is stable, the HZs of ϵ Eridani and HD 114729 are *marginally stable* meaning that they are filled with several MMRs

6.3.1.3 Stability of Terrestrial Planets in the Habitable Zones

In this section, we show how to use the catalogue to determine the stability of hypothetical Earth-like planets in exoplanetary systems. As an example, we consider the case of HD10697, where $a_1 = 2.13$ AU, $e_1 = 0.11$ and $\mu = 0.0055$. Figure 6.16 shows a stability map, calculated for $\mu = 0.005$ for inner orbits of the test planet. This corresponds to the minimum mass of the giant planet (minimum masses are used throughout). The stability of a small planet (starting with $e = 0$) in the system HD10697 can be studied along the line $e_1 = 0.11$. One can see that for small semimajor axes, $a < 0.33a_1 = 0.729$ AU the parameter space is very stable.

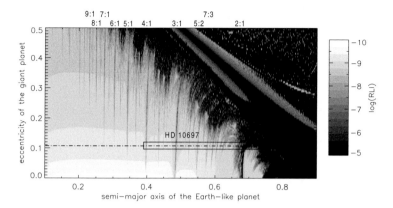

Fig. 6.16 Stability map for inner orbits of an Earth-like planet, when $\mu = 0.005$ and $M_1 = 0°$. The line $e_1 = 0.11$, corresponds to the system HD 10697. The scaled HZ of this system is between $0.39 < a < 0.77$. Its inner part is stable (containing only a few weakly chaotic MMRs), while the outer part is in the strongly chaotic regions

When $a > 0.33a_1$ several resonances appear, among which the most important are the 5:1, 4:1, 3:1 and 2:1 MMRs. For $a > 0.73a_1 = 1.55$ AU, a strongly chaotic region appears. The classical HZ of this system is between 0.85 and 1.65 AU therefore in Fig. 6.16 the scaled classical HZ is located between $0.85/a_1 = 0.39$ and $1.65/a_1 = 0.77$ (shown as a rectangle, elongated in horizontal direction). One can see that the inner part of the classical HZ contains ordered regions, but stripes of certain resonances are also present. The outer part of the classical HZ is in the strongly chaotic region.

6.3.2 Stability of Resonant Exoplanetary Systems

A significant amount of multiple extrasolar planetary systems contain pairs of giant planets which are orbiting in MMRs. The in situ formation of resonant planetary systems is very unlikely, since each resonance requires certain ratio of the semi-major axes. The more favorable scenario is the *type II migration* of giant planets being embedded in the still gas rich protoplanetary disc. Type II migration appears when a massive giant planet carves a gap in the gaseous protoplanetary disc practically inhibiting the gas flow through the gap. In that case the planet's semi-major axis is changing according to the viscous evolution of the protoplanetary disc, see more about the topic in [3].

If the migration of two giant planets is convergent (e.g. the difference between their semi-major axes is decreasing) the phenomenon of the *resonant capture* will occur between them, and the two planets can migrate very close to their host star. The efficiency of migration is excellently demonstrated by hydrodynamic simulations modeling the formation of the resonant system around the star GJ 876 [24]. There are other planetary systems in which the giant planets reached the 2:1 MMR through type II migration such as HD 128311 [35], and HD 73526 [39].

The detection of giant planets is based on the radial velocity method. To calculate the orbital elements of the planets of a multiple system is not an easy task. Although there are well-known and widely used algorithms to provide reliable orbital fits, the orbital elements obtained not always result in a stable configuration for the planetary system. Regarding the resonant system of giant planets around HD 73526, [47] published orbital elements and planetary/stellar masses which resulted in *stable* orbits of the giant planets over 1 million years. On the other hand, the orbits are *chaotic*, as was clearly visible from numerical integrations of the three-body problem using the given elements as initial conditions. Since chaotic behavior may be uncommon among the resonant extrasolar planetary systems, and may not guarantee the stability of the giant planets for the whole lifetime of the system (being certainly longer than 1 million years), we searched for regular orbital solutions for the giant planets as well [39]. As a first attempt to study the degree of the chaoticity, we mapped the parameter space around the solution of [47]. We have calculated the stability properties of the $a_1 - a_2$, $e_1 - e_2$, $M_1 - M_2$, and $\varpi_1 - \varpi_2$ parameter planes, where a is the semi-major axis, e is the eccentricity, M is the mean anomaly and ϖ is the longitude of periastron of one of the giant planets.

In Fig. 6.17 the stability structures of the parameter planes for the semi-major axis and the eccentricities are displayed. During the calculation of a particular parameter plane the other orbital data have been kept fixed to their original values. On each parameter plane the stable regions are displayed by white, the weakly chaotic regions by grey, and the strongly chaotic regions by black colors. The values of the corresponding orbital data are marked on each parameter plane. By studying the stability maps, one can see that the orbital elements given by Tinney et al. [47]

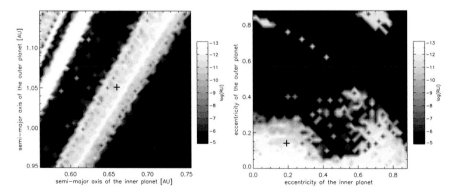

Fig. 6.17 Stability maps calculated around the orbital elements of [47]. The structure of the stability maps indicates that the orbital elements (marked by "+") are in a weakly chaotic region. Here *white colour* refers to ordered, *lighter grey shades* to weakly chaotic, and *darker shades* to unstable regions

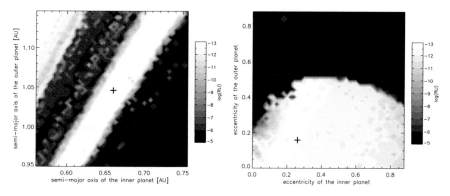

Fig. 6.18 Stability maps calculated around one set of orbital elements given in [39] It can be seen from the stability maps that the orbital elements (marked by "+") are embedded well in the stable region marked by *white colour*

are located in a weakly chaotic region, which explains the irregular behavior of the planetary eccentricities. We again stress that this does not automatically imply the instability of their fit, however by using these orbital data the system yields chaotic behavior and can be destabilized in longer timescales. Studying the stability maps it can also be concluded that the fit cannot be easily improved by the simple change of one of the orbital elements. The parameter plane is almost entirely weakly chaotic, there exists only a narrow strip of ordered behavior. After obtaining completely new sets of orbital elements using the Systemic Console [30] the stability maps around these fits were recalculated. In Fig. 6.18 parameter planes for the semi-major axis and the eccentricities are displayed, now around one of the stable orbital solutions. It can be clearly seen that the new orbital solution is well inside the region for ordered motion, which provides stability for the whole lifetime of the system. Based on the above example, it can be concluded that the RLI performs well in stability investigations of extrasolar planetary systems.

6.3.3   Application of the RLI to Study Libration Inside High Order MMRs

The most recent application of the RLI in dynamical astronomy has been presented by Érdi et al. [15] investigating the dynamical structure of high order resonances in the elliptic restricted three-body problem. This study has relevance in dynamics of Kuiper-belt objects. Moreover, it proved through numerical integration of a large set of orbits that the RLI excellently indicates the libration of the resonant angle inside a MMR. In what follows, first we present the results of Érdi et al. [15] obtained for

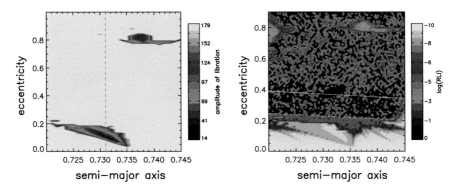

Fig. 6.19 The 8:5 inner MMR. *Left panel*: the two regions of libration as displayed by the libration amplitude of the resonant variable. The exact location of the MMR is at $a_r = 0.7310$ in normalized units. *Right panel*: the RLI map around the resonance, in which the two regions of libration can be clearly seen. The *continuous curve* denotes the unit apocentre distance, while the *dotted* and *dashed curves* the unit apocentre distance decreased by 1, and 1.5 Hill's radius, respectively

the 3rd order 8 : 5 inner resonance, in which case according to Eq. (6.5) the resonant variable is

$$\theta = 8\lambda' - 5\lambda - 3\varpi, \qquad (6.7)$$

where λ and ϖ are the mean orbital longitude, and longitude of the pericenter of the inner body, while λ' is the mean orbital longitude of the outer body.

A simple, but computationally demanding way to map the neighbourhood a MMR on the $a - e$ plane is to calculate the libration amplitude of θ for orbits whose initial semi-major axis and eccentricity values are taken from a grid, and the test particle has been started from pericentre. Such calculation can be seen on the left panel of Fig. 6.19. Using the same grid resolution and initial conditions the RLI values were also calculated, see the right panel of Fig. 6.19. The exact location of the resonance (when $\theta = 0°$) is marked with the vertical dashed line in the left panel of Fig. 6.19. The colour code corresponds to the value of the libration's amplitude: the darker the colour, the smaller the amplitude.

The 8 : 5 MMR has an interesting structure, there is a libration for low values of the eccentricities $e < 0.2$, and also for very high values $0.75 < e < 0.85$. Although the picture obtained by the RLI is more detailed, also showing some other neighbour MMRs, gives back the region of libration very well.

The RLI has also been applied to study the high order outer MMRs, in which case the massless body's orbit is outside the massive planet's orbit. In the study of Érdi et al. [15] the outer resonances of the Sun-Neptune system have been studied in two different models. These are the restricted three-body problem, in which only the gravitational effects of the Sun and Neptune were included, and also in a model in which the four giant planets were also taken into account. We will present here the results obtained for two different MMRs, namely the 8:5 and 7:3 outer resonances

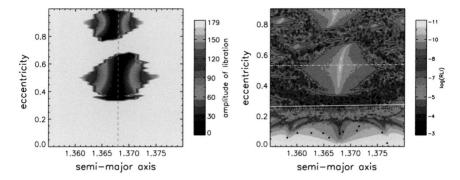

Fig. 6.20 The 8:5 outer MMR. *Left panel*: the two regions of libration displayed by the libration amplitude of the resonant variable. The exact location of the MMR is at $a_r = 1.3680$ in normalized units. *Right panel*: the RLI map around the resonance, in which the regions of libration are clearly visible. The *continuous curve* indicates the unit pericentre distance, while the *dotted curve* the unit pericentre distance increased by one Hill's radius. The *dash-dotted curve* shows the place of Uranus-crossing. The *filled circles* mark circulating TNOs

in the model of the restricted three-body problem. We note that the resonant variable of an outer $(p + q) : p$ MMR is given by Eq. (6.6).

Similarly to the $8 : 5$ inner MMR, the neighbourhood of the exact $8 : 5$ outer MMR ($a_r = 1.3680$ in normalized units) has been mapped on a dense grid of the $a - e$ plane (the test particle has been started from apocentre) and the libration regions are marked with different shades corresponding to the amplitude of θ. There are two regions of libration in this resonance located at relatively high eccentricities ($0.35 < e < 0.7$, $0.75 < e$), see the left panel of Fig. 6.20. Using the same $a - e$ grid the RLI values have also been calculated, see the right panel of Fig. 6.20. At the location of the exact resonance there is a nearly vertical dark strip indicating weak chaotic behaviour, and also the fact that in this configuration the $8 : 5$ MMR is not protective for low values of the test particle's eccentricity. The origin of the strong chaotic behaviour is due to the crossing of the orbits of the test particle and the perturbing body (Neptune in this case). There are different lines plotted to the RLI map. The continuous curve is the pericentre distance, the dotted curve is the pericentre distance increased by the Hill's radius of Neptune, and finally, the dash-dotted curve is the place of Uranus crossing. This latter curve may indicate that the high eccentricity libration regions of Fig. 6.20 being present in the model of the restricted three-body problem, might vanish in more advanced models including the planet Uranus, for instance. Finally, we roughly compare the $8 : 5$ outer resonance with the behaviour of some of the existing TNOs having inclination $i < 25°$, see the black filled circles for the corresponding a and e values of these objects. All of the TNOs displayed are circulating, and really they occupy the low eccentricity regions of the $a - e$ plane. We note, however, that the a and e values of the TNOs have different epochs, so their positions does not reflect the actual state of the system. On

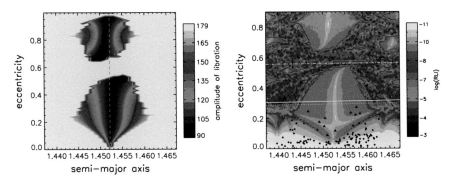

Fig. 6.21 The 7:4 MMR. *Left panel*: the two regions of libration displayed by the libration of the resonant variable. The exact location of the resonance is at $a_r = 1.4522$. *Right panel*: the RLI map around the resonance. The *triangles* mark librating, while the *filled circles* circulating TNOs. For the description of the curves see the previous figures

the other hand, it can be clearly seen in Fig. 6.20 that the TNOs at the 8 : 5 MMR have circulating resonance variable.

In order to have a more complete picture of the high order outer resonances, we also summarize the case when a MMR has a protective character, and the resonant variable of bodies lying in its vicinity can both librate and circulate. A good candidate for this purpose is the 7 : 4 3rd order outer MMR. The exact resonance is at $a_r = 1.4522$ (normalized units). Similarly to the 8 : 5 outer MMR, this resonance also has two regions of libration, but in this case the lower region of libration allows libration of test particles having low eccentricities, see Fig. 6.21. The right panel shows the dynamical structure of the resonance, and also TNOs found at this resonance. Most of them have circulating θ, but there are a few of them in the librating region, too. Studying the right panel of Fig. 6.21, one can see that the librating TNOs are clearly below the Neptune-crossing line, while the region of libration extends above it.

As an overall conclusion we can state that the RLI is a very reliable tool in detecting the positions of the lower and higher order MMRs being present in various planetary systems.

6.4   Discussion and Summary

In this chapter we summarize the basic properties of the recently introduced chaos detection method, the RLI. The RLI is based on the time evolution of the infinitesimally small tangent vector to the orbit, which is provided by solving numerically the variational equations. Hence, the RLI belongs to the family of the so–called variational indicators. Although the definition of the RLI is based on that of the LI, in this review we give evidence that the distinction between regular and chaotic motion is much clearer with the RLI, which makes it a more reliable alternative than the LI.

According to the comparative study with some wide–spread variational indicators, the RLI shows convincing performances in the experiments and considerably improves the performances of the classical LI. In generality, indicators like the FLI/OFLI or the MEGNO (actually there is a strong relationship between both indicators, see [31] for further details) are usually believed to be better options for a general analysis of the structure of the phase space. Therefore, our study reinforces the fact that the RLI can also be used as an alternative technique, which operates with reasonable computing times to make conclusive pictures of the dynamics despite the complexity of the problem.

Based on the comparative work presented in Sect. 6.2, we can summarize both the advantages and disadvantages of the RLI. In what follows we list its favorable properties/advantages, and also add that in which dynamical systems have been done the corresponding simulations.

- Having determined a reliable threshold in the Hénon–Heiles model, the RLI (and the FLI/OFLI) shows the best approximation rates in the ordered/chaotic regions. We note that the methods SALI and GALI also estimate the true percentage of the ordered/chaotic orbits but with a slight slower way.
- Comparing to the other chaos indicators also in the Hénon–Heiles model, the RLI detects much faster the orbits from the large chaotic sea (e.g. the "c-cs" orbit), than the other indicators.
- Studying a 4D symplectic mapping the RLI have been compared to the indicator MEGNO(2,0). In this case both the RLI and the MEGNO(2,0) reveal the fine structure of the phase space very accurately (much better than the LI, for instance). As a result of this experiment we also conclude that the RLI is a very effective tool in the characterization of a large array of initial conditions.
- In a complex 3D potential resembling a triaxial galactic halo (the so called NFW model), the RLI (together with the MEGNO, the OFLI, and the $GALI_5$) identifies the chaotic orbits within a Hubble time (~ 13 Gyrs). These indicators show that chaotic orbits can be identified within a physically meaningful time (i.e. the age of the Universe), which is important when studying the dynamics of a galaxy.

On the other hand, when applying the RLI one should be aware that:

- Among the presented CIs, in the Hénon–Heiles model the RLI identifies the so called "c-sl" orbit in the slowest way.
- In the study performed in the 4D symplectic mapping, the RLI cannot really distinguish between chaotic and sticky orbits. This is a disadvantage if one is interested in detecting the sticky orbits. On the other hand, if we are interested in detecting *all* chaotic orbits (including sticky orbits, as well), the application of the RLI might be useful.
- Regarding the time needed to calculate the RLI, we can conclude that it is not the least time consuming indicator. The FLI/OFLI, the MEGNO, the SALI and the SD might be more desirable options if the computation time really matters.

We note, however, that with the current generation of fast computers this option became less important.

In the last section of the current work we summarize the application of the RLI to planetary systems, which is its major application area. These studies include the development of a stability catalogue of hypothetical terrestrial exoplanets in extrasolar planetary systems, stability studies of resonant planetary systems and the investigation of high order mean motion resonances having relevance in studying the dynamics of the Kuiper-belt objects. We find that the RLI is an efficient and reliable numerical tool to map and characterize the dynamical structure of various mean motion resonances.

We note that the RLI (together with the SALI) has also been applied to map the stability regions of the Caledonian symmetric four-body problem, [46]. Since the preceding studies are mainly related to detecting the chaotic behaviour restricted four-body problem does not really belong to this line, thus to keep the length of the present study tractable, we omitted its presentation here.

We would also like to remark that the very simple computation of the RLI from the widespread well-known LI and its better performances reported in very different scenarios, make the RLI a serious candidate to replace the LI in a variety of fields, and not only in dynamical astronomy. For instance, in a paper published in a journal of Chemical Physics, the RLI is used as the default chaos indicator in the Lyapunov weighted path ensemble method. One of the capabilities of the method is to identify pathways connecting stable states which are relevant in the context of activated chemical reactions (see [18]).

Finally, the reliability of the RLI as a chaos indicator has been strongly demonstrated throughout this study, and as a result, the choice of the RLI to analyze a general dynamical system is well-founded.

Acknowledgements The authors thank the invitation and support of the scientific coordinators of the international workshop on Methods of Chaos Detection and Predictability: Theory and Applications: Georg Gottwald, Jacques Laskar and Haris Skokos and the hospitality of the Max Planck Institute for the Physics of Complex Systems where the meeting took place. ZsS is supported by the János Bolyai Research Scholarship of the Hungarian Academy of Sciences. NM is supported with grants from the Consejo Nacional de Investigaciones Científicas y Técnicas de la República Argentina (CCT - La Plata) and the Universidad Nacional de La Plata.

References

1. Barrio, R.: Sensitivity tools vs. Poincaré sections. Chaos Solitons Fractals **25**, 711–726 (2005)
2. Barrio, R., Borczyk, W., Breiter, S.: Spurious structures in chaos indicators maps. Chaos Solitons Fractals **40**, 1697–1714 (2009)
3. Baruteau, C., Masset, F.: Recent development in Planet Migration Theory. In: Souchay, J. et al. (eds.) Tides in Astronomy and Astrophysics. Lecture Notes in Physics, vol. 861, pp. 201–253. Springer, Berlin, Heidelberg (2013)
4. Belokurov, V. et al.: An orphan in the 'Field of Streams'. Astrophys. J. **658**, 337–344 (2007)

5. Benettin, G., Galgani L., Giorgilli, A., Strelcyn, J.: Lyapunov characteristic exponents for smooth dynamical systems; a method for computing all of them. Meccanica **15**, Part I: theory, 9–20; Part II: Numerical Applications, 21–30 (1980)
6. Carpintero, D., Maffione, N., Darriba, L.: LP–VIcode: a program to compute a suite of variational chaos indicators. Astron. Comput. **5**, 19–27 (2014)
7. Celletti, A., Kotoulas, T., Voyatzis, G., Hadjidemetriou, J.: The dynamical stability of a Kuiper Belt-like region. Mon. Not. R. Astron. Soc. **378**, 1153–1164 (2007)
8. Christodoulidi, H., Bountis, T.: Low-dimensional quasiperiodic motion in Hamiltonian systems. ROMAI J. **2**, 37–44 (2006)
9. Cincotta, P., Simó, C.: Simple tools to study global dynamics in non-axisymmetric galactic potentials – I. Astron. Astrophys. **147**, 205–228 (2000)
10. Cincotta, P., Giordano, C., Simó, C.: Phase space structure of multidimensional systems by means of the mean exponential growth factor of nearby orbits. Physica D **182**, 151–178 (2003)
11. Contopoulos, G., Voglis, N.: A fast method for distinguishing between ordered and chaotic orbits. Astron. Astrophys. **317**, 73–81 (1997)
12. Cooper, A.P., et al.: Galactic stellar haloes in the CDM model. Mon. Not. R. Astron. Soc. **406**, 744–766 (2010)
13. Darriba, L.A., Maffione, N.P., Cincotta, P.M., Giordano, C.M.: Comparative study of variational chaos indicators and ODEs' numerical integrators. Int. J. Bifurcat. Chaos **22**, 1230033 (2012)
14. Darriba, L., Maffione, N., Cincotta, P., Giordano, C.: Chaos detection tools: the LP–Vicode and its applications. In: Cincotta, P., Giordano, C., Efthymiopoulos, C. (eds.) Chaos, Diffusion and Non-integrability in Hamiltonian Systems-Application to Astronomy, pp. 345–366. Universidad Nacional de La Plata and Asociación Argentina de Astronomía Publishers, La Plata (2012)
15. Érdi, B., Rajnai, R., Sándor, Z., Forgács-Dajka, E.: Stability of higher order resonances in the restricted three-body problem. Celest. Mech. Dyn. Astron. **113**, 95–112 (2012)
16. Fouchard, M., Lega, E., Froeschlé, Ch., Froeschlé, Cl.: On the relationship between fast lyapunov indicator and periodic orbits for continuous flows. Celest. Mech. Dyn. Astron. **83**, 205–222 (2002)
17. Froeschlé, Cl., Gonczi, R., Lega, E.: The fast lyapunov indicator: a simple tool to detect weak chaos. Application to the structure of the main asteroidal belt. Planet. Space Sci. **45**, 881–886 (1997)
18. Geiger, P., Dellago, C.: Identifying rare chaotic and regular trajectories in dynamical systems with Lyapunov weighted path sampling. Chem. Phys. **375**, 309–315 (2010)
19. Gómez, F., Helmi, A., Cooper A., Frenk, C., Navarro, J., White, S.: Streams in the Aquarius stellar haloes. Mon. Not. R. Astron. Soc. **436**, 3602–3613 (2013)
20. Guzzo, M., Lega, E., Froeschl'e, C.: On the numerical detection of the effective stability of chaotic motions in quasi-integrable systems. Physica D **163**, 1–25 (2002)
21. Hénon, M., Heiles, C.: The applicability of the third integral of motion: some numerical experiments. Astrophys. J. **69**, 73–79 (1964)
22. Ibata, R.A., Irwin, M.J., Lewis, G.F., Stolte, A.: Galactic halo substructure in the sloan digital sky survey: the ancient tidal stream from the Sagittarius Dwarf Galaxy. Astrophys. J. **547**, 133–136 (2001)
23. Kasting, J.F., Whitmire, D.P., Reynolds, R.T.: Habitable zones around main sequence stars. Icarus **101**, 108–128 (1993)
24. Kley, W., Lee, M.H., Murray, N., Peale, S.J.: Modeling the resonant planetary system GJ 876. Astron. Astrophys. **437**, 727–742 (2005)
25. Kopparapu, R.K., Ramirez, R., Kasting, J.F., et al.: Habitable zones around main-sequence stars: New estimates. Astrophys. J. **765**, (2013), article id. 131, 16 pp.
26. Laskar, J.: Frequency analysis of a dynamical system. Celest. Mech. Dyn. Astron. **56**, 191–196 (1993)
27. Maffione, N., Darriba, L., Cincotta, P., Giordano, C.: A comparison of different indicators of chaos based on the deviation vectors: application to symplectic mappings. Celest. Mech. Dyn. Astron. **111**, 285–307 (2011)

28. Majewski, S.R., Skrutskie, M.F., Weinberg, M.D., Ostheimer, J.C.: A two micron all sky survey view of the Sagittarius dwarf galaxy. I. Morphology of the Sagittarius core and tidal arms. Astrophys. J. **599**, 1082–1115 (2003)
29. Manos, T., Skokos, Ch., Antonopoulos, Ch.: Probing the local dynamics of periodic orbits by the generalized alignment index (GALI) method. Int. J. Bifurcat. Chaos **22**, 1250218 (2012)
30. Meschiari, S., Wolf, A.S., Rivera, E., Laughlin, G., Vogt, S., Butler, P.: Systemic: a testbed for characterizing the detection of extrasolar planets. I. The systemic console package. Publ. Astron. Soc. Pac. **121**, 1016–1027 (2009)
31. Mestre, M., Cincotta, P., Giordano, C.: Analytical relation between two chaos indicators: FLI and MEGNO. Mon. Not. R. Astron. Soc. Lett. **414**, 100–103 (2011)
32. Murray, C.D., Dermott, S.F.: Solar System Dynamics. Cambridge University Press, Cambridge (1999)
33. Navarro, J., Frenk, C., White, S.: The structure of cold dark matter halos. Astrophys. J. **462**, 563–575 (1996)
34. Navarro, J., Frenk, C., White, S.: A universal density profile from hierarchical clustering. Astrophys. J. **490**, 493–508 (1997)
35. Sándor, Z., Kley, W.: On the evolution of the resonant planetary system HD 128311. Astron. Astrophys. **451**, L31–L34 (2006)
36. Sándor, Z., Érdi, B., Efthymiopoulos, C.: The phase space structure around L_4 in the restricted three-body problem. Celest. Mech. Dyn. Astron. **78**, 113–123 (2000)
37. Sándor, Z., Érdi, B., Széll, A., Funk, B.: The relative lyapunov indicator: an efficient method of chaos detection. Celest. Mech. Dyn. Astron. **90**, 127–138 (2004)
38. Sándor, Z., Süli, Á., Érdi, B., Pilat-Lohinger, E., Dvorak, R.: A stability catalogue of the habitable zones in extrasolar planetary systems. Mon. Not. R. Astron. Soc. **375**, 1495–1502 (2007)
39. Sándor, Z., Kley, W., Klagyivik, P.: Stability and formation of the resonant system HD 73526. Astron. Astrophys. **472**, 981–992 (2007)
40. Skokos, Ch.: Alignment indices: a new, simple method to for determining the ordered or chaotic nature of orbits. J. Phys. A Math. Gen. **34**, 10029–10043 (2001)
41. Skokos, Ch.: The Lyapunov Characteristic Exponents and Their Computation. Lecture Notes in Physics, vol. 790, pp. 63–135. Springer, Berlin, Heidelberg (2010)
42. Skokos, Ch., Antonopoulos, Ch., Bountis, T., Vrahatis, M.: Detecting order and chaos in Hamiltonian systems by the SALI method. J. Phys. A Math. Gen. **37**, 6269–6284 (2004)
43. Skokos, Ch., Bountis, T., Antonopoulos, Ch.: Geometrical properties of local dynamics in Hamiltonian systems: the Generalized Alignment Index (GALI) method. Physica D **231**, 30–54 (2007)
44. Skokos, Ch., Bountis, T., Antonopoulos, Ch.: Detecting chaos, determining the dimensions of tori and predicting slow diffusion in Fermi-Pasta-Ulam lattices by the Generalized Alignment Index method. Eur. Phys. J. Spec. Top. **165**, 5–14 (2008)
45. Szebehely, V.: Theory of Orbits. The Restricted Problem of Three Bodies. Academic, New York (1967)
46. Széll, A., Érdi, B., Sándor, Z., Steves, B.: Chaotic and stable behaviour in the Caledonian symmetric four-body problem. Mon. Not. R. Astron. Soc. **347**, 380–388 (2004)
47. Tinney, C.G., Butler, R.P., Marcy, G.W., et al.: The 2:1 resonant exoplanetary system orbiting HD 73526. Astrophys. J. **647**, 594–599 (2006)
48. Vogelsberger, M., White, S., Helmi, A., Springel, V.: The fine-grained phase-space structure of cold dark matter haloes. Mon. Not. R. Astron. Soc. **385**, 236–254 (2008)
49. Voglis, N., Contopoulos, G., Efthymiopoulos, C.: Detection of ordered and chaotic motion using the dynamical spectra. Celest. Mech. Dyn. Astron. **73**, 211–220 (1999)
50. Voyatzis, G.: Chaos, order, and periodic orbits in 3:1 resonant planetary dynamics. Astrophys. J. **675**, 802–816 (2008)

Chapter 7
The 0-1 Test for Chaos: A Review

Georg A. Gottwald and Ian Melbourne

Abstract We review here theoretical as well as practical aspects of the 0-1 test for chaos for deterministic dynamical systems. The test is designed to distinguish between regular, i.e. periodic or quasi-periodic, dynamics and chaotic dynamics. It works directly with the time series and does not require any phase space reconstruction. This makes the test suitable for the analysis of discrete maps, ordinary differential equations, delay differential equations, partial differential equations and real world time series. To illustrate the range of applicability we apply the test to examples of discrete dynamics such as the logistic map, Pomeau–Manneville intermittency maps with both summable and nonsummable autocorrelation functions, and the Hamiltonian standard map exhibiting weak chaos. We also consider examples of continuous time dynamics such as the Lorenz-96 system and a driven and damped nonlinear Schrödinger equation. Finally, we show the applicability of the 0-1 test for time series contaminated with noise as found in real world applications.

7.1 Introduction

The 0-1 test for chaos was developed in a series of papers [19, 20, 22] to distinguish between regular and chaotic dynamics in deterministic dynamical systems. Rather than requiring phase space reconstruction which is necessary to apply standard Lyapunov exponent methods to the analysis of discretely sampled data, the test works directly with the time series and does not involve any preprocessing of the data. The test requires only a minimal computational effort independent of the dimension of the underlying dynamical system under investigation.

G.A. Gottwald (✉)
School of Mathematics & Statistics, The University of Sydney, Sydney, 2006 NSW, Australia
e-mail: georg.gottwald@sydney.edu.au

I. Melbourne
Mathematics Institute, University of Warwick, Coventry CV4 7AL, UK
e-mail: I.Melbourne@warwick.ac.uk

© Springer-Verlag Berlin Heidelberg 2016   221
Ch. Skokos et al. (eds.), *Chaos Detection and Predictability*, Lecture Notes
in Physics 915, DOI 10.1007/978-3-662-48410-4_7

The test has found applications in a wide range of fields. Besides general studies of dissipative [12, 35, 67] and Hamiltonian [72] dynamical systems and multi-agent systems [39], the test has found its way into as disparate areas as engineering [42, 43, 55], electronics [65], finance and economics [28, 36–38, 68–70], geophysical applications [7, 47, 48, 60], hydrology [32, 40], epidemology [8, 9, 50] and traffic dynamics [34]. In particular its application to non-smooth processes [2, 42, 43], to systems with fractional derivatives and delays [3, 5, 71], and to nonchaotic strange attractors [18] are notable as those are not amenable to standard methods employing Lyapunov exponents. The test has also been used to analyse systems with non-local operators in integro-differential equations [62] and integro-partial differential equations [10]. Moreover, it has been used to analyse experimental data and observations [13, 33, 34, 37, 38].

The remainder is organised as follows. In Sect. 7.2 we briefly describe the test. The algorithm is then presented in Sect. 7.3 where we discuss several implementations of the test. The theoretical underpinning of our test is explained in Sect. 7.4. This is followed by numerical results in Sect. 7.5 illustrating the efficiency of our test to deal with intermittent maps, chaos in thin separatrix layers in Hamiltonian systems, partial differential equations and data contaminated by observational noise. We conclude with a summary in Sect. 7.6.

7.2 Description of the Test

The input of the test is a one-dimensional time series $\phi(n)$ for $n = 1, 2, \ldots$ We use the data $\phi(n)$ to drive the 2-dimensional system

$$p(n + 1) = p(n) + \phi(n) \cos cn,$$
$$q(n + 1) = q(n) + \phi(n) \sin cn, \tag{7.1}$$

where $c \in (0, 2\pi)$ is fixed. Define the (time-averaged) mean square displacement

$$M(n) = \lim_{N \to \infty} \frac{1}{N} \sum_{j=1}^{N} \left([p(j + n) - p(j)]^2 + [q(j + n) - q(j)]^2 \right), \quad n = 1, 2, 3, \ldots$$

and its growth rate

$$K = \lim_{n \to \infty} \frac{\log M(n)}{\log n}.$$

Under general conditions, the limits $M(n)$ and K can be shown to exist, and K takes either the value $K = 0$ signifying regular dynamics or the value $K = 1$ signifying chaotic dynamics.

A brief explanation of the rationale behind the test is as follows. (The mathematics is described more carefully in Sect. 7.4.) In the regular case (periodic or quasiperiodic dynamics) the trajectories for the system (7.1) are typically bounded, whereas in the chaotic case the trajectories for (7.1) typically behave approximately like a two-dimensional Brownian motion with zero drift and hence evolve diffusively (i.e. with growth rate $\sqrt{n}$). A convenient method for distinguishing these growth rates, bounded or diffusive, is via the mean square displacement $M(n)$ which accordingly is either bounded or grows linearly. The diagnostic $K \in \{0, 1\}$ captures this growth rate.

To summarise, we have the following two scenarios:

Underlying dynamics	Dynamics of $p(n)$ and $q(n)$	$M(n)$	K
Regular	Bounded	Bounded	0
Chaotic	Diffusive	Linear	1

In the following section we describe the test in more detail focusing on the practical issues in the implementation of the 0-1 test.

7.3 Description of the Algorithm

The test can be readily implemented in a few lines of code. We briefly describe its implementation and refer the reader to [22] for more details. Given a time series $\phi(j)$ for $j = 1, \ldots, N$ we perform the following sequence of steps:

1. For $c \in (0, \pi)$, we solve the system (7.1) to obtain

$$p_c(n) = \sum_{j=1}^{n} \phi(j) \cos jc, \quad q_c(n) = \sum_{j=1}^{n} \phi(j) \sin jc \qquad (7.2)$$

for $n = 1, 2, \ldots, N$. Typical plots of p and q for regular and chaotic dynamics are given in Fig. 7.1 which clearly illustrates the bounded motion of p and q for underlying regular dynamics and asymptotic Brownian motion for underlying chaotic dynamics.

2. To analyse the diffusive (or non-diffusive) behaviour of p_c and q_c we compute

$$M_c(n) = \frac{1}{N} \sum_{j=1}^{N} ([p_c(j+n) - p_c(j)]^2 + [q_c(j+n) - q_c(j)]^2) . \qquad (7.3)$$

To assure the limit $N \to \infty$ we require $n \ll N$. Hence we calculate $M_c(n)$ only for $n \leq N_0$ where $N_0 \ll N$. In practice we find that N_0 should not be chosen much larger than $N/10$.

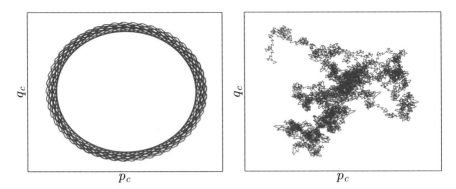

Fig. 7.1 Plot of p versus q for the logistic map $x_{n+1} = \mu x_n(1 - x_n)$. *Left*: Regular dynamics at $\mu = 3.55$; *Right*: Chaotic dynamics at $\mu = 3.97$. We used $N = 5000$ data points

In [23] a modified mean square displacement

$$D_c(n) = M_c(n) - V_{\text{osc}}(c, n) \tag{7.4}$$

was derived which exhibits the same asymptotic growth rate as $M_c(n)$ but with better convergence properties. The correction term

$$V_{\text{osc}}(c, n) = (E\phi)^2 \frac{1 - \cos nc}{1 - \cos c}$$

is readily estimated from the time average of the observable

$$E\phi = \lim_{N \to \infty} \frac{1}{N} \sum_{j=1}^{N} \phi(j) \ .$$

Note that the asymptotic growth rates of $M_c(n)$ and $D_c(n)$ are the same.

In Fig. 7.2 we show the two mean square displacements $M_c(n)$ and $D_c(n)$ for the logistic map $x_{n+1} = \mu x_n(1-x_n)$ with $\mu = 3.97$ (which corresponds to chaotic dynamics) and an arbitrary value of $c = 0.9$. The subtraction of the oscillatory term $V_{\text{osc}}(c, n)$ clearly regularizes the linear behaviour of $M_c(n)$. This allows for a much better determination of the asymptotic growth rate K_c of the mean square displacement which is described in the next step.

3. The contrasting behaviour of the translation variables p_c and q_c as seen in Fig. 7.1 can be distinguished by the asymptotic growth rate K_c of the mean square displacement (or of the modified mean square displacement $D_c(n)$). We present here two different methods to compute K_c, namely the *regression method* and the *correlation method*.

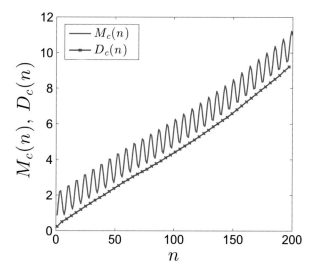

Fig. 7.2 Plot of the mean square displacement versus n for the logistic map with $\mu = 3.97$ corresponding to chaotic dynamics. We used $N = 2000$ data points and computed $M_c(n)$ and $D_c(n)$ for $n = 1, \ldots, 200$ and an arbitrary value of $c = 0.9$

Regression method The regression method consists of linear regression for the log-log plot of the mean square displacement (cf. Fig. 7.2). For the original mean square displacement $M_c(n)$, the asymptotic growth rate K_c is given by the definition

$$K_c = \lim_{n \to \infty} \frac{\log M_c(n)}{\log n} \ .$$

Numerically, K_c is determined by fitting a straight line to the graph of $\log M_c(n)$ versus $\log n$ through minimizing the absolute deviation.[1] It is recommended to minimize the absolute deviation rather than employing the usual least square method as the latter assigns a higher weight to outliers. Outliers are typical for small values of n since the linear behaviour of the mean square displacement is only given asymptotically.

As seen in Fig. 7.2, $D_c(n)$ exhibits far less variance than $M_c(n)$ so it is natural to apply the regression method to $D_c(n)$. However, since $D_c(n)$ may be negative due to the subtraction of the oscillatory term $V_{\text{osc}}(c, n)$, we need to set

$$\tilde{D}_c(n) = D_c(n) + a \min_{1 \le n \le N_0} |D_c(n)| \ ,$$

---

[1]One may either use off the shelf routines provided for example in *Numerical Recipes* [61] or build-in routines in MATLAB [49].

where $a > 1$ (in the simulations presented here we chose $a = 1.1$) to obtain the asymptotic growth rate

$$K_c = \lim_{n\to\infty} \frac{\log \tilde{D}_c(n)}{\log n} \ .$$

Again, K_c can be determined numerically by regression (minimizing the absolute deviation) for the graph of $\log \tilde{D}_c(n)$ versus $\log n$.

Correlation method In the correlation method, we form vectors $\xi = (1, 2, \ldots, N_0)$ and $\Delta = (D_c(1), D_c(2), \ldots, D_c(N_0))$ (alternatively, $M_c(n)$ could be used instead of $D_c(n)$). Recalling the definition of covariance and variance of given vectors x, y of length q

$$\text{cov}(x, y) = \frac{1}{q}\sum_{j=1}^{q}(x(j) - \bar{x})(y(j) - \bar{y}), \quad \text{where } \bar{x} = \frac{1}{q}\sum_{j=1}^{q}x(j) \ ,$$

$$\text{var}(x) = \text{cov}(x, x) \ ,$$

we define the correlation coefficient

$$K_c = \text{corr}(\xi, \Delta) = \frac{\text{cov}(\xi, \Delta)}{\sqrt{\text{var}(\xi)\text{var}(\Delta)}} \in [-1, 1] \ .$$

This quantity measures the strength of the correlation of $D_c(n)$ with linear growth. The correlation method greatly outperforms the regression method (see Figs. 7.3 and 7.4 below), but assumes that the dynamics is such that with probability 1 we have $K_c = 0$ or $K_c = 1$. This is justified for large classes of dynamical systems [23].

4. Steps 1–3 need to be executed for various values of c. In practice, 100 choices of c is sufficient. We then compute the median of these values of K_c to compute the final result $K = \text{median}(K_c)$. The values of c are chosen randomly in the interval $c \in (\pi/5, 4\pi/5)$ to avoid resonances. Resonances occur when the dynamics involves a periodic component with frequency ω implying a term in the Fourier decomposition of the observable ϕ proportional to $\exp(-i\omega k)$. In this case there is a resonance at $c = \omega$ leading to $p_c(n) \sim n$ and $q_c(n) \sim n$ and hence $M_c(n) \sim n^2$ (and $D_c(n) \sim n^2$) implying $K_c = 2$ for the regression method and $K_c = 1$ for the correlation method. Note that for $c = 0$ the test would yield a resonance irrespective of the underlying dynamics (which is why this value should be excluded). See [19, 22] for more details on resonances. In Fig. 7.3 we show K_c versus c for the logistic map for regular and chaotic dynamics. Resonances are clearly visible for the periodic case, with $K_c = 2$ for the regression method and $K_c \approx 1$ for the correlation method.

   Our test states that a value of $K \approx 0$ indicates regular dynamics, and $K \approx 1$ indicates chaotic dynamics. This is exemplified in Fig. 7.4 where K is shown as a function of the parameter μ of the logistic map.

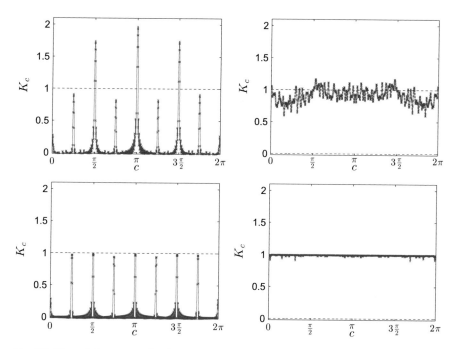

Fig. 7.3 Plot of K_c versus c for the logistic map calculated using the regression method (*top*) and correlation method (*bottom*). We used here $N = 5000$ data points, and 1000 equally spaced values for c. *Left*: $\mu = 3.55$ corresponding to regular dynamics; *Right*: $\mu = 3.97$ corresponding to chaotic dynamics

Remark on Finite Size Problems There are finite size issues that are inherent to all methods for chaos detection, namely that the length of the time series is sufficiently long to capture the dynamics across the whole of the attractor. Specifically, for the 0-1 test the determination of the mean square displacement requires $n \leq N_0 \ll N$, and the test relies on asymptotic behaviour of the (non)-diffusive behaviour of p and q which for too small time series data length may not yet be dominant. Concerning the latter point it is pertinent to mention that even in cases of time series which are too short to allow for convergence of K to either 0 or 1, strong indications for the presence or absence of chaos can be found by looking at the behaviour of K with the length of the time series used to determine K. Figure 7.5 shows typical decreasing/increasing behaviour of K near parameter values of the logistic map at the so called *edge of chaos*, indicating regular or chaotic dynamics, respectively. This was discussed at length in [21, 22].

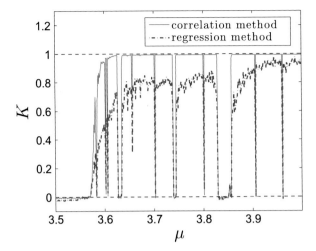

Fig. 7.4 Plot of K versus μ for the logistic map with $3.5 \leq \mu \leq 4$ increased in increments of 0.001. We used $N = 2000$ data points. Shown are results when K is calculated via the regression method (*dashed line, blue*) and when K is calculated via the correlation method (*continuous line, red*). The *horizontal lines* indicate the limiting cases $K = 0$ and $K = 1$. We used 100 randomly distributed values of c, and the mean square displacement was determined using $D_c(n)$

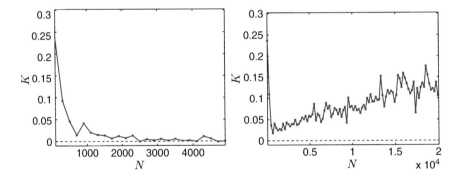

Fig. 7.5 Plot of K versus the length of the time series N for the logistic map near the edge of chaos. We used 100 randomly distributed values of c, and the mean square displacement was determined using $D_c(n)$ with $N_0 = N/10$, and the correlation method was used to determine K_c. *Left*: $\mu = 3.569$ corresponding to regular dynamics; *Right*: $\mu = 3.571$ corresponding to chaotic dynamics

7.4 Theoretical Framework for the 0-1 Test

Systems of the type (7.1) were studied extensively in [1, 14, 52, 53, 56]. The motivation there was to understand growth rates of trajectories in systems with Euclidean symmetry. A large class of discrete time systems with planar Euclidean symmetry are given by skew product equations of the form

$$x(n+1) = f(x(n)),$$
$$\vartheta(n+1) = \vartheta(n) + h(x(n)), \tag{7.5}$$
$$p(n+1) = p(n) + \Phi(x(n))\cos(\vartheta(n)) - \Psi(x(n))\sin(\vartheta(n)),$$
$$q(n+1) = q(n) + \Phi(x(n))\sin(\vartheta(n)) + \Psi(x(n))\cos(\vartheta(n)).$$

Here $f : X \to X$ defines the base dynamics (perpendicular to the symmetry variables) while $\vartheta(n)$ represents two-dimensional rotations and $(p(n), q(n))$ represent planar translations. It is assumed that the functions $h, \Phi, \Psi : X \to R$ are smooth. In [56], it was shown that if the dynamics on X is periodic or quasiperiodic, then typically the translation variables $p(n), q(n)$ remain bounded. However, sufficiently chaotic dynamics on X leads to diffusive behaviour in the translation variables. (See [14, 53] for the case of uniformly hyperbolic dynamics, and more recently [25] for nonuniformly hyperbolic dynamics.) Using these results, and computing the growth rate K of the mean square displacement as described in Sect. 7.2, we obtain with probability one the growth rates $K = 0$ and $K = 1$ respectively in these two situations. It is pertinent to stress that the test does not rely in anyway on a possible exponential decay of the auto-correlation function; the test holds for polynomial decay as well and even does not require summability of the auto-correlation function [24, 25] (see also Sect. 7.5.1.1). Also it is irrelevant whether the dynamics is mixing. For example, in the case of the logistic map mentioned in Sect. 7.5.1, the attractor is almost always a periodic sink or a strongly chaotic attractor consisting of a finite union of intervals permuted cyclically by the dynamics. In both cases, the attractor is mixing up to a finite cycle but is generally nonmixing. More importantly, in the case of continuous time dynamics, it is rarely the case that mixing can be established but the 0-1 test is still valid.

It should be noted that a similar dichotomy holds in the absence of the rotation variables, except that the bounded/diffusive behaviour is superimposed on a linear drift. The rotation symmetry kills off the linear drift [15, 56], rendering the bounded/diffusive dichotomy more readily detectable.

The idea behind the 0-1 test is to adjoin rotation and translation variables ϑ, p, q to a given (but unknown) dynamical system $f : X \to X$ generating data $\phi(n)$, thus producing a system with Euclidean symmetry to which the above theoretical results apply. Note that choosing $h \equiv c, \Psi \equiv 0$ and making the identification $\Phi(x(n)) = \phi(n)$ reduces the skew product system (7.5) to the 2-dimensional system (7.1) used in the 0-1 test.

The original version of the test [19] used "generic" choices of h so that certain theoretical results of Field et al. [15] could be applied in justifying the test. The current version is much more effective for noisy data [20] but the original theoretical justification for the test no longer applies. Nevertheless, it transpires that the simplified nature of the equations in (7.1) enables certain improvements to the theoretical underpinnings for the test, as described in [23]. The structure of the simplified equations means that they are amenable to techniques from Fourier analysis. In particular, there are connections with power spectra as described in the next subsection. In the aforementioned cases of periodic/quasiperiodic dynamics and uniformly/nonuniformly hyperbolic dynamics, we typically obtain $K = 0$ and $K = 1$, respectively. Here "typically" is in the sense of probability one: for almost every choice of c. As mentioned previously, we take the median value of K, computed with 100 randomly chosen choices of c, to circumvent the issue regarding bad choices of c. Moreover, the considerations in [23] lead directly to the modified mean square displacement $D_c(n)$ which we have seen leads to improved results (there is no analogue of this modification for the original test).

7.4.1 Connection with the Power Spectrum

Consider a discrete dynamical system $f : X \rightarrow X$ with ergodic invariant measure μ. Given a square-integrable observable $v : X \rightarrow \mathbf{R}$, the power spectrum S is defined to be the square of the Fourier amplitudes of $v \circ f^j$ per unit time[2]

$$S(\omega) = \lim_{n \rightarrow \infty} \frac{1}{n} \int_X \left| \sum_{j=0}^{n-1} e^{ij\omega} v \circ f^j \right|^2 d\mu, \quad \omega \in [0, 2\pi]. \tag{7.6}$$

It was proven in [51] that the power spectrum has a broadband nature and is nowhere zero for a large class of dynamical systems, including slowly mixing systems such as Pomeau–Manneville maps provided the auto-correlation function is summable. (For a discussion of recent results in the case of nonsummable autocorrelations, see Sect. 7.5.1.1.)

A simple short calculation shows that

$$S(c) = \lim_{n \rightarrow \infty} \frac{1}{n} \int_X \left| \sum_{j=0}^{n-1} e^{ijc} v \circ f^j \right|^2 d\mu = \lim_{n \rightarrow \infty} \frac{1}{n} M_c(n), \tag{7.7}$$

implying that

$$M_c(n) = S(c)n + o(n). \tag{7.8}$$

This may give the wrong impression that the 0-1 test for chaos is simply evaluating the power spectrum. From (7.8) one can only conclude that if the power spectrum is nowhere nonzero ($S(c) \neq 0$ for all c), then the asymptotic growth rate of the mean square displacement becomes $K_c = 1$ for all c. On the other hand, if $S(c) = 0$ for all c, it does not automatically follow that $K_c = 0$ (for example, the $o(n)$ term could be

---

[2] One may use $e^{2\pi ij\omega/n}$ rather than $e^{ij\omega}$ for a rescaled domain.

of the form $n/\log(n)$ implying $K_c = 1$). However in [23] it was rigorously proven that for a large class of dynamical systems, the $o(n)$ terms are such that for chaotic dynamics one obtains $K_c = 1$ and for regular dynamics $K_c = 0$.

It is pertinent to mention the computational advantage of the 0-1 test which extracts in a single number K the property of the power spectrum which is relevant for underlying chaotic or regular dynamics, i.e. whether it is everywhere or nowhere nonzero. The test completely bypasses the explicit computation of the power spectrum which would require considerably more data. Moreover, K can be plotted against a parameter of the system as in Fig. 7.4 and the convergence of K can be seen against the number of iterates N as in Fig. 7.5. There do not exist analogous plots for the power spectrum.

7.5 Numerical Examples for the 0-1 Test

We now illustrate the applicability of our test to be able to distinguish regular dynamics from chaotic dynamics in discrete and continuous time systems, dissipative and Hamiltonian systems, noise free and noise contaminated data.

7.5.1 Discrete Time Systems

One of the simplest families of dynamical systems that exhibits regular and chaotic dynamics is the logistic map $f : [0, 1] \rightarrow [0, 1]$ given by $f(x) = \mu x(1 - x)$. Here, $\mu \in [0, 4]$ is a parameter. It is well-known that there is a unique attractor for each value of μ and that the basin of attraction is of full measure in $[0, 1]$. For almost every value of μ, the attractor is either a periodic orbit or a strongly chaotic attractor. Throughout the earlier sections, this family of maps was used as an illustrative example for various features of the 0-1 test, see Figs. 7.1, 7.2, 7.3, 7.4 and 7.5.

We now proceed to explore two further families of discrete time systems.

7.5.1.1 Pomeau–Manneville Map

A prototypical family of maps exhibiting intermittency and weakly chaotic dynamical systems with "sticky" equilibria are Pomeau–Manneville intermittency maps $x_{n+1} = f(x_n)$ with $f : [0, 1] \rightarrow [0, 1]$ given by

$$f(x) = \begin{cases} x(1 + 2^{\gamma}x^{\gamma}) & 0 \leq x \leq \frac{1}{2} \\ 2x - 1 & \frac{1}{2} \leq x \leq 1 \end{cases} \tag{7.9}$$

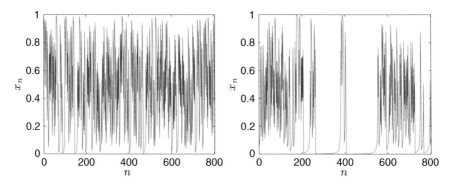

Fig. 7.6 Time series of the Pomeau–Manneville map (7.9). *Left*: Strongly chaotic case with $\gamma = 0.2$. *Right*: Weakly chaotic case with $\gamma = 0.7$

where γ is a parameter [44, 59]. For $\gamma \in [0, 1)$ there exists a unique absolutely continuous invariant probability measure (SRB measure). When $\gamma = 0$ the map reduces to the doubling map with exponential decay of correlations. For $\gamma \in (0, 1)$ the decay of correlations is polynomial with rate $1/n^{(1/\gamma)-1}$ which is summable for $\gamma < \frac{1}{2}$ and nonsummable for $\gamma \in [\frac{1}{2}, 1)$ [30]. For $\gamma > 0$ the fixed point at 0 is indifferent ($f'(0) = 1$) and plays the role of the "sticky" regular dynamics leading to laminar behaviour interspersed with intermittent chaotic bursts. This is illustrated in Fig. 7.6, where we show a trajectory of the Pomeau–Manneville map in the strongly chaotic case with $\gamma = 0.2$, where the correlations are summable, and in the intermittent weakly chaotic case with $\gamma = 0.7$, where the correlations are nonsummable.

It is well-known [17] that for such intermittent systems the usual central limit theorem breaks down for $\gamma \in (\frac{1}{2}, 1)$ leading to fluctuations of Lévy type rather than of Gaussian type. For mathematically rigorous results on this, see [27, 54, 74]. Despite this, our test for chaos is still able to detect chaos in the weakly chaotic case with nonsummable correlations. A proof of this statement is currently work in progress, but the underlying reason, as discussed in related results in [24, 25], is that the anomalous diffusion is suppressed due to the rotation symmetry induced by the presence of c in (7.2). As a result of this, when the dynamics is trapped near the indifferent fixed point, the dynamics appears regular and therefore leads to bounded dynamics of the translation variables p and q as seen in Fig. 7.7 for $\phi_n = 1 + x_n$. In Fig. 7.8 we show the asymptotic growth rate as a function of γ.

7.5.1.2 Standard Map

We now consider the area-preserving Standard map [6, 41]

$$\pi_{n+1} = \pi_n + \kappa \sin(\theta_n) \tag{7.10}$$

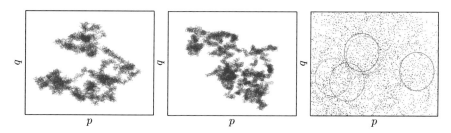

Fig. 7.7 Typical plots of the translation variables p and q driven by an observable $\phi_n = 1 + x_n$ of the Pomeau–Manneville map (7.9) for $c = 2.1375$. *Left*: Strongly chaotic case with $\gamma = 0.2$. *Middle*: Weakly chaotic case with $\gamma = 0.7$. *Right*: Zoom for weakly chaotic case with $\gamma = 0.7$ showing the trace of the laminar phases of the Pomeau–Manneville dynamics in form of bounded *circles*

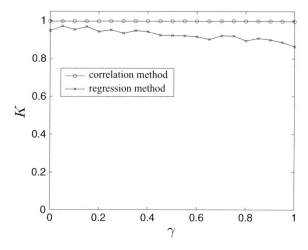

Fig. 7.8 Plot of K as a function of γ for an observable $\phi_n = 1 + x_n$ of the Pomeau–Manneville map (7.9), using the regression method (*crosses*) and the correlation method (*circles*). We used $N = 10,000$ data points and 100 randomly distributed values of c

$$\theta_{n+1} = \theta_n + \pi_{n+1} . \tag{7.11}$$

Figure 7.9 shows the trajectories of 100 randomly chosen initial conditions after a transient of 10,000 iterates for $\kappa = 0.9$. The phase space consists of regular islands embedded in chaotic layers. In contrast, at $\kappa = -0.3$ there is a hyperbolic fixed point at the origin and the asymptotic dynamics occurs in a thin separatrix layer as seen in Fig. 7.10 for 100 randomly chosen initial conditions $\pi_0 \in [0, 0.03]$ and $\theta_0 \in [-0.03, 0.03]$. This thin separatrix layer contains complex structures with many tiny islands embedded within a chaotic sea [29]. This case exhibits *weak* chaos in the sense of [73] with small Lyapunov exponents which may be difficult to distinguish from those corresponding to regular orbits.

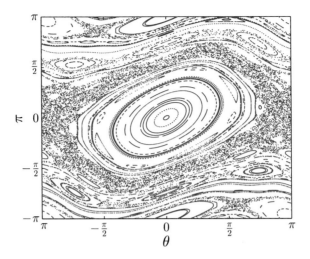

Fig. 7.9 Standard map (7.11) exhibiting chaos with $\kappa = 0.9$

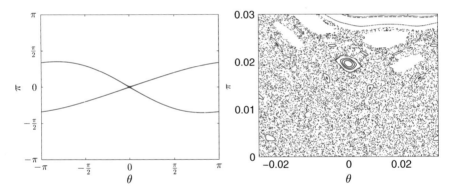

Fig. 7.10 Standard map (7.11) exhibiting weak chaos in a small separatrix layer with $\kappa = -0.3$. The *right figure* is a zoom near the hyperbolic point showing the enlarged stochastic layer

In Figs. 7.11 and 7.12 we show how the 0-1 test is able to detect regular and chaotic orbits, even in the weakly chaotic case. We have chosen 1000^2 initial conditions and run them for 10,000 steps. We used the correlation method and applied it to the modified mean square displacement $D_c(n)$.

7.5.2   Continuous Time Systems

We have so far formulated the 0-1 test for discrete time systems. For continuous time series $\phi(t)$, we obtain a discrete time series $\phi(t_1)$, $\phi(t_2)$, $\phi(t_3)$, ... for given discrete times $0 < t_1 < t_2 < t_3 < \cdots$ to which the test for chaos may be

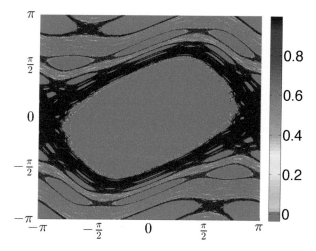

Fig. 7.11 Contour plot of K for the standard map (7.11) exhibiting chaos with $\kappa = 0.9$. We used 1000^2 equally spaced initial conditions and calculated K via the correlation method from D_c

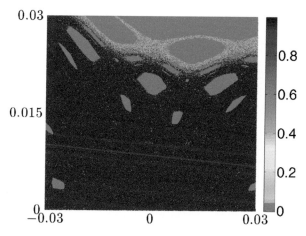

Fig. 7.12 Contour plot of K for the standard map (7.11) exhibiting chaos with $\kappa = -0.3$. We used 1000^2 equally spaced initial conditions and calculated K via the correlation method from D_c

applied as in previous sections. The sequence t_j, $j \geq 1$, has to be chosen in a deterministic manner to assure that the time series $\phi(t_j)$ is deterministic. One may choose the t_j as the intersection times with a cross-section. In this case the time series $\phi(t_j)$ corresponds to observing a Poincaré map. A second, perhaps more usual, approach is to take $t_j = j\tau_s$ where $\tau_s > 0$ is the sampling time. The time series $\phi(t_j) = \phi(j\tau_s)$ corresponds to observing the "time-τ_s" map associated with the underlying continuous time system. Contrary to the case of observing a Poincaré map, in the latter approach one is faced with a well-known oversampling issue: If τ_s is too small, then the system is *oversampled* and this often leads to incorrect results

[22]. Although oversampling is a practical problem for data series of finite size, it should be emphasized that theoretically the test works for all sampling times τ_s in the limit $N \rightarrow \infty$. We now present numerical results for an ordinary differential equation and a partial differential equation where care has to be taken to overcome the issue of oversampling for the realistic case of finite data series.

7.5.2.1 Rössler Equations

To illustrate how the issue of oversampling manifests itself in the 0-1 test for chaos and how to overcome it, as proposed in [22], we consider here the 3-dimensional Rössler system [64]

$$\dot{x} = -y - z$$
$$\dot{y} = x + ay$$
$$\dot{z} = b + z(x - d) . \tag{7.12}$$

For the values $a = 0.432$, $b = 2$ and $d = 4$, the system exhibits chaos with a maximal Lyapunov exponent of about $\lambda_{max} \approx 0.1$ (we use the natural logarithm). We have integrated this system with a fourth-order Runge–Kutta scheme with variable step-size and recorded 100,000 data points each $\Delta t = 0.01$ (i.e. 1000 time units) after disregarding a transient behaviour of 50 time units to allow for the dynamics to settle on the attractor. A plot of the dynamic in the x-y-plane is provided in Fig. 7.13.

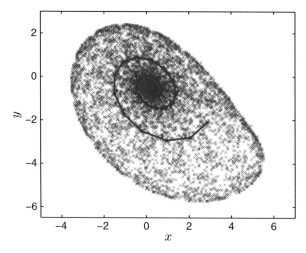

Fig. 7.13 Phase portrait for the Rössler system (7.12). The short trajectory segment (*blue*) was sampled at $\tau_s = 0.5$

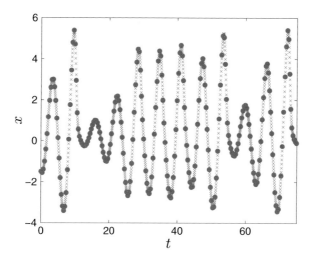

Fig. 7.14 Plot of the observable $\phi(t) = x(t)$ for the Rössler system (7.12). The finely sampled data (*crosses*) are sampled at $\tau_s = 0.05$ time units. The coarsely sampled data (*filled circles*) are sampled at $\tau_s = 0.35$ time units

Figure 7.14 shows an oversampled and a sufficiently coarsely sampled observable for the Rössler system (7.12). The finely sampled time series ($\tau_s = 0.05$) yields $K \approx 0$ whereas the coarsely sampled data ($\tau_s = 0.35$) yields $K \approx 1$ already despite using only 1/7th of the data.

A good choice of the sampling time τ_s can often be obtained by visual inspection as in Fig. 7.14. A more quantitative method is to use the e-folding time of the autocorrelation function or to use the first minimum of the mutual information [16, 31]. For the data depicted in Fig. 7.14 these method yield $\tau_s = 1.15$ and $\tau_s = 1.50$, respectively. However, we observed here that the smaller sampling time $\tau_s = 0.35$ already yields $K \approx 1$ with the advantage of being a longer data set. In general, the optimal sampling time will depend on the dynamical system and the time series under consideration. We refer the reader to [31] for a discussion on optimal time delays in the context of phase space reconstruction.

In the following we show how the issue of oversampling arises in the 0-1 test. For continuous time systems, the (time-averaged) mean square displacement is defined as

$$M_c(t) = \lim_{T \to \infty} \frac{1}{T} \int_0^T (p(t + \tau) - p(\tau))^2 + (q(t + \tau) - q(\tau))^2 \, d\tau \, ,$$

which, for a time series sampled with sample time τ_s, is approximated by

$$M_c(n) = \lim_{N \to \infty} \frac{1}{N} \sum_{j=1}^{N} \left([p_{\tau_s}(j + n) - p_{\tau_s}(j)]^2 + [q_{\tau_s}(j + n) - q_{\tau_s}(j)]^2 \right) \tau_s^2 \, .$$

Similarly the power spectrum for the time-continuous case discretizes to

$$S(\nu) = \lim_{n \to \infty} \frac{1}{n} E \left| \sum_{j=0}^{n-1} e^{2\pi i \frac{\nu}{\nu_s} j} \phi(j) \right|^2 \tau_s^2, \qquad (7.13)$$

where $\nu_s = 1/\tau_s$ is the sample frequency. For chaotic systems the power spectrum decays for large frequencies ν, and so for frequencies larger than some $\nu_{\max}$ the power spectrum is zero for all practical purposes.

Comparing (7.13) with the power spectrum (7.6) for discrete-time data, we identify

$$c = 2\pi \frac{\nu}{\nu_s}, \quad \nu \in [0, \nu_{\max}] . \qquad (7.14)$$

Sampling at the Nyquist rate with $\nu_s^\star = 2\nu_{\max}$ corresponds to $c \in (0, \pi)$ as before. However, oversampling at a higher frequency $\nu_s > \nu_s^\star$, restricts the effective choices of c to $c \in (0, c^\star)$ where $c^\star = \frac{\nu_s^\star}{\nu_s} \pi < \pi$. In this case, the test for chaos may incorrectly classify the dynamics of a chaotic system as regular, since it is possible that more than half of the randomly chosen values of $c \in (0, \pi)$ will lie in $(c^\star, \pi)$ yielding a median $K = 0$. Note that once a sampling time $\tau_s \neq 0$ is fixed, the problem of oversampling cannot be alleviated by increasing the length N of the time series.

We illustrate this using the Rössler system (7.12) sampled with τ_s ranging from $\tau_s = 0.05$ up to $\tau_s = 1$. In Fig. 7.15 the median of the asymptotic growth rate

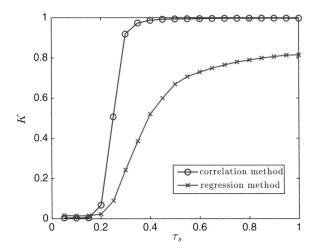

Fig. 7.15 Plot of K as a function of the sample time τ_s for the Rössler system (7.12). At the finest sampling rate $\tau_s = 0.05$ we recorded $N = 100,000$ data points. Results are shown for the correlation method (*circles*) and the regression method (*crosses*)

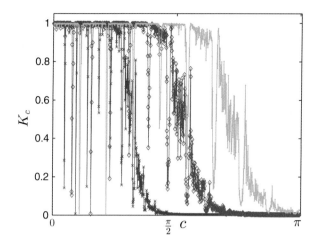

Fig. 7.16 Plot of K_c as a function of the frequency c for the Rössler system (7.12). *From left to right* we used $\tau_s = 0.15$, $\tau_s = 0.25$ and $\tau_s = 0.35$. The corresponding values of the growth rate (calculated using the correlation method) are $K = 0.005$, $K = 0.5$ and $K = 0.97$, respectively. At the sampling rate $\tau_s = 0.05$ we recorded $N = 100,000$ data points

K is shown as a function of the sample time. For data that is too finely sampled, we obtain $K = 0$ although the dynamics is actually chaotic. Figure 7.16 illustrates how the range of effective values of c depends on the sampling time τ_s. The linear scaling of the range of c for which $K_c \approx 1$ as suggested by (7.14) is clearly seen. The pronounced dips of K_c for certain values of c are caused by near resonances which occur in the chaotic Rössler system for our parameter values caused by regularly appearing revolutions of the dynamics as illustrated in Fig. 7.13. Note that the presence of resonances does not affect the value of the median K as seen in Fig. 7.15.

7.5.2.2 Partial Differential Equations

We apply now our test to the driven and damped nonlinear Schrödinger equation

$$\imath q_t + q_{xx} + 2|q|^2 q = -\imath \gamma q + \varepsilon e^{\imath(\omega t + \sigma)},$$

which describes a plasma resonantly driven by a capacitor with frequency ω and damped via collisions [4, 11, 57]. For $q = Q \exp(\imath(\omega t + \sigma))$ we solve

$$\imath Q_t + Q_{xx} + 2|Q|^2 Q = \omega Q - \imath \gamma Q + \varepsilon . \tag{7.15}$$

It is well known that for given system length L the system (7.15) undergoes a period doubling bifurcation route into chaos [57] for increasing values of the driving amplitude ε. We present here results for a system with length $L = 80$,

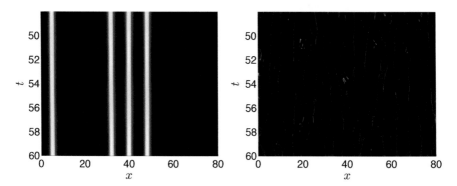

Fig. 7.17 Hovmöller diagram of $|Q(x,t)|$ for regular dynamics with $\varepsilon = 0.095$ (*left*) and chaotic dynamics with $\varepsilon = 0.2$ (*right*) for the driven and damped Schrödinger equation (7.15)

$\gamma = 0.11$, $\omega = 1$ for $\varepsilon = 0.095$ and $\varepsilon = 0.2$ for regular and chaotic dynamics, respectively. The system (7.15) is integrated with a second-order in space and time finite difference Crank–Nicolson solver where the nonlinear term is treated with an Adams–Bashforth scheme. We use $n_x = 256$ grid points and an integration time step of $dt = 0.0001$ and evolve from an initial condition $q = -\iota\sqrt{2} + 0.1\cos(kx)$ with $k = 15$ with reflective (von Neumann) boundary conditions. In Fig. 7.17 we show Hovmöller diagrams of $|Q(x,t)|$ for regular and chaotic dynamics. We construct observables by evaluating the field $Q(x,t)$ at spatial locations $x_j = j\,dx$ with $dx = L/n_x$ and $j = 1, \cdots, n_x$. In particular, we consider the following observables

$$\phi_1(t) = \sum_{j=1}^{n_x} |Q(x_j, t)|,$$

$$\phi_2(t) = |Q(L/2, t)|,$$

$$\phi_3(t) = \sum_{j=1}^{5} |Q(x_{j^\star}, t)|.$$

For the last observable $\phi_3(t)$ we randomly choose five locations $x_{j^\star}$ from the $n_x = 256$ spatial gridpoints at time $t = 0$.

The observables $\phi_{1,2,3}$ are sampled time every 0.3 time units with a total of 10,000 snapshots taken. This sampling time is sufficiently large to avoid the oversampling effects for continuous time systems which would lead to $K \approx 0$ irrespective of the underlying dynamics, as discussed in Sect. 7.5.2.1 (see also [22]). Using the correlation method on $D_c(n)$, for each of the three observables we obtain values of K smaller than 0.0017 in the regular case with $\varepsilon = 0.095$, and values of K within 0.003 from $K = 1$ for the chaotic case.

We note that although the nonlinear Schrödinger equation (7.15) is formally infinite-dimensional, its dynamics evolves on a finite dimensional attractor [11].

7.5.3 Data Contaminated by Noise

For a test for chaos to be able to analyse real world data one needs to show its capability to be able to process observations contaminated by noise. In the following we revisit the example of measurement noise in an 8-dimensional Lorenz-96 model studied in our previous work [20]. There it was shown that our 0-1 test for chaos is far superior to traditional methods using phase space reconstruction and Lyapunov exponents [63, 66], without preprocessing the data with standard noise reduction methods [31].

7.5.3.1 Lorenz-96 System

We revisit the tough test case of analysing quasi-periodic dynamics with measurement noise [20]. In particular we study the Lorenz-96 system [45]

$$\dot{z}_i = z_{i-1}(z_{i+1} - z_{i-2}) - z_i + F \qquad i = 1, \cdots, D \qquad (7.16)$$

with periodic $z_{i+D} = z_i$. This system is a toy-model for midlatitude atmospheric dynamics, incorporating linear damping, forcing and nonlinear transport. In the atmospheric context one usually uses $D = 40$. This particular value for D is chosen such that the spacing between adjacent grid points z_i roughly equals the Rossby radius of deformation at midlatitudes where the circumference of the earth is roughly 30,000 km. The dynamical properties of the Lorenz-96 system have been investigated, for example, in [46, 58]. We use $D = 8$ modes where quasi-periodic windows were found to alternate with chaotic dynamics [58].

  In the previous examples, the test was able to distinguish sharply between regular and chaotic dynamics in noise free data. However such good performance of the test for noise free data is detrimental for noise contaminated data—even small amounts of noise would be detected and noisy regular dynamics would be falsely classified as chaotic. The sensitivity of our test was enhanced by the subtraction of the oscillatory term (cf. Eq. (7.4) and Fig. 7.2), and the application of the correlation method. In [20] our test employed directly the mean square displacement $M_c(n)$ including the oscillatory term rather than the modified version $D_c(n)$. Furthermore, it used the regression method rather than the correlation method to calculate K_c from the mean square displacement $M_c(n)$. We showed that in this case our test greatly outperforms methods involving phase space reconstruction and Lyapunov exponents in distinguishing quasi-periodic dynamics from chaotic dynamics when 10 % measurement noise was added to the observations.

  Here we will revisit our test including a method to deal with measurement noise proposed in [22]: If we assume that the noise is independent of the dynamics, i.e. pure measurement noise, and also independent of the forcing F, we can decompose the linear part of the modified mean square displacement as

$$D_c(n) = (S_{\mathrm{dyn}}(c) + S_{\mathrm{noise}}(c))n + o(n) . \qquad (7.17)$$

Here $S_{\mathrm{dyn}}(c)$ is the variance associated with the underlying deterministic dynamics to be analysed and $S_{\mathrm{noise}}(c)$ is the variance associated with the measurement noise. We can estimate the variance associated with the measurement noise $S_{\mathrm{noise}}(c)$ by estimating $D_c(n)$ from the noisy observations at a parameter F where the dynamics is known to be regular with $S_{\mathrm{dyn}}(c) = 0$. We then estimate $S_{\mathrm{noise}}(c)$ to compensate for the linear growth of $D_c(n)$ due to the noise. We remark that this is not always possible and requires (at least) that gauge experiments can be performed. For example, this method cannot help with studying the regularity of planetary motion for noisy observations. If gauge experiments are not possible, our test can still be used to analyse noise-contaminated experimental data using the formulation proposed in reference [20] as done, for example, in [13, 33, 34, 37, 38]. Once $S_{\mathrm{noise}}(c)$ is estimated the test can proceed with

$$\hat{D}_c(n) = D_c(n) - S_{\mathrm{noise}}(c)\, n \qquad (7.18)$$

as described in the previous sections. In particular we can employ the more sensitive correlation method. Note that the linear growth term of the mean-square displacement will not be entirely eliminated (unless by chance $S_{\mathrm{noise}}(c)$ is correctly estimated from the data) but the proposed scheme controls its magnitude. This allows the test to analyse observational data of finite length; for unlimited noise-contaminated data, one would, of course, obtain $K = 1$, irrespective of the underlying deterministic dynamics.

In Fig. 7.18 we show results of our test using first $M_c(n)$ and the regression method as in [20] and second using $\hat{D}_c(n)$ and the correlation method. We use a fourth-order Runge–Kutta scheme with a time step of $dt = 0.05$ to generate observations

$$\phi(t) = (1 + \zeta)(z_2 + z_3 + z_4). \qquad (7.19)$$

The measurement noise $\zeta = \eta u$ is drawn from a uniform distribution $u \sim \mathscr{U}[-1, 1]$. In Fig. 7.18 we show results for noise-free observations with $\eta = 0$, and for measurement noise levels of 10 % and 20 % with $\eta = 0.1$ and $\eta = 0.2$, respectively. Observations are taken every 2.5 time units and a total of $N = 100{,}000$ observations are taken.

We gauge the variance due to the noise at $F = 5.25$. We obtain $S_{\mathrm{noise}} = 0.3$ and $S_{\mathrm{noise}} = 0.62$ for a noise level η of 10 % and 20 % observational noise, respectively. Both methods detect the quasi-periodic windows well for 10 % measurement noise (we remark that methods relying on phase space reconstruction and maximal Lyapunov exponents were not able to accurately distinguish quasi-periodic dynamics from chaotic dynamics [20]). The distinction between quasi-periodic dynamics and chaotic dynamics with a noise level of 20 % is less clear in the test employing $M_c(n)$ but still remarkably good when using $\hat{D}_c(n)$. We propose this example which involves noise contaminated quasi-periodic dynamics as a challenge for other tests.

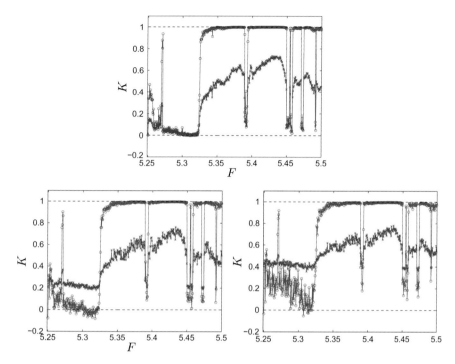

Fig. 7.18 Plot of K versus F for the Lorenz-96 system (7.16) for $5.25 \leq F \leq 5.5$ increased in increments of 0.005. We used $N = 100,000$ data points sampled at 2.5 time units. *Top*: Noise free data; *Bottom left*: 10 % measurement noise; *Bottom right*: 20 % measurement noise. K is calculated via the regression method for $M_c(n)$ (*crosses, blue*) and for the correlation method for $D_c(n)$ with subtracted noise variance (*circles, magenta*). The *horizontal lines* indicate the limiting cases $K = 0$ and $K = 1$. We used 100 randomly distributed values of c

7.6 Summary

We have described the 0-1 test for chaos, focusing on its implementation and several practical issues as well as on the theoretical justifications outlining the realm of validity of the test. We have illustrated the versatility and efficiency of our method by treating the notoriously difficult case of weakly chaotic separatrix layers in the standard map as well as analysing measurement noise contaminated data.

The advantage of our method lies in (a) its computational low cost and ease of implementation, (b) its generality of applicability independent of the nature of the dynamical system and its dimension, (c) its working directly with the time series without the need for phase space reconstruction, (d) its ability to detect weak chaos and (e) its ability to detect regular behaviour within noisy data. In particular, we mention the 8-dimensional Lorenz-96 model contaminated by noise (see Sect. 7.5.3). We are not aware of any other method that comes close to matching the effectiveness of our test for this example.

The theoretical justification of the 0-1 test depends on the nature of attractors for general smooth (or piecewise smooth) dynamical systems. In [23] we challenged the skeptical reader to construct a robust smooth example where the test fails. So far, no such example has come to light. This was explored further in [26] where we formulated a conjecture which, roughly speaking, states that for typical smooth dynamics, either $K_c = 1$ for almost every c or $K_c = 0$ for almost every c.

Acknowledgements GAG acknowledges support from the Australian Research Council. The research of IM was supported in part by the European Advanced Grant StochExtHomog (ERC AdG 320977).

References

1. Ashwin, P., Melbourne, I., Nicol, M.: Hypermeander of spirals; local bifurcations and statistical properties. Physica D **156**, 364–382 (2001)
2. Bernardini, D., Rega, G., Litak, G., Syta, A.: Identification of regular and chaotic isothermal trajectories of a shape memory oscillator using the 0–1 test. Proc. Inst. Mech. Eng. K: J. Multibody Dyn. **227**(1), 17–22 (2013)
3. Cafagna, D., Grassi, G.: An effective method for detecting chaos in fractional-order systems. Int. J. Bifurc. Chaos **20**(3), 669–678 (2010)
4. Cai, D., McLaughlin, D.W.: Chaotic and turbulent behaviour of unstable one-dimensional nonlinear dispersive waves. J. Math. Phys. **41**(6), 4125–4153 (2000)
5. Cao, J., Syta, A., Litak, G., Zhou, S., Inman, D., Chen, Y.: Regular and chaotic vibration in a piezoelectric energy harvester with fractional damping. Eur. Phys. J. Plus **130**(6) (2015)
6. Chirikov, B.V.: A universal instability of many-dimensional oscillator systems. Phys. Rep. **52**, 263 (1979)
7. Chowdhury, D.R., Iyengar, A.N.S., Lahiri, S.: Gottwald Melborune (0–1) test for chaos in a plasma. Nonlinear Process. Geophys. **19**(1), 53–56 (2012)
8. Dafilis, M., Frascoli, F., McVernon, J., Heffernan, J.M., McCaw, J.M.: The dynamical consequences of seasonal forcing, immune boosting and demographic change in a model of disease transmission. J. Theor. Biol. **361**, 124–132 (2014)
9. Dafilis, M., Frascoli, F., McVernon, J., Heffernan, J., McCaw, J.: Dynamical crises, multistability and the influence of the duration of immunity in a seasonally-forced model of disease transmission. Theor. Biol. Med. Model. **11**(1), 43 (2014)
10. Diddens, C., Linz, S.J.: Continuum modeling of particle redeposition during ion-beam erosion. Eur. Phys. J. B **86**(9), 1–13 (2013)
11. Eickermann, T., Grauer, R., Spatschek, K.H.: Identification of mass capturing structures in a perturbed nonlinear Schrödinger equation. Phys. Lett. A **198**, 383–388 (1995)
12. Erzgräber, H., Wieczorek, S., Krauskopf, B.: Dynamics of two semiconductor lasers coupled by a passive resonator. Phys. Rev. E **81**, 056201 (2010)
13. Falconer, I., Gottwald, G.A., Melbourne, I., Wormnes, K.: Application of the 0–1 Test for chaos to experimental data. SIAM J. Appl. Dyn. **6**, 395–402 (2007)
14. Field, M., Melbourne, I., Török, A.: Decay of correlations, central limit theorems and approximation by Brownian motion for compact Lie group extensions. Ergodic Theor. Dyn. Syst. **23**, 87–110 (2003)
15. Field, M., Melbourne, I., Török, A.: Stable ergodicity for smooth compact Lie group extensions of hyperbolic basic sets. Ergodic Theor. Dyn. Syst. **25**, 517–551 (2005)
16. Fraser, A.M., Swinney, H.L.: Independent coordinates for strange attractors from mutual information. Phys. Rev. A **33**, 1134–1140 (1986)

17. Gaspard, P., Wang, X.-J.: Sporadicity: between periodic and chaotic dynamical behaviours. Proc. Natl. Acad. Sci. USA **85**, 4591–4595 (1988)
18. Gopal, R., Venkatesan, A., Lakshmanan, M.: Applicability of 0–1 test for strange nonchaotic attractors. Interdiscip. J. Nonlinear Sci. **23**(2), 023123 (2013)
19. Gottwald, G.A., Melbourne, I.: A new test for chaos in deterministic systems. Proc. R. Soc. A **460**, 603–611 (2004)
20. Gottwald, G.A., Melbourne, I.: Testing for chaos in deterministic systems with noise. Physica D **212**(1–2), 100–110 (2005)
21. Gottwald, G.A., Melbourne, I.: Comment on "Reliability of the 0–1 test for chaos". Phys. Rev. E **77**, 028201 (2008)
22. Gottwald, G.A., Melbourne, I.: On the implementation of the 0–1 test for chaos. SIAM J. Appl. Dyn. **8**, 129–145 (2009)
23. Gottwald, G.A., Melbourne, I.: On the validity of the 0–1 test for chaos. Nonlinearity **22**, 1367–1382 (2009)
24. Gottwald, G.A., Melbourne, I.: A Huygens principle for diffusion and anomalous diffusion in spatially extended systems. Proc. Natl. Acad. Sci. USA **110**, 8411–8416 (2013)
25. Gottwald, G.A., Melbourne, I.: Central limit theorems and suppression of anomalous diffusion for systems with symmetry (2013, submitted)
26. Gottwald, G.A., Melbourne, I.: A test for a conjecture on the nature of attractors for smooth dynamical systems. Chaos **24**, 024403 (2014)
27. Gouëzel, S.: Central limit theorem and stable laws for intermittent maps. Probab. Theor. Relat. Fields **128**, 82–122 (2004)
28. He, K., Xu, Y., Zou, Y., Tang, L.: Electricity price forecasts using a Curvelet denoising based approach. Phys. A Stat. Mech. Appl. **425**, 1–9 (2015)
29. Howard, J.: Discrete virial theorem. Celest. Mech. Dyn. Astron. **92**(1–3), 219–241 (2005)
30. Hu, H.: Decay of correlations for piecewise smooth maps with indifferent fixed points. Ergodic Theor. Dyn. Syst. **24**, 495–524 (2004)
31. Kantz, H., Schreiber, T.: Nonlinear Time Series Analysis. Cambridge University Press, Cambridge (1997)
32. Kędra, M.: Deterministic chaotic dynamics of Raba River flow (Polish Carpathian Mountains). J. Hydrol. **509**, 474–503 (2014)
33. Krese, B., Govekar, E.: Nonlinear analysis of laser droplet generation by means of 0–1 test for chaos. Nonlinear Dyn. **67**, 2101–2109 (2012)
34. Krese, B., Govekar, E.: Analysis of traffic dynamics on a ring road-based transportation network by means of 0–1 test for chaos and Lyapunov spectrum. Transp. Res. Part C Emerg. Technol. **36**, 27–34 (2013)
35. Kulp, C.W., Smith, S.: Characterization of noisy symbolic time series. Phys. Rev. E **83**, 026201 (2011)
36. Kříž, R.: Chaotic analysis of the GDP time series. In: Zelinka, I., Chen, G., Rössler, O.E., Snasel, V., Abraham, A. (eds.) Nostradamus 2013: Prediction, Modeling and Analysis of Complex Systems. Advances in Intelligent Systems and Computing, vol. 210, pp. 353–362. Springer International Publishing, Berlin (2013)
37. Kříž, R.: Finding chaos in finnish gdp. Int. J. Autom. Comput. **11**(3), 231–240 (2014)
38. Kříž, R., Kratochvíl, Š.: Analyses of the chaotic behavior of the electricity price series. In: Sanayei, A., Zelinka, I., Rössler, O.E. (eds.) ISCS 2013: Interdisciplinary Symposium on Complex Systems. Emergence, Complexity and Computation, vol. 8, pp. 215–226. Springer, Berlin, Heidelberg (2014)
39. Leon, F.: Design and evaluation of a multiagent interaction protocol generating behaviours with different levels of complexity. Neurocomputing **146**, 173–186 (2014)
40. Li, X., Gao, G., Hu, T., Ma, H., Li, T.: Multiple time scales analysis of runoff series based on the Chaos theory. Desalin. Water Treat. **52**(13–15), 2741–2749 (2015)
41. Lichtenberg, A., Lieberman, M.: Regular and Chaotic Dynamics. Applied Mathematical Sciences. Springer, New York (1992)

42. Litak, G., Syta, A., Wiercigroch, M.: Identification of chaos in a cutting process by the 0-1 test. Chaos Solitons Fractals **40**, 2095–2101 (2009)
43. Litak, G., Radons, G., Schubert, S.: Identification of chaos in a regenerative cutting process by the 0-1 test. Proc. Appl. Math. Mech. **9**(1), 299–300 (2009)
44. Liverani, C., Saussol, B., Vaienti, S.: A probabilistic approach to intermittency. Ergodic Theor. Dyn. Syst. **19**, 671–685 (1999)
45. Lorenz, E.N.: Predictability - a problem partly solved. In: Palmer, T. (ed.) Predictability. European Centre for Medium-Range Weather Forecast, Shinfield Park, Reading (1996)
46. Lorenz, E.N., Emanuel, K.A.: Optimal sites for supplementary weather observations: simulation with a small model. J. Atmos. Sci. **55**(3), 399–414 (1998)
47. Lugo-Fernández, A.: Is the loop current a chaotic oscillator? J. Phys. Oceanogr. **37**(6), 1455–1469 (2007)
48. Martinsen-Burrell, N., Julien, K., Petersen, M.R., Weiss, J.B.: Merger and alignment in a reduced model for three-dimensional quasigeostrophic ellipsoidal vortices. Phys. Fluids **18**, 057101 (2006)
49. MATLAB: version 7.10.0 (R2010a). The MathWorks Inc., Natick, MA (2010)
50. McLennan-Smith, T.A., Mercer, G.N.: Complex behaviour in a dengue model with a seasonally varying vector population. Math. Biosci. **248**, 22–30 (2014)
51. Melbourne, I., Gottwald, G.A.: Power spectra for deterministic chaotic dynamical systems. Nonlinearity **21**, 179–189 (2008)
52. Melbourne, I., Nicol, M.: Statistical properties of endomorphisms and compact group extensions. J. Lond. Math. Soc. **70**, 427–446 (2004)
53. Melbourne, I., Török, A.: Statistical limit theorems for suspension flows. Israel J. Math. **144**, 191–209 (2004)
54. Melbourne, I., Zweimüller, R.: Weak convergence to stable Lévy processes for nonuniformly hyperbolic dynamical systems. Ann. Inst. H. Poincaré (B) Probab. Stat. **51**, 545–556 (2015)
55. Nair, V., Sujith, R.: A reduced-order model for the onset of combustion instability: physical mechanisms for intermittency and precursors. Proc. Combust. Inst. **35**(3), 3193–3200 (2015)
56. Nicol, M., Melbourne, I., Ashwin, P.: Euclidean extensions for dynamical systems. Nonlinearity **14**, 275–300 (2001)
57. Nozaki, K., Bekki, N.: Low-dimensional chaos in a driven damped nonlinear Schrödinger equation. Physica D **21**, 381–393 (1986)
58. Orrell, D., Smith, L.: Visualising bifurcations in high dimensional systems: the spectral bifurcation diagram. Int. J. Bifurcat. Chaos **13**(10), 3015–3028 (2003)
59. Pomeau, Y., Manneville, P.: Intermittent transition to turbulence in dissipative dynamical systems. Commun. Math. Phys. **74**, 189–197 (1980)
60. Prabin Devi, S., Singh, S.B., Surjalal Sharma, A.: Deterministic dynamics of the magnetosphere: results of the 0-1 test. Nonlinear Process. Geophys. **20**(1), 11–18 (2013)
61. Press, W.H., Teukolsky, S.A., Vetterling, W.T., Flannery, B.P.: Numerical Recipes 3rd Edition: The Art of Scientific Computing, 3rd edn. Cambridge University Press, New York (2007)
62. Radons, G., Zienert, A.: Nonlinear dynamics of complex hysteretic systems: oscillator in a magnetic field. Eur. Phys. J. Spec. Top. **222**(7), 1675–1684 (2013)
63. Rosenstein, M.T., Collins, J.J., De Luca, C.J.: A practical method for calculating largest Lyapunov exponents from small data sets. Physica D **65**, 117–134 (1993)
64. Rössler, O.: An equation for continuous chaos. Phys. Lett. A **57**(5), 397–398 (1976)
65. Swathy, P.S., Thamilmaran, K.: Dynamics of SC-CNN based variant of MLC circuit: an experimental study. Int. J. Bifurcat. Chaos **24**(02), 1430008 (2014)
66. Takens, F.: Detecting strange attractors in turbulence. In: Dynamical Systems and Turbulence, Warwick 1980 (Coventry 1979/1980). Lecture Notes in Mathematics, vol. 898, pp. 366–381. Springer, Berlin (1981)
67. Tsai, T.-L., Dawes, J.H.: Dynamics near a periodically-perturbed robust heteroclinic cycle. Physica D **262**, 14–34 (2013)
68. Webel, K.: Chaos in German stock returns - new evidence from the 0–1 test. Econ. Lett. **115**(3), 487–489 (2012)

69. Xin, B., Li, Y.: 0-1 test for chaos in a fractional order financial system with investment incentive. Abstr. Appl. Anal. **2013**, 876298 (2013)
70. Xin, B., Zhang, J.: Finite-time stabilizing a fractional-order chaotic financial system with market confidence. Nonlinear Dyn. **79**(2), 1399–1409 (2014)
71. Yuan, L., Yang, Q., Zeng, C.: Chaos detection and parameter identification in fractional-order chaotic systems with delay. Nonlinear Dyn. **73**(1–2), 439–448 (2013)
72. Zachilas, L., Psarianos, I.N.: Examining the chaotic behavior in dynamical systems by means of the 0–1 test. J. Appl. Math. **2012**, 681296 (2012)
73. Zaslavskii, G.M., Sagdeev, R.Z., Usikov, D.A., Chernikov, A.A., Sagdeev, A.R.: Chaos and Quasi-Regular Patterns. Cambridge University Press, Cambridge (1992)
74. Zweimüller, R.: Stable limits for probability preserving maps with indifferent fixed points. Stoch. Dyn. **3**, 83–99 (2003)

Chapter 8
Prediction of Complex Dynamics: Who Cares About Chaos?

Stefan Siegert and Holger Kantz

Abstract We compile knowledge on limitations to prediction of the time evolution of complex systems. Although such systems are typically highly chaotic, the inverse of the maximal Lyapunov exponent, the Lyapunov time, is not the time scale beyond which predictions fail. Instead, as the example of weather forecasting will show, predictions can be successful on lead times which are several orders of magnitude longer. We analyze the reasons which prevent errors from growing exponentially fast with a rate related to the maximal Lyapunov exponent. Moreover, we advocate that standard practices from weather forecasting should be transferred to other fields of complex systems' predictions, which includes a statement about the uncertainty related to the actual prediction and a performance measure on past predictions so that a decision maker can assess the potential quality of a forecasting scheme.

8.1 What Do We Want to Predict?

Throughout human history we find reports about attempts to predict the future. The targets of prediction were typically quite complex, such as future personal fate, the outcome of a war, or about the richness of harvest. Most ancient prediction methods (e.g., oracles of Greek and Roman antiquity) are nowadays not any more considered as scientifically sound. The skeptic, however, might think similarly about, e.g., the statements of economic research institutes which regularly publish predictions, precise up to tenths of percentages, of economic growth or tax volume change without providing any estimates of prediction uncertainty and without explicitly mentioning the scenarios for which such predictions were made [1].

S. Siegert (✉)
College of Engineering, Mathematics and Physical Sciences, University of Exeter, Laver Building, North Park Road, Exeter EX4 4QE, UK
e-mail: s.siegert@exeter.ac.uk

H. Kantz
Max Planck Institute for the Physics of Complex Systems, Nöthnitzer Str. 44, 01187 Dresden, Germany
e-mail: kantz@pks.mpg.de

© Springer-Verlag Berlin Heidelberg 2016   249
Ch. Skokos et al. (eds.), *Chaos Detection and Predictability*, Lecture Notes
in Physics 915, DOI 10.1007/978-3-662-48410-4_8

It is, however, fact that a large part of our daily decisions are based on or at least influenced by predictions. It is another fact that most of these predictions concern very complex systems where our understanding is limited. In this article, we want to discuss the issue of predictions and predictability of complex systems, analyzing the sources of prediction errors, suggesting optimal prediction strategies, and arguing for a fair presentation of not only the prediction result, but also of its uncertainty.

We will use weather forecasts as our reference, which allows us to illustrate most of our statements. Weather forecasts are the most sophisticated and most developed forecasts known to us, based on profound physical understanding of the system and its dynamics, on an excellent network of observations of the current state, and on a huge number of people and computers generating the daily forecasts and further improving the forecast quality. Also, weather forecasters have done pioneering work in assessing the uncertainty of their forecasts (e.g., by ensemble forecasts [2]), in developing scoring schemes (e.g., [3, 4]), and improving predictions by postprocessing model output and by blending new measurements and past forecasts into new initial conditions (data assimilation).

Weather forecasts are omnipresent. Actually, weather forecasting has become a considerable business. Many sectors of economy, not only agriculture and traffic, but also renewable energy production and energy consumption are strongly dependent on weather and therefore benefit from accurate predictions [5]. Even if for most of us it would not be a real problem to be mislead by an inaccurate forecast, we are aware of, and often like to complain about lack of forecast accuracy. Considering huge effort spent for weather prediction, in terms of persons, measurement stations, computer power, we conclude that forecasting the weather is very difficult. The most evident reason for this seems to be the underlying chaoticity or the turbulent nature of air flow.

Weather hence is a typical complex phenomenon, and its complexity has several aspects. First of all, as said, the global hydrodynamic transport equations for air, the Navier Stokes equations, yield turbulent solutions with positive Lyapunov exponents and hence with sensitive dependence on initial conditions. Secondly, turbulence is a multi-scale problem, related to the fact that the atmosphere is a continuum system in 3-d space, with a very small cut-off scale where molecular dissipation sets in. Thirdly, beyond that, weather is a consequence of the interaction of many different physical (and chemical and biological) processes which take place on very different time and length scales. On the other hand, weather is a natural system of which we believe that in principle we can set up model equations for all these sub-processes.

Economy and finance, to mention another field where predictions are highly desirable and part of the daily business, suffers from the lack of first-principles models and from the lack of "observation" data to assess the current state of the system: many relevant economic data are acquired with some time lag (such as unemployment rates, tax revenue, economic growth), and many relevant data are not publicly available. But in addition, these systems are controlled by humans. This has two striking consequences: Firstly, humans can introduce innovations and hence are able to change the rule of the game by their knowledge and activities. This means that model equations might become invalid in the course of time, and even new degrees of freedom might be introduced. Secondly, humans are responding to forecasts. Whereas the weather does its thing regardless of what we predict about it, economy and finance will react to forecasts. This makes it hard to verify forecasts, and even harder to control the system. Nonetheless, it is plausible that also economy shares relevant aspects of complexity with weather: The sensitive dependence on tiny details in initial conditions, the existence of micro- and macro-scales, the interaction of subsystems.

In the first part of this article, where we will be concerned with model based forecast, we clearly have to require the existence of reasonable mathematical models of the phenomenon to forecast. In the second part, where we discuss data based forecasts, more general situations of complex systems are addressed. For the atmosphere, *chaos* is considered to be one of the main limitation of predictability [6]: Chaos is defined by the mixing property and hence is apparent through temporal irregularity of the solutions. The lack of predictability is evident from the linear instability with respect to tiny errors in initial conditions: as it is well known, such errors can grow exponentially fast in time, if there is at least one positive Lyapunov exponent. This exponential error growth is so fast that it always will win over an improved precision by which the initial condition is determined. For long and inspired by Lorenz' pioneering work [6], this was considered to be the main limitation for accurate predictions in complex systems.

When we speak of complex systems in this article, we mean really complex systems such as the atmosphere, large economic or social systems, or technical systems such as the power grid or telecommunication networks. But also biological systems such as ecosystem dynamics and the irregular fluctuations of physiological parameters such as blood pressure or heart rate have been discussed in the context of chaos and have been subject to prediction attempts (e.g., [7]). Such systems are not only chaotic, but in addition they have a huge dimensional phase space and are coupled to some environment, which often cannot be explicitly modelled or observed. They are composed of different sub-systems, which introduces a strong inhomogeneity among the different degrees of freedom, and they typically also act on a wide range of different time and length scales. All this together introduces types of instability and limitations to successful predictions which go far beyond the consequences of chaos.

In this article, we will pick up arguments showing that the issue of whether or not such a complex system is chaotic, and in particular the value of its largest Lyapunov exponent, is not really relevant for the accuracy of many types of predictions: Both the difficulties in forecasting and the methods to perform good forecasts do not really care about linear instability. We will present both, obstacles against good forecasts, and the current answer to this, where we will distinguish between data based forecasts and model based forecasts. Most of our considerations are not at all new—an accessible and more comprehensive discussion of some of them can be found in the book by L. Smith [8].

8.2 Sources of Unpredictability

8.2.1 Chaos

As is well known, chaotic dynamics suffers from sensitive dependence on initial conditions. This creates an evident source of unpredictability due to the exponential divergence of nearby trajectories: Given a chaotic dynamical system f and two initial conditions x_0 and y_0 with a very small Euclidean distance, $||x_0 - y_0|| = \epsilon \ll 1$, we consider the two trajectories emerging from x_0 and y_0 as $x(t)$ and $y(t)$. If a system has at least one positive Lyapunov exponent $\lambda_{max} > 0$, then $||x(t) - y(t)|| \simeq \epsilon \exp(\lambda_{max} t)$ for large t and for Lebesgue-almost all x_0 and y_0 with $||x_0 - y_0|| = \epsilon \ll 1$. (Remark: In order to be chaotic, a system with continuous time must possess an at least 3-dimensional phase space, i.e., x and y are vectors.) There is, however, a transient time during which the distance between $x(t)$ and $y(t)$ might grow faster or slower than the asymptotic exponential growth, depending on the direction of the initial perturbation and on the structure of the local Jacobians. The exponential growth is a limiting behavior in the double and non-exchangeable limit $||\epsilon|| \rightarrow 0$ and $t \rightarrow \infty$.

It is evident that even with perfect model equations the prediction of every natural system will suffer from exponential error growth: We can never assume to know the "true" state of the observed system with arbitrary precision, so that our initial condition inserted into the model equations will always deviate from the true state. Whatever tiny this errors is, after some time the exponential factor $\exp(\lambda_{max} t)$ will be large enough to make it visible and to eventually make our forecast trajectory diverge fully from the true one. Actually, the time $1/\lambda_{max}$ is often called the *Lyapunov time* and determines the order of magnitude of the time when this error growth is dominant. If we assume an initial error of size ϵ, then it will be of size unity after $-\ln \epsilon$ times the Lyapunov time. This illustrates that when increasing the precision of the knowledge of the initial condition by orders of magnitudes, the time till the error reaches order one increases only linearly [9].

In weather forecasting, the error in initial conditions is not a simple consequence of measurement inaccuracies. It is much more determined by the lack of suitable observations: Whereas a weather model, let it be global or regional, due to limitations of computer power currently represents the continuum of the atmosphere by up to 10^{10} degrees of freedom[1] located on a 3-dimensional grid in space. The World Meteorological Organization lists only of the order of 10^5 measurement stations which provide direct observations, in addition there are satellite data which are often projections along the vertical axis. From these observed data, a model state has to be constructed which has a much higher dimension. The logic behind this is, see below, the belief that the true attractor dimension, and hence the effective number of degrees of freedom, is much lower than 10^{10}, so that much less observations

---

[1] For example, the ECMWF model in the current 40th cycle is a T1279 model with 91 vertical layers.

might suffice in order to determine where on the attractor the new initial condition should be located. How does one constrain the initial condition to the attractor? This is implicitly done by techniques summarized as *data assimilation* [10]: Data assimilation tries to merge the information obtained from the new, most recent set of measurement data with the model trajectory of the past few days in order to create a model vector which is both close to the model attractor and is close to the observed data. Despite great successes, this inherent lack of observation data (and its very inhomogeneous distribution across the Earth) causes considerable errors of initial conditions, which, since they are constructed using the model dynamics, also are affected by model errors.

An additional problem lies exactly in these model equations: We can not assume that they are perfect. In the best case, there is simply a small error in parameter values contained in these equations. Model errors would first (say, after one time step) introduce a tiny error in our forecast even if the initial condition were perfectly known. Then, this deviation between forecast trajectory and true trajectory will grow again exponentially due to chaos irrespectively of the model error. Hence, the effect of small model errors is similar to the effect of errors in the initial conditions, if the motion is chaotic.

There is an additional and potentially very severe problem with model errors: Lack of structural stability. In physicists' words, the system dynamics is structurally stable, if a small change of model parameters can be compensated by a coordinate transform which should be close to identity. That is, the properties of the attractor such as invariant measure, Lyapunov exponent, KS-entropy do not change. Almost all studied low-dimensional models are non-hyperbolic, which means that they are *not* structurally stable. Hence, a tiny perturbation of some model parameter might result in a very different dynamical behavior, e.g., the "wrong" model might possess an attracting periodic orbit, whereas the "true" system is chaotic. A prominent example for such behavior is the logistic equation and the Hénon map, which have tiny periodic windows all-over their parameter space. There is common believe (and this is formalized in the "chaotic hypothesis" [11]), that high dimensional systems, even if not structurally stable in the strict mathematical sense, behave practically like hyperbolic systems. And indeed, free running GCMs seem not to lock into periodic behavior when parameters are slightly detuned.

Aren't the last few paragraphs a demonstration of how relevant chaos is in limiting predictions of complex systems? Yes, they are, and the reasoning above has been stressed many times during the past 50 years.

However, when we aim at predictions of really complex systems, i.e., systems much more complex than a simple chaotic few-degrees of freedom system, then several additional aspects arise which might be more relevant to forecasts than the system's chaoticity, and these are essentially all consequences of the phase space being very high dimensional.

8.3   Beyond Chaos

In this section we will discuss why very often chaos, although present in the system for which one intends to make forecasts, is not the limiting aspect. Most of these arguments are not new, but we are not aware that they have been put together before as we do it here.

8.3.1   Outside the Linear Regime

The exponential divergence and thereby exponential error growth is only present when errors are tiny, in the so called linear regime. We argue that in many real world prediction tasks, one works outside this regime, and we refer here again to our favorite example, weather forecasts.

In Fig. 8.1 we show root mean squared (rms) prediction errors of temperature forecasts for Hanover, Germany, for four different forecast schemes, as a function of lead time. The benchmark of forecasting is the so called *climatology* $c(d)$, $d = 1, \ldots, 365$. Climatology is the average value at a given day d in the year, averaging over many such days d from past years. As a function of the date d, climatology $c(d)$ shows the seasonal cycle. It is the systematic part of temperatures, whereas the deviation of the actual temperature on some day, $T(d)$, from $c(d)$ is due to the specific weather at the particular time. Actually, $T(d) - c(d)$ is called *temperature*

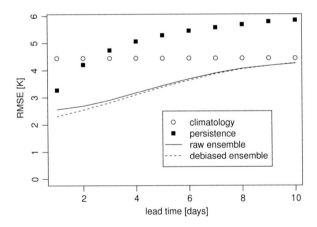

Fig. 8.1 Forecast error of temperature as a function of lead time: Shown are the root mean squared prediction errors of the three forecast schemes "persistence" (future temperature is identical to today's temperature), "climatology" (future temperature is the multi-annual mean temperature for that day of the year), and a model forecast obtained from the NCEP/NCAR re-forecast project, for temperature predictions in Hanover (Germany) averaged over the years 1981–2010 with and without re-calibration of the model output

anomaly and its prediction is the nontrivial part of a temperature forecast. Therefore, the rms error of climatology represents the standard deviation of the distribution of true temperature values around the long year average, averaged over every day in the year. Roughly, this standard deviation is constant over the year for the Hanover station, so that the rms error of climatology represents the uncertainty of what we need to forecast: Knowing the date, one knows that the true temperature will be distributed around $c(d)$ with a distribution whose standard deviation is, as we see from Fig. 8.1, about 4 K. Actually, this distribution is in rather good approximation symmetric and Gaussian (not shown here, see [12]). In other words, if we subtract the seasonal cycle, the task of weather forecasting is to predict which value from this approximately Gaussian distribution will be realized. Its standard deviation therefore sets the scale for the dynamical range of the quantity to be predicted. This range is much smaller than the range of the temperatures proper, since the latter can be seen as the superposition of a specific weather related value over the seasonal cycle.

The rms error of the calibrated weather model, i.e., a weather model output with adjustments to compensate for systematic errors observed in the past, is about 2 K, i.e., one half of the rms error of climatology. In the context of time series analysis [13], one normalizes the rms prediction error of a forecast scheme by the standard deviation of the quantity to be predicted, and in this sense the average error of the re-calibrated weather model output is about 50 %. Is that good or bad? That is not our concern here, although we should say that there are better models: We use data from a global weather model from a reanalysis/reforecast project [14], which provides us with the excellent statistics of more than 40 years of forecast/analysis pairs with the very same model. The local prediction suffers from the fact that this model is not perfectly up to date, and that it is a global model. State-of-the-art regional weather models predict temperature anomalies with about 25 % error. Our concern is that if at the shortest lead time of 24 h the rms error is already 50 % (or 25 %) of the standard deviation of the signal, then we are definitely out of the linear regime of the underlying dynamics. Trajectories are such far away from the assumed true trajectory that we are already closer to the saturation regime than we are to the exponential divergence regime. Hence, chaos is not the main cause of further error growth when we increase lead time up to 10 days.

We can illustrate the fact that typical distances between trajectories are so large that exponential divergence is not any more relevant in two more ways. Both are related to the fact that for state-of-the-art weather predictions, ensembles forecasts are performed [2]. This means that a deterministic model of the atmosphere is run with several slightly different initial conditions. Why this is done will be discussed more thoroughly in Sect. 8.4. Here we want to stress that although the number of ensemble members is limited to 10–50 by the demand of saving computation time to do real time predictions, such ensembles often include for every perturbed initial condition $x + \epsilon u$ the perturbation into the opposite direction, $x - \epsilon u$ (here, x is the high-dimensional state vector, u is a unit vector in state space, and ϵ is the amplitude of the perturbation). If errors stayed in the linear regime, then one of the two trajectories emerging from these two initial conditions would be redundant, and in particular their average would exactly yield the unperturbed solution. That they are both included in the ensemble shows that generally they do not produce redundant information, simply because they soon leave the linear regime around the unperturbed solution.

Error growth as a function of error magnitude has been characterized by the finite size Lyapunov exponent of Boffetta et al. [15]. There it was shown for much less complex systems than weather that scale dependent measures of instability can be employed to characterize the growth of finite errors and that indeed such error growth might depend strongly on the magnitude of the error, as opposed to the mathematical limit of infinitesimal errors.

That ensemble members do not diverge from each other exponentially fast can also be visualized directly. In Fig. 8.2 we show 13 traces of temperature over a lead time of 10 days, together with the observed value. These traces are the temperature values of 13 different initial conditions of a weather model, and on visual inspection there is no evidence that they diverge exponentially from each other.

One could argue that local stretching rates or *finite time Lyapunov exponents* (FTLE) [16] do fluctuate along a trajectory, as a consequence of the fact that instability varies over the phase space. Indeed, this is a relevant aspect also for forecasting: There are situations, where a forecast can be more precise, and others where it is less precise, as a function of the current state of the system. However, if one averages FTLEs over many points in phase space, their average is usually close to the maximum Lyapunov exponent (they are identical to the maximum Lyapunov exponent only if the most expanding direction is not a function of the phase space point), whereas the average over the spread of many ensembles does not yield an

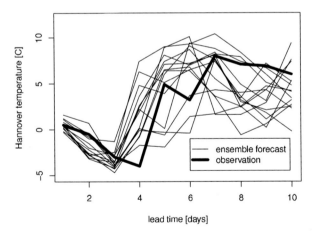

Fig. 8.2 Sketch of an ensemble forecast: Shown are temperature forecasts for the city of Hanover, initialized at December 31, 2007, up to 10 days into the future. The *bold black line* represents the actual measurements, whereas the 13 *thin lines* are model forecasts with 13 slightly different initial conditions, as they are produced by the NCEP/NCAR ensemble forecast within the reanalysis/reforecast project [14]. Notice that the divergence of nearby trajectories is in no stage exponential in time but instead very irregular

estimate of the maximal exponent, simply because they are not reflecting the growth of infinitesimal errors.

In summary, the uncertainty of the initial condition is so big that two initial conditions which are, within these uncertainties, both candidates for the true atmospheric state, depart from each other not exponentially fast but with some very complicated behavior which is slower than exponential. Without any doubt, this complicated behavior has its origin in complicated, nonlinear dynamics, or even in the aperiodic nature of chaotic flows. Nonetheless, exponential growth of infinitesimal perturbations due to chaos is irrelevant, and the error growth in a complex nonlinear stochastic process would look very much alike.

8.3.2 Dilute Phase Space

For the simulation of a PDE system on a computer, one needs some coarse graining, i.e., one projects the equations of motion from an infinite dimensional phase space to a finite dimensional one. It has been shown in many studies [17, 18] that the attractor dimension of such approximations as, e.g., represented by the Kaplan–Yorke dimension, approach a constant when spatial resolution of the discrete system is improved beyond a certain level. In other words, the positive Lyapunov exponents remain essentially unchanged and only negative ones are added, if one increases the phase space dimension by better spatial resolution beyond a physically motivated minimum. A typical complex system such as the dynamics of the atmosphere has an attractor dimension which is indeed large, but finite, as to be compared to the in principle infinite phase space dimension (see, e.g., [19]). The study in [20] might serve as an example of a quantitative result: a quasi-geostrophic model with 1449 degrees of freedom is studied, its Kaplan–Yorke dimension is about 300. It is expected that this discrepancy between phase space dimension and attractor dimension is much larger for more realistic models. As mentioned before, operational weather models today have of the order of 10^{10} degrees of freedom. Nobody has been able to compute the Kaplan–Yorke dimension of such a model, but it seems to be consensus that it is by several orders of magnitude smaller than 10^{10}.

In order to illustrate the effect of such sparseness of the unstable phase space directions among all possible directions, we introduce here a very simple model in discrete space and time (coupled map lattice) with the following properties: Its phase space dimension can be adjusted as the parameter N of coupled maps, it is chaotic, its maximal Lyapunov exponent is independent of N, and so is its very small attractor dimension. The model consists of one chaotic map (Ulam map) at lattice position $k = 1$, and of $N - 1$ linear maps (most of them being stable, some are unstable) at the other lattice points. We introduce a kind of convective transport by a skew coupling. The dynamics hence reads as follows:

$$x_{1,n+1} = 1 - 2x_{1,n}$$

$$x_{k,n+1} = a_k x_{k-1,n} \quad \text{for } k = 2, \ldots, N, \tag{8.1}$$

where the parameters a_k are uniformly distributed random numbers from the interval [0.87, 1.07]. The maximal Lyapunov exponent is $\lambda_1 = \ln 2$ from the chaotic driving at lattice site $k = 1$. Along with the trajectory in phase space, we iterate tangent space vectors by the linearized dynamics. We start a perturbation at "$n = -\infty$", i.e., far in the past, which, in the long time limit, provides a numerical estimate of the maximal Lyapunov exponent, but in addition yields the instantaneous stretching rate from one time step to the next. We additionally start ensembles of random perturbation vectors at different points along the reference trajectory. Initially, these random perturbations grow much more slowly than with the instantaneous stretching rate. Notice that this is in contrast to observations of "super-exponential growth" of such transients in low dimensional systems [20]. Only after a transient are these random perturbations dominated by the most unstable direction in tangent space. Their growth is then governed by the same instantaneous stretching rate as given by the perturbation started far in the past. In Fig. 8.3 we illustrate this: We show the average of the logarithm of the local stretching rates as a function of the number of time steps after initialization of an ensemble of perturbations, where the average is taken over many ensemble members and over very many points along the chaotic trajectory. As expected, after several time steps, it converges to the maximal Lyapunov exponent. However, and this is the main result of this exercise, during the first few iterations, the growth is *much slower* and can even be negative (shrinking of the perturbation vectors). Moreover, the transient time needed till the growth rate approaches its asymptotic value (for this system: $\ln 2$) increases with system size N. Hence this is an illustration of what is commonly known from ensemble weather

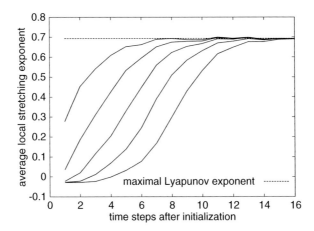

Fig. 8.3 Averaged over ensemble members and many starting points for perturbations, the *curves* represent the instantaneous stretching rate of infinitesimal randomly initialized perturbations as a function of iterations past initialization for the system (8.1). Asymptotically, the growth rate coincides with the maximum Lyapunov exponent $\lambda_{max} = \ln 2$, but during the very first iterations there might even be shrinkage. The *curves* represent system sizes $N = 10, 100, 1000, 10,000, 100,000$ *from left to right*. Hence, the effect is the more pronounced, the higher the dimension of phase space

predictions: If one wants to have perturbations growing from the very first moment, one has to chose the initial perturbation vectors in a very special way, namely they should be aligned with the most unstable directions in phase space.

This system, Eq. (8.1) is, admittedly, very specific, even though we argue above that it shares essential features with PDE systems. However, it has been observed in many studies on specific model PDEs or coupled ODEs that Lyapunov vectors localize in real space [21, 22], i.e., that at a given time, only a few components of a Lyapunov vector are considerably different from zero. Hence, the projection of a vector with random components onto such a localized vector is small, more precisely, is the smaller the larger the phase space dimension. A simple estimate which considers the normalization of the vector in N-dimensional space shows that on average, the projection onto a localized vector has a magnitude of $1/\sqrt{N}$, which means that it takes a time $t = \ln N/2\lambda_{max}$ till this perturbation reaches unity. This $\ln N$ behavior can also be observed in Fig. 8.3: Every order of magnitude by which we increase N leads to a constant shift of the curves to the right. This is plausible since in our system Eq. (8.1) the Lyapunov vector of the maximal exponent is trivially localized, namely it is the vector $(1, 0, \ldots, 0)$.

Back to the issue of chaos: In a complex system with very many degrees of freedom, a random error on the initial condition will not grow exponentially fast during the first few time steps. Only asymptotically, infinitesimal errors will be governed by the maximal Lyapunov exponent. Therefore, if we assume that a typical initial condition used for forecasting is a random perturbation of the unknown true state of the system, the prediction error should grow less fast than expected if the maximum Lyapunov exponent is known.

8.3.3 Multi-Scale Dynamics

The other reason why chaos is not the true limitation of weather predictions lies in the multi-scale aspect of weather. The most unstable structures of atmospheric dynamics are eddies of boundary layer turbulence, i.e., the gusty motion of air in interaction with the rough surface of the Earth, driven by shear and by the heating from the ground. Their instability determines the value of the largest Lyapunov exponent of this dynamical system "atmosphere", and there are estimates that it is in the range of 1/s to 100/s [23]. On the other hand, our weather instead is determined by large scale atmospheric conditions such as low and high pressure systems, whose lifetime is of the order of several days and which typically move not faster than about a few hundred kilometers per day. Hence, these structures are much more stable, such that weather forecasts are meaningful on lead times which are by orders of magnitude larger than the Lyapunov time of atmospheric dynamics. Indeed, these features of multi-scale dynamics in weather were first discussed in a quantitative way by Lorenz [24].

It is worthwhile to reformulate this issue: The maximal Lyapunov exponent is a well defined mathematical concept, involving two limits. The initial perturbation should be infinitesimal, and the growth of errors is studied in the infinite time limit. By this procedure, the linear subspace with maximal linear instability is identified and characterized. In a chaotic system with a few degrees of freedom, this maximally unstable subspace will after short time dominate the growth of perturbations. In a very high dimensional but homogeneous system such as the coupled map lattice Eq.(8.1), the transient time till a tiny random perturbation will grow with the maximal Lyapunov exponent might be quite long as shown before. The dynamics of the atmosphere is, in addition to living in an infinite dimensional phase space, spatially inhomogeneous, in particular in vertical direction. Hence, strongly chaotic motion of many small scale degrees of freedom seems to have only a weak effect on the large scale dynamics. It has not been investigated so far but it is plausible that Lyapunov vectors corresponding to the largest exponents are localized close to the ground. Hence, the multi-scale nature of atmospheric processes is another reason why the Lyapunov time is not a limitation to the prediction of those quantities in which we are interested in weather forecasts.

8.3.4 Model Errors and Noise

The more complex a natural system is, the less probable is it that the model equations used for its mathematical description are perfect. For models of the atmosphere it is very well known that they are imperfect in many respects. As in every model, relevant parameters in the model equations are known only with finite accuracy, and some of them might be even poorly known. In addition, many parameters of the model might vary slowly in time. In weather models, this is, e.g., change of albedo (the reflectivity of the Earth's surface) and of the hydrology (water storage and evaporation of the soil) due to change of land use. Hence, model parameters need regular updates, which are not always available.

A severe shortcoming of weather models is the fact that due to lack of computational resources one is forced to represent the 3-dimensional continuum of the atmosphere by a set of grid-points in space. None of the physical processes living on scales which are smaller than the grid spacing can be resolved in such a model. Grid spacings of operational models vary between about 1 km to hundreds of kilometers, depending on whether the model represents a small region or the whole globe. That is, even models with the currently achievable best resolution are hardly able to represent individual clouds, whereas the global models will even skip large orographic structures such as narrow mountain ranges and islands. All physical processes on the smaller scales have to be parameterized. This means that they are not represented by their own dynamical variables, but that instead their action on the resolved variables is approximated as a function of the latter themselves (also known as *closure*). Hence, the model variable at a grid-point lacks the interaction with potentially fast fluctuating quantities; only their average effect can be captured by deterministic parametrization. The missing fluctuations can, in the simplest approximation, be understood as some kind of noise which acts on the true, natural variable at a grid point and which is missing in the model. It is, however, evident that introducing a noise term in the model equations, as, e.g., by stochastic parametrization, would not solve the prediction problem: In order to make accurate predictions, the model noise should have exactly the same realization as the natural fluctuations have. An illustration of this statement is the prediction of a small particle in water: We know that this undergoes Brownian motion. But every single simulated Brownian path would be a bad prediction for the observed Brownian path, since in the prediction we are unable to guess the correct realization of the noise which would be needed to generate the observed path. The best prediction with least rms error would be to predict the particle being at rest. Hence, stochastic terms in atmospheric models are useful for long term modelling (one might get much better statistics of all sorts of events), but every single stochastically driven trajectory has no evident predictive power.

However, even if the individual prediction cannot become better by noise, one can use the concept of stochastic driving to explore the uncertainty of a prediction induced by unresolved degrees of freedom. The benefit from stochastic terms modelling the action of unresolved variables is in ensembles of forecasts [2, 25]: If one has a good estimate of the magnitude and the temporal correlations of these noise terms, one can use an ensemble of stochastically driven trajectories where each ensemble member has a different noise realization, and thereby construct a forecast distribution. This distribution then reflects the uncertainty of the forecasts due to unresolved degrees of freedom.

It is inevitable that models resolve only a part of all relevant degrees of freedom, so that the unresolved degrees act as some kind of noise on the resolved ones. This is true for many fields of science, such as weather, economy, traffic, ecology, human health, or biology. In addition, in many cases there are influences which cannot be modelled deterministically. In economics as well as in traffic, one might model the unpredictability of human behavior stochastically. Also the impact of natural disasters on both could be considered as a stochastic processes. These noises as well as those from unresolved degrees of freedom further limit predictability.

Actually, it is the natural tendency of researchers to make models as realistic as possible, resolving as many degrees of freedom as can be handled (from the numerical point of view). However, simply making one part of a model more realistic does not always improve predictions, as new dynamical instabilities might be introduced. Although we currently do not have a good concept for this problem, with our experience in dynamics we can state that every refinement of an existing model needs to be checked for its improved performance, one must not take that improvement for granted.

The conclusion from these considerations is that ignored degrees of freedom and noises in the real system cause additional prediction errors. These will grow slower than exponentially fast in time; for a pure random walk approximated by a fixed point they would grow like square root of time.

8.4 Best Practice Predictions

Given these many uncertainties, one should follow some best practice in predictions. As we complained earlier in this text, it is bad practice that when forecast results are communicated to the public, including political decision makers, usually only numbers are given as if they were unquestionable and precise. Decision makers have to use these numbers without any information about their reliability. This is intolerable and also unnecessary, as the example of weather forecasting shows.

So firstly, every individual prediction should not only report the best guess of the value of an entity, but also the best guess for the uncertainty of this value. In an ideal case, one could issue a probability distribution of which value is how probable, the *forecast probability*. This is the probability for the future to attain a certain value, given our rough knowledge about the system and its current state. So even in perfectly deterministic settings, such a forecast probability is different from a δ-distribution. For the purpose of giving simplified messages to the end user, an error bar representing the standard deviation of this probability might be sufficient.

In model based forecasts, the forecast probability can be constructed from ensemble forecasts, as it was outlined in Sect. 8.3. The ensemble of different forecast values then yields, after applying a technique called *kernel dressing*, a forecast distribution: A probability distribution of the prediction target, conditioned on our lack of knowledge about the system. Kernel dressing means that every forecast value of an individual ensemble member is replaced (or "dressed") by a probability distribution which has its maximum at the predicted value. The shape and the width of this kernel has to be chosen appropriately [26]. The forecast distribution is the normalized sum of all of these probability distributions.

Model errors are also partially represented by this method, but can also partly be removed by post-processing: If predictions are made with sufficiently high frequency, one can study the prediction-observation pairs of past predictions in a statistical way and correct, by re-calibration of the model output, for systematic deviations: One desirable property of ensemble forecast is unbiasedness, i.e. having a mean-zero prediction error. Another desirable property is a flat Talagrand histogram [27]: If one considers the rank of the observation among an ensemble of forecasts, this rank should be uniformly distributed, leading to a flat Talagrand histogram. Flatness of the histogram is a necessary condition for the forecast to be *reliable* [28]. A systematic bias of the ensemble which is reflected by an imbalance of outliers to the larger and outliers to the smaller values (tilted diagram), can be removed directly by a seasonal shift of the model forecasts. A systematic violation of reliability, as indicated by a non-flat, ∪-shaped or ∩-shaped Talagrand histogram, can partly be corrected by adjusting the kernel width in ensemble dressing, thereby adjusting the ensemble spread. Clearly, from the physical point of view it would be nicer to improve the model itself in order to avoid systematic forecast errors.

In situations where debatable values for certain model parameters have strong impact on the prediction result, also these must be mentioned together with the prediction. In climate projections, this is done by simulations for different scenarios of greenhouse gas emissions. But also in every forecast of economic growth some assumptions about, e.g., the oil price have to be made, and these assumptions must be communicated to forecast users.

Finally, a decision maker who has competing forecasts at disposal wants to know the performance of each forecaster in the past. Hence, generally accepted skill scores for predictions such as the rms prediction error for deterministic predictions, or the Brier score [3] for probabilistic predictions, should be computed and published by the forecaster.

In particular in weather forecasting, where at least one prediction per day is issued, every weather service could report, e.g., a score for its predictions as a function of the past years. We are only aware of the ECMWF providing information on performances [29], whereas many public weather services and commercial weather platforms in the internet do not. But also repeated economic forecasts such as of the expected growth of GNP or of tax revenue could be scored and would, after a decade or more, present a nice picture of how well forecasts have been in the past.

8.5   Data Driven Forecasts

In this section, we want to tackle the issue "who cares about chaos" from the point of view of data based forecasts. It is a common expectation that a mathematical model of a complex system will usually outperform any purely data based prediction scheme. The reason why this is usually indeed the case lies in the additional information contained in the model equations, if they are constructed on the basis of first principles, independent of a training data set. Nonetheless, a model based prediction of a really complex system requires a huge effort. In weather prediction, this is a whole industry developing the models, running a supercomputer, setting up and operating an observation network. Instead, data based prediction requires much less effort and relies essentially on a few input data sets in order to produce a forecast, and often can be optimized by a single person on a small computer. Moreover, there are many forecast issues where detailed mathematical models are lacking, so that the only chance to predict the future is based on recordings of past data.

In Fig. 8.1 we presented the performance of two data driven forecasts, which were both not particularly successful: *Climatology* was called a prediction which learns the seasonal cycle from many past observations and hence makes a prediction which depends on the calendar day for which the prediction is made, but not on the actual weather situation. It serves as a benchmark in Fig. 8.1. We also show the performance of *persistence*. Persistence means that the system has the tendency to remain in the same state as it is. Hence, a predictor exploiting persistence simply predicts for the future the value of the last available observation. As we see, this is not too bad 1 day ahead, but after 3 days this prediction is worse than climatology. Although being a function of the current state (represented by the last observation), the persistence predictor ignores dynamics. In the following we present a predictor exclusively derived from data which depends on the state and includes dynamics.

Much more promising than persistence and climatology are data driven forecasts which rely on a principle which in atmospheric sciences is called the "Lorenz method of analogues". It was introduced by Edward Lorenz [30] and suggests that if we find an atmospheric state in the past which is similar to the present state, then the future evolution of the atmosphere will follow approximately the evolution of this similar past state. Hence, one can read off from the past how the future will be. We should remark that this is actually how experience works, and that this was the way of weather forecasting in times before numerical weather models. But here we (as most surely also E. Lorenz) mean a much more detailed analogue than one on synoptic scales, which is found by a computer comparing the actual situation to a very large data base of past situations, and hence can also cope with sensitive dependence on initial conditions.

This method has been specified in many different ways, and it is worth to mention Farmer and Sidorowich [31] who formalized this for chaotic deterministic dynamics. However, using the simplest version, called "zeroth order" by Casdagli [32], is at the same time also the optimal predictor for stochastic processes, provided one is able to reconstruct their Markov property. Therefore, the issue of a chaotic or a stochastic origin of observed irregularities in data is not an issue. In the following two subsections we will discuss this in more detail.

8.5.1 Data Driven Forecasts for Deterministic Dynamics

Determinism means that the initial condition uniquely determines the trajectory which evolves from it. As a consequence of this, deterministic trajectories of the same dynamics cannot intersect each other: If they intersect, then their further evolution must be identical (because the intersection point must have a unique future), and if the dynamics is time invertible, then also the past of the intersection point must be unique, hence the two trajectories are identical for all times.

Data driven prediction for deterministic systems then consists of two parts: First, prepare the data in such a way that they represent uniquely the state of the dynamical system in the above sense. Second, find an algorithm which identifies the unique future of a new initial condition on the basis of already observed pairs of initial condition and future.

In practice, both parts can only be achieved approximately. For the identification of the state vector, Takens time delay embedding [33] is a mathematically rigorous concept for low dimensional systems. For high dimensional systems, it still serves as a useful guideline, despite practical limitations. Given a time series of observations equidistant in time, $x_1, \ldots, x_N$, and choosing a dimension m of the phase space to be reconstructed, one defines *delay vectors* $s_n := (x_n, x_{n-d}, x_{n-2d}, \ldots, x_{n-(m-1)d})$ with the time lag d in an overlapping way. Hence, the observed time series $x_1, \ldots, x_N$ is represented by a sequence of $N - md$ vectors s_n. The embedding dimension m and the time lag d are parameters by whose variation the prediction performance can be optimized, for guidelines for their proper choice see, e.g., [13]. An independent test to assure the suitability of the embedding parameters m and d consists of computing the fraction of false nearest neighbors [34]: If two points in phase space are close to each other (i.e., they are neighbors), then, due to continuity of the dynamics, their near future under the dynamical evolution should be neighbors as well. If the embedding dimension is too small, points which are quite distant from each other in the true phase space might appear to be close due to projection. However, it is improbable that their future values are also close under the very same projection. Hence, the method of false nearest neighbors counts how many pairs of nearby points s_n, $s_{n'}$ have future points s_{n+1}, $s_{n'+1}$ which are not close, in order to detect the effects of insufficient embedding dimension. If additional input variables exist, a mixed multivariate/time delay vector might serve best. This has to be optimized through minimization of the prediction error.

After one has decided on the form of the state vector, the algorithm for actual prediction following the concept of analogues can be approached: Given a state vector as input to the prediction scheme, one searches in a training set of historic recordings for similar state vectors. *Similar* here means that one defines a measure of distance in embedding space such as the Euclidean norm of the difference vector and collects all states from the data base where this distance towards the new input is smaller than a threshold ϵ. The smaller ϵ the more similar are these states, but the lesser are found in the finite data base. Hence, ϵ is a parameter which balances statistical robustness (many neighbors) with accuracy (good analogues) and has to be adjusted *a posteriori*.

Since the state vectors are constructed from a time series, one can read, from this time series, a future value of this observable at arbitrary lead time. Given the lead time, we call the corresponding time series value the "future" of this state vector. The prediction is then a suitable function of the futures of all the similar states. In the simplest case, this function is simply the mean of these futures, which was called "0th-order model" in [32]. Instead, these futures can be seen as a sample of a forecast distribution, similar to the ensemble from Sect. 8.4. Their empirical standard deviation is a measure for the prediction error to be expected and quantifies the uncertainty of this forecast, and by kernel dressing the set of futures can be converted into a forecast distribution, so that this data driven forecast complies with our requirements for state of the art forecasts in Sect. 8.4.

Such a local prediction is very flexible. The collection of all input/forecast pairs can model an arbitrary nonlinear relationship between input and output. The parameters which are embedding dimension m, time lag d, neighborhood diameter ϵ, can be intuitively understood and therefore they can be adjusted using intuition. The local approach has the drawback that it fails when data are sparse, because it cannot extrapolate. The consequence of sparse data is that either the number of neighbors is so small that the forecast suffers from statistical fluctuations, or that one has to enlarge the size of the neighborhood ϵ to unsuitably large values. This leads to using "analogues" which are not really analogues, so that this might replace statistical errors by systematic errors.

8.5.2 Data Driven Forecasts for Stochastic Data

If we assume that the observed data stem from a stochastic process, a precise prediction is impossible. Nonetheless, there exists an optimal predictor, and even the prediction of the uncertainty of the forecast is possible. Let us assume that the stochastic process is Markovian. This means that knowing the current state of the process, x_t, the conditional probability $p(x', t + \delta t | x_t, t)$ to find any other state x' at $t + \delta t$ is well defined through the transition probability from x_t to x' over a small time step δt. For a Markov process, the set of all transition probabilities defines completely its dynamics (e.g., correlation functions) and its asymptotics (e.g., stationary distribution) [35].

Hence, knowing the current state, $p(x', t + \delta t | x_t, t)$ gives us the full information about what happens at time $t + \delta$. In a stochastic setting, no further information exists. If we intend to issue a sharp forecast value and we measure the forecast quality by the rms prediction error, then the best forecast is the mean value of $p(x', t + \delta t | x_t, t)$, which can be easily verified by minimization of the rms error on a large sample of forecast trials. The uncertainty of this forecast can be readily quantified by the standard deviation of this distribution. But as well, one can use $p(x', t + \delta t | x_t, t)$ for a probabilistic forecast. Hence, knowing $p(x', t + \delta t | x_t, t)$, one can make a forecast which complies with Sect. 8.4.

How can we determine $p(x', t + \delta t | x_t, t)$ from time series data? One possible approach is again by the principle of analogues. First we have to find a good approximation to the state vector x_t. A good approach in practice is to use a time delay embedding as in the deterministic case, i.e., in the case of a scalar time series we use $s_t = (x_t, x_{t-\delta t}.x_{t-2\delta t}, \ldots)$, where the optimal value of the embedding dimension is found by minimizing the prediction error with respect to this parameter. Then, the principle of analogues assumes that all state vectors which are similar to s_t will be subject to a very similar transition probability. I.e., $p(x', t + \delta t | s_t, t) \approx p(x', t' + \delta t | s_{t'}, t')$, if $|s_t - s_{t'}| \ll 1$. This includes the assumption of stationarity, since s_t and $s_{t'}$ are not just different state vectors, but also observed at different times. The analogue-concept together with the stationarity assumption then yields the following result: The futures of all sufficiently close neighbors of s_t form a random sample according to $p(x', t + \delta t | s_t, t)$. The empirical mean of this sample is an estimator of the mean of $p(x', t + \delta t | s_t, t)$, and the sample variance is an estimator of its variance. And as in the deterministic model based forecast, where an ensemble was interpreted as a random sample of a forecast distribution, we can exploit the sample of future values here in a probabilistic way. A variant of this concept was first used in [36] for modelling, and the predictor was introduced in [37].

Now back to the title of this article: When we perform data based forecasts, the algorithm to perform the forecast is independent of whether we believe that the underlying dynamics is deterministic or stochastic, since the forecast schemes of Sects. 8.5.1 and 8.5.2 are identical, even if their motivation is quite different.

The formal reason behind this lies in the fact that a deterministic system is a limiting case of a Markovian stochastic process: the transition probability is a Dirac-δ distribution.

8.6 Conclusions

There are many sources for prediction errors, if we try to predict the future of complex systems. Very different from low-dimensional systems, prediction errors in real prediction tasks, i.e., starting from estimated model states, are usually not dominated by the exponential growth of errors on the initial condition, even if the dynamics of the system is chaotic. We discussed the sources for prediction errors and also showed several reasons why the exponential error growth with a rate given by the maximal Lyapunov exponent is usually not relevant. In weather forecasting, current models possess predictive skill beyond the annual cycle up to about 10–14 days into the future, which is by many orders of magnitude larger than the Lyapunov time, which is determined by the instability of air flow on very small spatial scales. Nonetheless, the nonlinearities of a complex system are essential for predictive skill: among others, they are responsible for the fact that prediction uncertainties are state dependent, i.e., that there are situations where predictions are more precise and other situations where they are less precise, in the very same system.

An essential message of this contribution is also that we advocate a best practice for predictions which should be routinely applied also outside the realm of weather predictions. It is highly desirable that together with every prediction, a quantitative assessment of its expected precision should be issued, alongside with results of scoring schemes applied to past predictions. Only then a decision maker can assess how reliable an individual forecast is and compare the average past performance of different forecasters.

References

1. See, e.g., the web-site http://www.pwc.co.uk/economic-services/global-economy-watch/gew-projections.jhtml of PricewaterhouseCoopers as an example of "economic forecasts" two years into the future.
2. Toth, Z., Kalnay, E.: Ensemble forcasting at NMC: the generation of perturbations. Bull. Am. Meteorol. Soc. **74**, 2317 (1993)
3. Brier, G.: Verification of forecasts expressed in terms of probability. Mon. Weather Rev. **78**, 1 (1950)
4. Christensen, H.M., Moroz, I.M., Palmer, T.N.: Evaluation of ensemble forecast uncertainty using a new proper score: application to medium-range and seasonal forecasts, Q. J. R. Meteorol. Soc. (2014). doi:10.1002/qj.2375
5. Mailier, P.J., Jolliffe, I.T., Stephenson, D.B.: Quality of weather forecasts. Online publication of the Royal Meteorological Society (2006). Available through the publication lists of its authors, e.g. http://empslocal.ex.ac.uk/people/staff/dbs202/publications/booksreports/mailier.pdf

6. Lorenz, E.N.: Deterministic nonperiodic flow. J. Atmos. Sci. **20**, 130 (1963)
7. Glass, L., Belair, J., Milton, J.J., an der Heiden, U. (eds.): Dynamical Disease: Mathematical Analysis of Human Illness. AIP, New York (1997)
8. Smith, L.A.: CHAOS A Very Short Introduction. Oxford University Press, Oxford (2007)
9. Shaw, R.: Strange Attractors, Chaotic Behavior, and Information Flow. Z. Naturforsch. **36a**, 80–112 (1981)
10. Kalnay, E.: Atmospheric Modeling, Data Assimilation, and Predictability. Cambridge University Press, Cambridge (2003)
11. Gallavotti, G., Cohen, E.G.D.: Dynamical ensembles in stationary states, J. Stat. Phys. **80**, 931–970 (1995)
12. Siegert, S., Bröcker, J., Kantz, H.: Skill of data based predictions versus dynamical models - case study on extreme temperature anomalies (2013). http://arxiv.org/abs/1312.4323
13. Kantz, H., Schreiber, T.: Nonlinear Time Series Analysis. Cambridge University Press, Cambridge (2004)
14. Hamil, T.M., Whitaker, J.S., Mullen, S.L.: Reforecasts, an important data set for improving weather predictions. Bull. Am. Meteorol. Soc. **87**(1), 33–46 (2005)
15. Boffetta, G., Giuliani, P., Paladin, G., Vulpiani, A.: An extension of the Lyapunov analysis for the predictability problem. J. Atmos. Sci. **55**, 3409–3416 (1998)
16. Paladin, G.: Intermittency and equilibrium measures in dynamical systems. In: Jullien, R., Peliti, L., Rammal, R., Boccara, N. (eds.) Universalities in Condensed Matter. Springer Proceedings in Physics, vol. 32. Springer, Berlin, Heidelberg (1988)
17. Doering, C.R.: Exact Lyapunov dimension of the universal attractor for the complex Ginzburg-Landau equation. Phys. Rev. Lett. **59**, 2911 (1097)
18. Yang, H.-L., Radons, G.: Geometry of inertial manifolds probed via a Lyapunov projection method. Phys. Rev. Lett. **108**, 154101 (2012)
19. Lions, J.L., Manley, O.P., Teman, R., Wang, S.: Physical interpretation of the attractor dimension for the primitive equations of atmospheric circulation. J. Atmos. Sci. **54**, 1137 (1997)
20. Vannitsem, S., Nicolis, C.: Lyapunov vectors and error growth patterns in a T21L3 quasi-geostrophic model. J. Atmos. Sci. **54**, 347 (1997)
21. Pikovsky, A., Politi, A.: Dynamic localization of Lyapunov vectors in spacetime chaos. Nonlinearity **11**, 1049 (1998)
22. Yang, H.-L., Radons, G.: Universal features of hydrodynamic Lyapunov modes in extended systems with continuous symmetries. Phys. Rev. Lett. **96**, 074101 (2006)
23. Basu, S., Foufoula-Georgiou, E., Porté-Angel, F.: Predictability of atmospheric boundary-layer flows as a function of scale. Geophys. Res. Lett. **29**, 2038 (2002)
24. Lorenz, E.N.: Predictability of a flow which possesses many scales of motion. Tellus **21**, 289 (1969)
25. Palmer, T.N.: A nonlinear dynamical perspective on model error: a proposal for non-local stochastic dynamic parametrization in weather and climate prediction models. Q. J. R. Meteorol. Soc. **127**, 279 (2001)
26. Silverman, B.W.: Density Estimation for Statistics and Data Analysis, 1st edn. Chapman and Hall, London (1986)
27. Hamill, T.M.: Interpretation of rank histograms for verifying ensemble forecasts. Mon. Weather Rev. **129**, 550–560 (2001)
28. Gneiting, T., Raftery, A.E.: Strictly proper scoring rules, prediction, and estimation. J. Am. Stat. Assoc. **102**, 359–378 (2007); Bröcker, J.: Reliability, sufficiency, and the decomposition of proper scores. Q. J. R. Meteorol. Soc. **135**, 1512–1519 (2009)
29. Buizza, R., Houtenkamer, P.L., Toth, Z., Pellerin, G., Wei, M., Zhu, Y.: A comparison of the ECMWF, MSC, and NCEP global ensemble prediction systems. Mon. Weather Rev. **133**, 1076 (2004)
30. Lorenz, E.N.: Atmospheric predictability as revealed by naturally occurring analogues. J. Atmos. Sci. **26**, 636 (1969)

31. Farmer, J.D., Sidorowich, J.J.: Predicting chaotic time series. Phys. Rev. Lett. **59**, 845–848 (1987)
32. Casdagli, M.: Nonlinear prediction of chaotic time series. Physica D **35**, 335 (1989)
33. Takens, F.: Detecting Strange Attractors in Turbulence. Lecture Notes in Mathematics, vol. 898. Springer, Berlin (1981); Sauer, T., Yorke, J.A., Casdagli, M.: Embedology. J. Stat. Phys. **65**, 579 (1991)
34. Kennel, M.B., Brown, R., Abarbanel, H.D.I.: Determining embedding dimension for phase-space reconstruction using a geometrical construction. Phys. Rev. A **45**, 3403 (1992)
35. Gardiner, C.W.: Handbook of Stochastic Methods. Springer, Berlin (2009)
36. Paparella, F., Provenzale, A., Smith, L.A., Taricco, C., Vio, R.: Local random analogue prediction of nonlinear processes. Phys. Lett. A **235**, 233 (1997)
37. Kantz, H., Holstein, D., Ragwitz, M., Vitanov, N.K.: Markov chain model for turbulent wind speed data. Physica A **342**, 315 (2004)

Erratum to: Chaos Detection and Predictability

Charalampos (Haris) Skokos, Georg A. Gottwald, Jacques Laskar

Erratum to:
Ch. Skokos et al. (eds.), *Chaos Detection and Predictability*,
Lecture Notes in Physics 915, DOI 10.1007/978-3-662-48410-4
© Springer-Verlag Berlin Heidelberg 2016

The Editor name "Charalampos Haris Skokos" was not correct. It is now corrected to Charalampos (Haris) Skokos throughout the book.

The publisher apologizes for having published an early draft edition of the preface. This has been updated to the final version.

The updated original online version for this book can be found at
DOI 10.1007/978-3-662-48410-4

Ch. (Haris) Skokos
Department of Mathematics and Applied Mathematics, University of Cape Town, Rondebosch, South Africa
e-mail: haris.skokos@uct.ac.za

G.A. Gottwald
School of Mathematics and Statistics, University of Sydney, Sydney, Australia

J. Laskar
Observatoire de Paris, IMCCE, Paris, France

© Springer-Verlag Berlin Heidelberg 2016   E1
Ch. Skokos et al. (eds.), *Chaos Detection and Predictability*, Lecture Notes
in Physics 915, DOI 10.1007/978-3-662-48410-4_9

Printed in the United States
By Bookmasters